Studies in Philosophy

Edited by
Robert Nozick
Pellegrino University Professor
at Harvard University

A Garland Series

Teleology and the Norms of Nature

William J. FitzPatrick

Garland Publishing, Inc.
A member of the Taylor & Francis Group
New York and London
2000

Published in 2000 by
Garland Publishing Inc.
A Member of the Taylor & Francis Group
2 Park Square, Milton Park, Abingdon, Oxfordshire OX14 4RN
711 Third Avenue, New York, NY 10017

First issued in paperback 2011

10 9 8 7 6 5 4 3 2 1

Library of Congress Cataloging-in-Publication Data
FitzPatrick, William Joseph, 1964–
Teleology and the norms of nature / William J. FitzPatrick.
p. cm. — (Studies in philosophy)
"A Garland series."
Includes bibliographical references and index.

1. Ethics. 2. Natural selection. 3. Teleology. 4. Naturalism. I. Title.
II. Studies in philosophy (New York, N.Y.)
BJ58.F58 2000
124—dc21 99-055659

ISBN13: 978-0-815-33602-0 (hbk)
ISBN13: 978-0-415-51567-2 (pbk)

Contents

Preface

The project represented in this book has a long and somewhat unusual history. I began thinking about natural teleology (the proper functions and ends associated with living things) while studying with Philippa Foot at UCLA in the late 1980's. There was at that time a great deal of interest in this topic at UCLA, primarily due to Foot's engaging pursuit of what she has come to call "teleological naturalism" in ethics: the attempt to shed light on the nature and source of ethical normativity by appealing to natural teleology in human life, and the objective norms it grounds concerning human deliberation and action. She wanted to show how ethical judgment could be objectively rooted in facts about human life by construing it as ultimately just a special case of the sort of evaluative judgment bound up with species-relative natural teleological judgments, which she in turn sought to understand in need-based terms, having to do with the characteristic ways in which creatures of a given species flourish.

To those of us attracted to both naturalism and objectivism in ethics, such a project was extremely intriguing, especially given the enthusiastic and supportive climate of the department. I, for one, was quick to get on board, and over the next several years devoted much of my graduate work to trying to defend and further articulate this view. Fortunately for me, however, Foot—having for several years run into substantial resistance from all sides—was not satisfied with my often vague attempts at developing the view, and encouraged me instead to focus on some of the *problems* that would likely arise for it. This advice turned out to be extremely sound. What it led me to do was finally to start reading biology, looking for a deeper knowledge of the living things we had previously spent so much time discussing from the perspective of a superficial armchair biology—apparently assuming (though it seems so naive in retrospect) that philosophical sophistication alone, aided by a few nature documentaries, would be sufficient for understanding the nature of teleological discourse about living things. My hope was that a more careful investigation of real cases in biology would give me problems to work on and solve, showing how they could best be understood according to a need-based or welfare-based account of natural teleology, which I would refine and develop in thinking through the examples; I could then go on to apply this account to human life in the manner suggested by Foot. In fact, something very different happened.

In the course of my reading—which included biology textbooks and books and articles by both biologists and philosophers of biology—I came across a copy of Richard Dawkins' *The Selfish Gene* (ironically, given to me a few years earlier by Foot, as part of a stack of books she was discarding). It seemed well suited to my purposes, containing a wealth of interesting and lucidly spelled out examples to work on. As I read, however, and continued on to his other books (especially his more rigorous development of his views in *The Extended Phenotype*), it became clear to me just how grossly inadequate the ahistorical approach we had taken to living things had been. In the past, I would have said that talk of genes and natural selection was quite beside the point when trying to understand present teleological facts about species, and I would have regarded an account of teleology that made heavy weather of such things as a case of scientism that misses the deeper philosophical insights. But reading Dawkins' treatment of numerous and varied biological examples in light of the plausible natural selection background behind them changed my mind: I soon became convinced that we cannot really begin to understand natural teleology until we understand more fully what living things are, in a way that is deeply informed by the *details* of the principles of evolution. And, as I recall, it was when I got to his discussion of dominance hierarchies among elephant seals that I first realized in particular that the welfare-based approach to natural teleology I had been defending for years—and likewise, the ethical project that relied on it—was in deep trouble.

It is worth keeping in mind, then, that I came to this project not as a philosopher of biology who happened to take an interest in refuting a certain metaethical theory, but as someone with a strong prior interest in that ethical theory, who ultimately had to delve deeply into the philosophy of biology in order to get to the bottom of it, eventually working to show it to be untenable. At the same time, however, as the project got rolling I became fascinated by the issues surrounding the metaphysics of teleology and theoretical biology in their own right. I have thus tried to examine a wide range of these issues in detail for their own sake, arguing for and developing a systematic account of natural teleology and exploring its implications in various areas, as well as tying it into the questions that originally motivated the project.

In addition to my debt to Philippa Foot for her original (if unintentional) push in the direction I've pursued, I owe particular thanks to Marc Lange for extensive and very helpful written comments on an earlier draft, and most of all to Robert Adams, whose thoughtful and insightful comments on every chapter benefitted the project immeasurably. I am also grateful to Torin Alter, Sean Foran, Barbara Herman, Gavin Lawrence, Mike Otsuka, Stefano Predelli, Dominik Sklenar, Houston Smit, and Carol Voeller for helpful conversations and comments at various points in the writing of an earlier draft—and especially to Melissa Weiner, for her indispensable support and encouragement at every stage.

Teleology and the Norms of Nature

CHAPTER I

Introduction

1. TELEOLOGY AND LIVING THINGS

Living things stand dramatically apart from other objects in the natural world, and this is due only in part to their unparalleled degree of complexity. Equally significantly, organisms—unlike non-living natural objects—seem to exhibit a *teleological* organization among their parts, features and activities. This is not to suggest that they were created for a purpose, or even that they were created at all; nor is it to suggest that the type of process that has given rise to organized complexity in living things—evolution through cumulative natural selection—is a purposive or end-directed one. The point is just that organisms appear to be integrated systems, the parts and features and activities of which can be said to have *functions* in a sense that implies existing or occurring *for the sake of* certain unifying biological *ends* of the organism—this being a teleological relation. If this is so, it is a remarkable fact about living things, and raises interesting questions about the nature and source of such natural teleological relations.[1]

It is useful to begin by recalling just how pervasive function ascriptions are in connection with living things. We naturally attribute functions to biological entities ranging from organs (e.g. the heart), tissues (e.g. cartilage), cells (e.g. red blood cells), subcellular organelles (e.g. chloroplasts), and genes (e.g. a hemoglobin gene), to processes (e.g. photosynthesis), secreted or ingested substances (e.g. hormones, vitamins and minerals), behavioral traits (e.g. the stalking of prey) and

---

[1] Throughout I shall use the expressions "natural teleology", "biological teleology" and "functional teleology in biology" interchangeably.

certain products of behavior (e.g. spider webs). The concept of function is clearly central to our thinking about organisms. It is important, then, to be clear about what exactly is being said when a biological function is ascribed to something, and what implications such judgments have.

First, we should notice that the concept of function employed in such ascriptions is not merely the broad and uninteresting one employed, for example, in the claim that the rain functions *as* a street cleaner, or functions *to* clean the streets. The notion of function employed in these claims about the rain is just the notion of a useful effect that something happens to have, as the rain happens to have the useful effect of cleaning the streets. By contrast, the heart does not merely function to pump the blood, as the rain functions to clean the streets: It is, we say, *the heart's function* to pump the blood. We cannot say that it is *the rain's function* to clean the streets, for while the rain *has the effect* of cleaning the streets, it cannot be said to *have the function* of cleaning the streets, i.e. to *have* the cleaning of the streets as *its* function. Similarly, while certain aberrant heart sounds may *function as* warning signs of heart trouble, proving very useful to physicians, it cannot be said that they *have the function* of indicating heart trouble to physicians—i.e. that this is *their function*, just as it is the heart's function to pump the blood. When we say that the heart's function is to pump the blood, we are employing the term "function" in a more restricted and interesting way than we are when merely speaking of useful effects.[2]

This more restricted use of the term "function"—which is also commonly found in connection with such things as artifacts, institutions and social roles—is basically equivalent to a certain familiar use of the term "purpose", where this is understood in a way that does not essentially have any psychological connotations.[3] Thus, we might say that the heart's *purpose* in the body—the biological *point* of the heart, as it were—is to pump the blood. By contrast, we would not say that it is in any sense the *purpose* of the heart to make certain noises, even if this proves useful as a diagnostic aid (i.e. where deviation from the

---

[2] Similar observations on these different uses of "function" are made by Wright (1976, p. 79), Millikan (1989, pp. 291–4) and Griffiths (1993, p. 411).

[3] Cf. Millikan (1989, pp. 291f.). To avoid begging any questions at this point, we might weaken the above to say that this use of "purpose" does not at any rate *obviously* have any psychological connotations, and is certainly commonly found in non-psychological contexts.

usual noise pattern would indicate trouble); nor would we say that it is the *purpose* of the rain to clean the streets. Often the expression "proper function" is used when speaking of *a thing's function* or *purpose*, as opposed to what it merely happens to do, or "functions to do" or "functions as", and this is illuminating. The term "proper" indicates the connection of this concept of function to the normative notion of proper functioning—a connection which is absent in the other case. A heart is the sort of thing that can be said to be functioning properly or malfunctioning, for example, unlike the rain or the clouds. And such judgments—along with related evaluative judgments about its being in good or poor condition, or being good of its kind or defective—are made in connection with the heart's proper function, not in connection with the various things it merely happens to do. A properly functioning heart is one that is performing its proper function well, i.e. pumping the blood properly, and a heart is in good condition if it is well-disposed with respect to the performance of its proper function.[4]

Finally, attributions of proper function—which I will henceforth refer to simply as "function" for convenience—are *teleological* in nature because this notion of function is bound up with the notion of an *end for the sake of which* the function is performed, as opposed to other effects which are merely incidental.[5] The proximate biological ends served by the pumping of the blood, for example, are its circulation

---

[4] On the significance of the term "proper" in "proper function", cf. Millikan (1989), Neander (1991a,b) and Plantinga (1993). It is worth noting that in Cummins' (1975) account of function, which focuses on contributions made by parts or features to the capacities of a containing system, there is no such connection to any non-arbitrary notion of proper functioning or to related notions of good/bad condition, excellence or defect. The application of such notions, if it occurs at all, would appear on his view to be arbitrary, dependent on the interests of the appraiser, thus not helping to provide any objective distinction between proper functions and mere effects. (Cf. also Searle, 1992, p. 238.) Likewise, the term "function" as Cummins understands it is not even close to being synonymous with "purpose" (in the non-psychological sense discussed above). This leads one to suspect, as Millikan (1989, p. 294) suggests, that he is not really attempting to give an account of *proper function* at all, but is concerned only with the much broader notion of whatever something "functions to do" within a containing system, which helps to explain some "interesting" capacity of that system. This would apply in principle even to the clouds' producing rain within the water system, which helps explain how the soil is kept moist, or to the heart's producing noises within the circulatory system, which helps explain the character of the overall sound produced. Cummins' view is further discussed briefly below and in more detail in chapter eight.

[5] Cf. Woodfield (1976, ch. 7), Achinstein (1983, pp. 280f.), Neander (1991b, p. 454), Griffiths (1993, p. 411).

throughout the body and the consequent distribution of nutrients and removal of wastes; the making of noise, by contrast, is recognized to be a mere side-effect, quite beside the point, as it were. If such talk of ends served by functions sounds stilted, we may notice that the same thoughts are regularly expressed more colloquially by using teleological expressions such as "in order to" or "for the sake of", as when it is said, for example, that it is *for the sake of* circulation that the blood is pumped by the heart—the circulation in turn being *for the sake of* nutrient distribution and waste removal throughout the body. By contrast, we would certainly not say that the heart beats *for the sake of* making noise, even though it does indeed make noise and its doing so proves to be useful. Nor would we say that the clouds make rain *for the sake of* cleaning the streets, even though the clouds' making rain *functions to* clean the streets. This again illustrates that the broad notion of "functioning to ϕ" or of "functioning as a ϕ-er" is not a teleological one (though it can of course be used that way), which further helps to isolate the notion of function we are after.

Teleological judgments are particularly interesting in that they appear to serve as explanations, answering a certain sort of "why?" question—i.e. one that concerns what something is *for*. We can ask, for example, why blood is circulated, meaning not "what causes it to be circulated?" (which may already be understood) but "what function is served by its being circulated—what's the physiological *point* of it?" And here the appropriate answer would be a functional teleological judgment: "It is circulated *for the sake of* nutrient distribution and waste removal"—and *not* for the sake of cooling, for example, as someone might wrongly have thought. Such teleological explanations are found throughout biology right alongside causal mechanistic explanations, and they appear to be genuine biological facts.

Despite fairly widespread agreement on the above points, however, there is still much disagreement over what it is that makes living things teleological systems, what exactly an entity's having a function—or a function's being *for the sake of* a certain end—consists in, and what kind of shape an organism's teleological organization can be expected to have. Of what relevance, if any, is the concept of organismic welfare, and on what level—individual, group, or species (or perhaps even ecosystem)? Do facts about the evolutionary history that has given shape to a given type of organism also determine the teleological profile of organisms of that type, and if so, how exactly do they do so? Different answers to these questions will yield different pictures of the shape of biological teleology—different accounts, for example, of the biological ends toward which a given organism's various proper functions are ultimately directed, as well as different accounts of the sort of explanation that is provided by appeals to functions or ends.

Much of the recent discussion has been shaped by the recognition that the clearest examples of functional characteristics in organisms—things like the heart's pumping of blood, or the spider's spinning of webs—are products of natural selection. In light of this connection it is only natural to wonder just how intimately the *historical facts* relevant to the presence of biological adaptations in current organisms are related to *present functional teleological facts* about them.[6] To some it does not seem obvious that the two issues are related at all, beyond the fact that processes of natural selection are often causally responsible for the presence of the properties that are directly relevant to functional teleology. The historical origins of a trait are one thing, it might be thought, and the trait's function in a species at a given time quite another. The latter might seem to have to do simply with the present possession of certain properties—such as dispositions to promote certain aspects of the organism's (or group's, etc.) welfare—regardless of how they came to be possessed. Why, after all, should a trait have to have any particular causal history in order to have a genuine function here and now? Could any discovery about natural selection history—or even the discovery that the theory of natural selection is false and that organisms are the result of intelligent design after all, or of something else altogether—possibly make a difference to our account of the present functions of such things as hearts or honeybee stings?[7]

---

[6] A biological *adaptation*, in the above sense, is a phenotypic trait of an organism resulting from the process of natural selection, where the trait (-type) has played a causal role in the selection process. (This last condition serves to rule out phenotypic effects that are simply neutral by-products of pleiotropic genes. The details of natural selection will be discussed in chapter three.) This use of the term "adaptation"—as in "the curved bill of the sickle-billed hummingbird is *an adaptation*"—may be contrasted with both its use to designate the *process of becoming adapted* (whether in the evolutionary sense of a species becoming adapted to a certain environment through natural selection over evolutionary time, or in the different sense in which an individual may adapt to a changed environment), and its use to designate the resulting *condition of adaptedness*. (Cf. Mayr, 1988, pp. 134–5.) I will use the term only in the first sense as defined above, to designate certain biological traits or trait-complexes (such as organs).

[7] Cf. Prior (1985, p. 317).

Someone who is thus inclined to ignore historical background in thinking about present function may not even see any reason to restrict natural or biological function to matters of survival and reproduction—a restriction that is primarily a product of historically-based views, with their focus on the process of natural selection. Instead, such functions may be thought to relate more generally to the *good* of the relevant organisms, i.e. to the ways in which the typical needs of members of a given species are regularly met, where such needs might in principle go beyond anything having to do with mere survival and reproduction (as they certainly do in the human case, for example).

I shall refer to this general approach to biological function—whether the focus is on contributions to organismic welfare narrowly conceived or as conceived in a broader way—as the *ahistorical, welfare-based approach*. The essence of such a view is the belief that we can arrive at an adequate understanding of the functional teleological facts about organisms just by considering the ways in which various traits presently tend to contribute to the organisms' welfare, without paying any attention to the causal history behind the development of those traits. There are a number of possible developments of such a view, though I shall focus critical attention on Philippa Foot's recent view, for reasons that will be explained below.[8]

At the other extreme, many will think it all too obvious that an understanding of the historical background of a biological trait bears directly on an understanding of the trait's present function: For they take the functionality of a trait just to *consist* at least largely in its possession of a certain kind of causal history. Wright, most notably, has

---

[8] For various examples of such an ahistorical approach, see Sorabji (1964), Hempel (1965), Boorse (1976), Prior (1985), Bigelow and Pargetter (1987), and especially Foot (1994, 1995). There are various differences among these views. Prior, for example, following Cummins, explicitly takes function ascriptions to be "theory and interest relative", relating in principle to contributions to *any* sort of capacity of an organism, depending on what we happen to be interested in; she identifies a subset of functions, "s-functions", that she thinks captures what we normally regard as biological functions, which are related to survival and reproduction—but again, there is on her view nothing special about survival and reproduction other than the interest we happen to take in them. By contrast, for Foot the notion of *good* (or of *need*) has a central place in biological teleology, and this is a conceptual matter, not simply a matter of what we happen to be interested in. Bigelow and Pargetter restrict functions to activities whose performance "confers a survival enhancing propensity" on the creature in question, rather than speaking more generally of the meeting of needs or the conferring of benefit on the creature or on others of the species, as do Sorabji and Foot.

argued that the essence of teleology is what he calls "consequence-etiology". A trait has a consequence-etiology if the causal explanation of how it came to be present in current organisms involves an appeal to certain causally relevant *consequences* that the trait (-type) has.[9] If the theory of natural selection is correct, then biological adaptations have consequence-etiologies, at least if they still have the effects that were selected for, since through natural selection they have come to be characteristic of populations or species—and hence have come to be widely present in current specimens—*because* of certain relevant *effects* they have, such as frightening away predators or attracting mates. On Wright's view, the possession of some such consequence-etiology, whether through natural selection or some other process (though he believes that it is in fact natural selection that plays this causal role in biology), is necessary and sufficient for the trait to be genuinely functional, the effect in question being its function; indeed, the trait's having a function just consists in its possessing a consequence-etiology.[10] Following Wright and others, we may call this sort of approach to biological function—whether characterized generally in terms of consequence-etiology, as in Wright's case, or formulated more narrowly to include only cases involving natural selection, as on other accounts—the *etiological approach*, since it ties function directly to causal history.

I believe that there is some truth to be found in each of these approaches, but that in the end both are flawed and must be rejected. The etiological approach is on the right track, I shall argue, insofar as it

---

[9] There is some unclarity in the use of the expression "consequence-etiology". In a broad sense, something might be said to have a consequence-etiology if its presence is causally explained in part by certain effects of that type of entity, whether or not the effects still occur; in a more restricted sense, something would be said to have a consequence-etiology only if that type of entity *still has* the effects that figure into the causal history leading to its presence. On Wright's view—which is discussed in detail in later chapters, especially chapter eight—it is only when something has a consequence-etiology in the more restricted sense that it can be said to have a present function.

[10] Ibid. Again, note that Wright requires that the trait (-type) still have the effect in question for that effect to count as its present function: This is how he attempts to get around the problem of vestigial traits, which are present because of effects the trait-types used to have, but no longer have. By contrast, Neander (1991a, pp. 173–4) rather surprisingly drops this condition without explicitly introducing anything else to take care of the problem of vestigials. As Prior (1985, p. 319) points out in connection with an earlier paper, this leaves her account to imply, for example, that the proper function of the appendix is the breakdown of cellulose. The consequence-etiological approach to functional teleology will be discussed and criticized in later chapters.

acknowledges the importance of the natural selection background to facts about biological function in actual living things. On some versions it even manages to yield a specification of the functional facts that is at least in large part extensionally accurate. It goes wrong, however, in its account of *why* historical facts pertaining to natural selection are relevant to facts about present function. In particular, it makes a fundamental mistake in just reducing facts about functional teleology directly and atomistically to facts about causal history (or to these together with the fact that the trait in question still has the effect that figures into the causal history). This is closely related to what I take to be the equally misguided reduction of functional teleological explanation to a kind of historical-causal explanation—namely, consequence-etiological explanation, which in this case would be the evolutionary explanation of how the trait in question came to be widely manifested in current specimens, involving an appeal to certain effects (or *consequences*) that type of trait has. The former mistake will be discussed in chapter seven, section two, and the latter is the topic of much of chapter eight.

In contrast to etiological views, the ahistorical welfare-based approach rightly avoids both of the above forms of reduction. It goes seriously wrong, however, in treating the natural selection background as something that can generally be ignored in seeking to understand facts about present biological function, as if the latter could be gleaned simply by looking at how members of a given species normally "get on", without regard for how they came to be the way they are. This, I shall argue, results in an inability to draw a non-arbitrary distinction between functional and merely incidental contributions to ends—or between genuine ends and merely useful effects—and thus leads to all sorts of difficulties once the biological world is considered in any real detail.

The truth, I believe, lies in a third alternative that has not yet been properly explored. According to this view, the facts pertaining to natural selection must be taken into account if we are to arrive at a proper general understanding of present biological function, and even if we are *fully* to understand the nature of familiar particular functions, such as the function of the heart. But facts about present biological function are nonetheless facts about *standard, non-incidental roles played by various entities in the working of organisms, understood as functional systems*; they are not directly or atomistically reducible to facts about the natural selection histories of the entities in question, as they are according to standard etiological accounts. The account I shall defend and develop might be called an *historically-informed system-oriented* account. I believe this account provides the most satisfactory basis for settling, in a non-arbitrary and comprehensive manner, the general questions raised earlier, revealing what teleology in living things is really all about, and thus telling us something important about what living things as such *are*. Among other things, it shows clearly why organismic welfare is of limited significance to natural teleology (as against welfare-based views); and it leads, in chapter eight, to an account of teleological explanation that shows it to be a distinct form of explanation, not reducible to any form of efficient-causal explanation (as against etiological views).

In the course of exploring this topic, I will also consider related questions about the parallels between natural and artificial teleology, the interactions among biological, psychological and artificial teleology in both animals and humans, and the levels of organization at which teleology is plausibly to be found in the biological world. But there is one related issue to which I want to draw attention in particular, since it is what lies behind my concern with undermining welfare-based views of natural teleology. Teleological claims are, as mentioned earlier, bound up with normative claims: Contexts allowing for attributions of proper functions and ends to certain kinds of things equally allow for related evaluative and normative claims about particulars belonging to those kinds. For example, the proper function of the mitral valve in the heart is to prevent the flow of blood back into the atrium during systole, and it is in relation to this that a condition known as mitral valve prolapse—which can result in the leaking of blood back into the atrium—counts as a *defect*; a prolapsed mitral valve, then, is said to be *defective*, or not doing what it *should* be doing, given the standards associated with the proper function of that kind of thing.

This means that the existence of objective functions and ends in nature implies the existence of objective norms in nature as well. Thus, if we can make sense of teleological attributions in biology as objective facts, then we have equally shown the existence of a range of objective evaluative and normative facts, obviating the need in at least one realm for familiar appeals to subjectivist or constructivist devices to make sense of such judgments. It would be almost a joke to suppose, for example, that the judgment that a particular oak tree has bad roots must be understood as an expression of some emotion or attitude or commitment of choice on the part of the judger. Someone might, of course, have a certain interest in connection with such a tree and mean only that these roots will not suit those interests (i.e. that they are *bad for his purposes*), perhaps thereby also expressing his disinclination to choose to buy such a tree. But he might equally be speaking simply of the condition of the roots with respect to their proper function in such an organism, observing that they are poorly disposed to carry it out, and hence defective of their kind—something that has nothing to do with his or anyone else's interests or choices.[11] Natural teleology, then, if there is indeed such a thing, seems to bring with it natural normativity of a certain kind.

Where this gets particularly interesting is in the application to human life. We are, after all, part of the natural world—living things belonging to a natural species—so that there ought to be natural teleology and related norms associated with human life no less than with other life forms. Indeed, this is perfectly familiar at the level of physiology, where we speak of good eyes or bad hearts in connection with the functional teleological standards for such things in human beings. But in principle we ought to expect such functional standards and norms to extend also to the psychological and behavioral levels of life, just as they surely do for other animals (as discussed below).[12] This then raises the question of what kinds of natural norms might pertain to human character and action, and what kind of significance they might have for us. Since, as described in the Preface, the present project was originally motivated in large part by a desire to answer just

---

[11] Cf. Foot (1978a, p. 145), (1994, p. 207), and Wallace (1978, pp. 25, 31–2).

[12] It is important not to conflate *functional teleology in the realm of psychology*, which is what is being referred to in the text, with *psychological teleology*. Psychological teleology is just the sort of teleology found in connection with purposeful behavior, i.e. goal-directedness rooted in desires and intentions. (My going to the store *in order to* get bread is an example.) Obviously this sort of teleology pervades human activity. The question being raised in the text, however, is whether there is *functional* teleology—the normative teleology of proper functions and ends—at this level of human life. This would be the case if it were part of proper human functioning to form and to act on certain desires and intentions in certain sorts of circumstances. Such a thought is familiar enough in ethology, where we commonly attribute proper functions to various psychological and behavioral dispositions in animals.

such questions—and in particular to show why recent appeals to natural human teleology in (certain versions of) neo-Aristotelian ethical theory are untenable—I shall digress to explain this broader issue in some detail before moving on.

2. ARISTOTLE AND THE HUMAN *ERGON*

In chapter seven of book one of the *Nicomachean Ethics*, Aristotle asks:

> Is [man] naturally functionless? Or as eye, hand, foot, and in general each of the parts evidently has a function (*ergon*), may one lay it down that man similarly has a function apart from all these?[13]

The implication is clear: Aristotle thinks we can infer from the fact that our organs have natural functions, that we ourselves, in some sense, have a natural function or *ergon*. But what might the latter mean, so as plausibly to follow from the former? One thing it *cannot* mean is that a human being has a natural function as a kind of "organ" within some larger natural functional entity, akin to the eye within the body. Whatever the merit of such a claim, it clearly wouldn't follow from the claim that our organs have functions. But this is plainly not what Aristotle has in mind here in any case—despite the admittedly misleading talk (at least in English) of a human being's having "a function". A better way to put the point is as follows:

> Just as (i) there are certain *characteristic activities* it *belongs* to any given human organ to carry out—as, for example, it belongs to the eye to see, or to the heart to pump the blood—so too (ii) there are characteristic activities it belongs to a human being as a whole to perform.[14]

Now why should we believe the first claim, and why might the inference to the second claim seem warranted? The former is again just meant to be another way of putting the familiar claim that organs (and

---

[13] *Nicomachean Ethics* (Revised Oxford Aristotle, 1984), 1097b30–2.

[14] Cf. McDowell (1980, p. 366). McDowell similarly defines the *ergon* of an F as "what it is the business of an F to do," stressing the normative nature of this concept, which is important to its purported connection with the notions of excellence and defect.

various other parts and features of an organism) have what I've referred to above as "proper functions". To say that the heart has the proper function of pumping the blood, for example, is to make a partly normative claim that might equally be put by saying that pumping the blood is an activity *constitutive of the heart's proper functioning*, which any *non-defective* heart will carry out; this is what it *belongs* to a heart, as such, to do—as opposed, for example, to cooling the blood, as someone might wrongly have thought. If we believe that it makes good physiological sense to speak of hearts as functioning properly or not, or as being healthy or defective, then we should be receptive to such a claim.

The point of putting things this way is that the same language can then be carried over to talk of the whole organism without the misleading implications of speaking of the organism's having "a proper function". Just as there are characteristic activities it belongs to the heart or eye to carry out, as constitutive of its proper functioning given the kind of thing it is, we might wonder whether similarly there are *characteristic activities it belongs to the whole organism to carry out, as constitutive of its proper functioning given the kind of thing it is*. The second claim (in my above reformulation of Aristotle's point) is that there are. And when it is understood in this way, we can see why it would seem to follow from the first. The inference from i to ii above may be seen simply as a conceptual point about hierarchically organized teleological systems such as organisms: In such a system, the proper functions of the parts are roughly the proper contributions they make or standard roles they play in the proper functioning of the system as a whole; thus, if it makes sense to speak of the parts as having proper functions, then it must equally make sense to speak of the proper functioning of the whole system as such, and thus to inquire into what it is that constitutes the system's proper functioning as such—the latter being the activities constitutive of its *ergon*.

This is in fact a familiar idea in ethology and sociobiology. A spider's spinning of webs or a lioness's teaching her cubs to hunt, for example, are naturally spoken of as constitutive of their proper functioning—web-spinning belonging to the *ergon* of a spider and the training of cubs or the stalking of prey belonging to the *ergon* of a lioness. And there might well be rather detailed species-relative standards for proper functioning in these areas, to be understood within the context of the characteristic life of the species. There are plausibly detailed natural norms, for example, for the golden orb spider regarding web-spinning (what kind of web to spin, under what conditions, etc.), which any given specimen must meet if it is to count as a non-defective golden orb spider, just as its eyes must meet certain species-relative standards of vision if they are to count as non-defective golden orb spider eyes.[15]

Now whether or not this is precisely the direction in which Aristotle meant to take his "function argument", it would at least explain and support his move from the claim that a human being's parts have natural functions to the claim that the whole human being has an *ergon*—understood as a set of activities characteristic of non-defective human beings and constitutive of their proper functioning, the standards for which are likewise species-relative and appropriate to the nature of human life. It would therefore also provide an argument for the possibility of objective, species-relative evaluations concerning the qualities and behavior of individuals. Just as we already make such judgments concerning parts, we could equally make similar judgments about human beings themselves. If a spider or a lioness can be objectively defective—bad of its kind—for failing to do what is constitutive of its species-typical *ergon*, then the same should be expected for an individual human being, if indeed there is likewise a human *ergon*. The question, of course, is how to account for the relevant standards, and whether this leads to anything of broader significance in the human case.

Aristotle certainly thought that the idea of a human *ergon* had broader significance: For him it was the key, at least in very general terms, to understanding the ultimate end of human life and the intellectual and moral virtues involved in its attainment. I shall now set Aristotle aside, however, and focus on neo-Aristotelian developments of this general line of thought.

3. TELEOLOGICAL NATURALISM IN ETHICS: FOOT'S RECENT WORK

In a famous passage, which inspired later developments by others, G.E.M. Anscombe has suggested that such natural function-related

---

[15] Cf. Foot (1994, 1995).

norms in the realm of human psychology and behavior might be nothing less than ethical norms:

> It might remain to look for [ethical] 'norms' in human virtues: Just as man has so many teeth, which is certainly not the average number of teeth men have, but is the number of teeth for the species, so perhaps the species man.....'has' such and such virtues, and this 'man' with the complete set of virtues is the 'norm', as 'man' with, e.g., a complete set of teeth is a norm.[16]

Teeth have a natural proper function in human life, related to the end of breaking up food, and because of this there are natural norms for teeth, governing such things as their size, structure and number in non-defective human beings. Anscombe's suggestion, then, is that something similar may be true of the qualities of character we call "virtues", which may likewise be properly related to the realization of important human ends. This line of thought was later developed by James Wallace, who argued that what makes certain character traits count as virtues are the functions they have in human life:

> Any study of living creatures as such, including modern biology, inevitably involves normative considerations. The view that life is a natural phenomenon leads to the conclusion that certain normative data are found in nature. Such normative data are properly studied by studying the lives of the appropriate organisms. On the basis of these considerations, I will defend two theses. (1) It is possible to find in human life itself an objective basis for normative theses. (2) Certain facts about the nature of human life do yield explanations of why certain traits are virtues.[17]

---

[16] Anscombe (1981c, p. 38).

[17] Wallace (1978, p. 18). This general idea is still very much alive—for example, in Gaut's (1997, pp. 184–5) recent claim that "value is a teleological, biological category", whether we're talking about non-defectiveness in the parts of plants or ethically relevant goodness in connection with human life. Thompson (1995) also develops closely related ideas. In fact, the contemporary appeal to natural teleology in ethical theory is not limited to neo-Aristotelians: Korsgaard (1996, p. 152), who is a neo-Kantian, claims rather strikingly that any living organism's natural end is "to preserve its own identity", and that therefore, given her account of obligation in terms of identity preservation, "life is a form

But the most intriguing and thoroughgoing appeal to natural teleology in ethical theory occurs in Philippa Foot's most recent work.[18] It is therefore on her approach that I shall focus, though I believe the problems I raise for it in following chapters will be seen to be equally damaging to other appeals to natural teleology in ethics.

At the heart of Foot's approach is a certain understanding of *life* according to which, as Wallace has put it, "life is a normative concept," and the norms bound up with any given form of life or species have to do ultimately with the *flourishing* of such organisms, understood as the meeting of their species-typical *needs*.[19] Foot's thought is that it belongs to the very nature of life that for any species of living organism, there is a set of facts about the typical needs of such creatures and about how those needs are properly met so that, all else being equal, such creatures flourish. The idea is then that it is in these welfare-oriented terms that the notion of a creature's natural proper functioning—both at the physiological level and at the behavioral level—is to be understood. The activities that are constitutive of proper functioning—from the beating of a heart to the stalking of prey—are those that make proper contributions to the meeting of such creatures' needs, and hence their flourishing, as part of the "way of life of the species." And it is in connection with this that derivative judgments about an individual's species-relative goodness or defectiveness are to be understood. This is what I have earlier referred to as a "welfare-based" view of natural teleology, according to which the notion of welfare or flourishing comes in as the final natural end that gives shape to the associated function-related norms.[20]

---

of morality. Or to put the point less strangely and in a way that has been made more familiar to us by Aristotle, morality is just the form that *human life* takes."

18 Foot (1994, 1995). These papers are developments of ideas presented in unpublished manuscripts and public lectures delivered over roughly the last fifteen years.

19 Wallace (1978, p. 16).

20 This is the most straightforward and charitable reconstruction of Foot's view. In her own presentation of it, she does not explicitly stress the concept of functional teleology, focusing instead on the concept of what she calls "autonomous, species-dependent goodness or defectiveness" in living things, which she connects with the concept of species-typical needs, as defined by the "life of the species" (Foot, 1994, pp. 208 f.). But she herself refers to her view as "teleological naturalism", and it is clear from her discussion and examples that evaluations of "species-dependent goodness or defectiveness" are precisely function-based evaluations of excellence or defect where the functions in question belong to species-relative, natural functional teleology, and that her

How exactly the notion of organismic welfare-promotion or flourishing comes in is a complicated matter. Foot recognizes that it would be too simple to relate all aspects of an organism's proper functioning to that individual's flourishing, since, for example, it is the welfare of offspring that is promoted by the proper function of the mammary glands or of parental nurturing behavior, and it is certainly not the welfare of the stinging bee that is promoted by the stinging of an intruder. Thus, a lioness that fails to teach her cubs to hunt, or a bee (of the appropriate caste) that fails to sting an intruder to protect the hive may be defective of their kind, though their behavior does not decrease their own welfare and may even promote it. Foot's claim, then, must be stated fairly generally, if vaguely: Proper functioning, at various levels, for members of a given species is to be understood in relation to the ways in which *they* typically meet *their* various needs and flourish, as part of the characteristic way of life of the species—where this will often involve certain individuals helping to meet the needs of others in the species.[21]

Now there will obviously be many differences across species in the *details* of proper functioning and in the nature of the needs served, and Foot recognizes this. Let me just note three points in this regard. First, she recognizes—and indeed, insists—that in the human case the notions of *need* and *flourishing* will take on a much richer content than in the case of plants and non-human animals. Human needs surely go well beyond needs associated ultimately with mere survival and reproduction.[22] Secondly, though she doesn't discuss this, it is certainly open to her to maintain that human needs are also special in that many of them cannot be understood from outside the living of a human life, in the way that the needs of a geranium can be understood from a

---

appeal to species-typical needs is meant precisely as her account of the *ends* of natural functional teleology, making possible such a structure of evaluation for living things.

[21] Foot (1994, p. 209). In other words, insofar as proper functioning is *instrumentally* related to the meeting of various needs, these needs may exist in different individuals from those whose proper functioning is at issue. Note that this general appeal to need-satisfaction in deriving facts about proper functioning is not incompatible with Aristotle's apparent claim that proper natural functioning is also *constitutively* related to *individual* flourishing—the idea being that an individual's proper natural functioning (whatever exactly that turns out to involve) is *constitutive* of *its* true flourishing (*Nicomachean Ethics* 1097b25–29, 1098a15–17). Foot does not address the latter issue in these papers, and I will set it aside here, except to note that if the argument in the following chapters is right, then Aristotle's apparent claim is no less problematic than Foot's claims.

[22] Foot (1995, p. 9), and (1994, p. 210).

botanist's external perspective. It may be that many relevant facts about human life can be appreciated only from the inside—without leaving the rich perspective of concrete human living.[23] Finally, while proper functioning for plants and non-human animals may simply be a matter of *doing* certain things, it is consistent with Foot's idea to recognize that for human beings, proper functioning often involves not merely doing certain things, but coming to such behavior by way of *taking* certain kinds of considerations *as* reasons for so acting. Acting on reasons is, after all, a natural part of human life, and it is plausible to suppose that this will be involved in proper human functioning.[24]

Returning to the main idea, then, Foot's claim is that in spite of these differences in detail across species, there is nonetheless a *single general framework of formal connections* among the notions of *life*, *species*, *proper functioning*, *needs* or *welfare*, and *excellence* or *defect*, as I've sketched.[25] Such a framework, she thinks, holds for any living species: The parts, features and activities of members of a given species are to be judged non-defective or defective in accordance with standards based on background facts about how the species-typical needs of such creatures are met within the characteristic way of life of the species, whereby they flourish. And her suggestion is that as applied to human life, at the level of will and character, this natural

[23] For a development of this idea, see Nagel (1979). Nagel rejects any "external" approach to ethics, as e.g. through a morally-neutral science of biology. See also McDowell (1980, pp. 367–371), Lawrence (1993, pp. 289–361), and Nussbaum (1995). McDowell, Lawrence and Nussbaum all argue against a reading of Aristotle that would commit him to having to justify a particular conception of the human *ergon* and excellences (or of proper human functioning) by appeal to a morally-prior investigation of human nature. One might have thought him so committed either because (i) such an investigation seems necessary for arriving at standards of individual happiness needed to show that living in accordance with the proposed excellences is indeed supremely desirable for a human being, as *eudaimonia* is supposed to be, or because (ii) he is taken to be trying to offer a *foundational* view of ethics, resting ethical truths on independently discoverable facts about human nature. McDowell, Lawrence and Nussbaum argue that these ideas are mistaken, and that Aristotle is plausibly not mixed up in them. Foot's view, as I understand it, needn't involve any commitment to such a justificatory method either, though her position on this issue is not entirely clear. Thus, my complaint against her will *not* be that she is wrongly involved in an external sort of foundationalist, teleological naturalism, which appears to be McDowell's main worry in McDowell (1995).

[24] Foot (1995, pp. 9–10).

[25] Cf. Lawrence (1993, esp. p. 321), where a similar claim is made in a discussion of Aristotle's "function argument".

framework makes sense of evaluative judgments belonging to *ethics*. That is, we should understand ethical evaluation "on the pattern of the [species-relative] evaluation of other operations in animals and in men":

> Nobody would, I think, take it as other than a plain matter of fact that there is something wrong with the hearing of a gull that cannot distinguish the cry of its own chick, as with the sight of an owl that cannot see in the dark. [Likewise with a lioness that does not teach her cubs to hunt, or a free-riding wolf who eats but does not take part in the hunt.] Similarly, it is obvious that there are objective, factual evaluations of such things as human sight, hearing, memory and concentration, based on the life-form of our own species. Why, then, [should] it seem so monstrous a suggestion that the evaluation of the human will should be determined by facts about the nature of human beings and the life of our own species?[26]
>
> It seems clear that in spite of the greater range of considerations that are relevant to the determination of absolute need in human beings, the evaluation of such things as locomotion, sight, and memory has the same conceptual structure in animals and men....And the tenor of this paper is to suggest that the same is true when the subject is the goodness or badness of human actions. If it is so, and if, moreover, action is evaluated in relation to human needs, as the things bees do are evaluated by what bees need, this will be highly relevant to the status of justice as a virtue. Is the behaviour of an unjust person defective, we should ask, for the same reason, broadly speaking, as is the behaviour of a free-riding non-dancing dancing bee, or a lioness who does not teach her cubs to hunt?[27]

Foot's suggestion, then, is that the objective norms of human character and action that we're after in ethics are really just a special case of the kind of species-relative norms found in connection with

---

[26] Foot (1995, p. 14). Cf. also Wallace (1978, p. 16): "Among the facts about living creatures are how they live normally, under what conditions they flourish or languish, and what the proper functioning is of their parts....There is no reason in principle why a study of human excellence based upon the nature of human life need be any less objective, well founded, or authoritative than the study of any sort of living creature."

[27] Foot (1994, p. 211).

living things generally. What's special is that they involve the human will, which is why they are relevant to ethics, unlike norms involving physiology or the behavior of non-human animals. In short, given her welfare-based view of natural teleology, it seems to her to be a plausible claim that ethical normativity is ultimately a species of natural teleological normativity. Following her, we may refer to this view as "teleological naturalism" in ethics.

Now given her welfare-based view of natural teleology, this is an understandably tempting move. Moral virtues are, after all, often thought to count as virtues precisely because of their importance for the attainment of various kinds of human good, and this idea would be neatly captured by such an approach. It might be argued, for example—following a suggestion from Anscombe—that since human beings have an important general need to be able to count on each other's promises, proper human functioning in the realm of action will involve general conformity to certain principles of honesty and fidelity; it's because of such human need that these character traits are moral virtues. On Foot's view, this could be understood precisely as an example of natural teleology as applied to the human will.[28] Such an approach might be particularly attractive to those looking for an account of ethical normativity that is both naturalistic and objective, as an alternative to subjectivist neo-Humean or constructivist neo-Kantian accounts. (There would, however, still be the difficulty of arguing for an objectivist account of claims of need or claims about goods insofar as these go beyond survival, reproduction and pleasure—something that is not settled by the objectivity of the function-related evaluative framework).

In summary, then, Foot's appeal to natural teleology offers us a general theoretical framework within which to understand intuitively plausible thoughts about virtues and human good. It provides a metaphysical explanation for the source of the connection between ethical normativity and human flourishing—something that is often just taken for granted by neo-Aristotelians. The appeal to conditions of human flourishing in an account of the virtues has been a constant theme in Foot's own work, going back to her well-known early papers

---

[28] Foot (1994, 1995). Cf. also G.E.M. Anscombe (1981d), and Wallace (1978) where virtues are understood in terms of their having certain proper functions in human life.

collected in *Virtues and Vices*.[29] But her current work goes further in offering an explanation for why ethical normativity is to be understood in terms of human flourishing: This is simply a consequence of the construal of ethical normativity as a species of natural teleological normativity, combined with her welfare-based view of natural teleology. Foot's teleological naturalism thus explains the characteristic structure of neo-Aristotelian ethical theory by appeal to the alleged metaphysical structure of life itself. So if it were successful, it would seem to offer a solid basis for a *neo-Aristotelian, naturalistic objectivism* in ethics—at least insofar as the relevant claims of need can be made out to be objective.

Unfortunately, as I shall bring out in the course of the present investigation into natural teleology, such an approach to ethics is fundamentally misguided, since the welfare-based view of natural teleology on which it relies is untenable. Some will no doubt think that this is obvious and hardly needs to be argued for—either because the very idea of natural teleology is nothing but an outmoded bit of Aristotelianism or creationism, or because even if there is such a thing as natural teleology in human life, it's *obviously* the wrong sort of thing to have any relevance to an understanding of ethical judgment, at least given what we now know about biology.[30] Both of these thoughts are mistaken. As to the first, the fact is that the existence of natural teleology is widely accepted by contemporary philosophers and biologists who have no truck with creationism and little sympathy for or concern with Aristotelian metaphysics. Moreover, as pointed out in section one, the pervasiveness of teleological concepts and language in even our most scientifically informed discourse about living things strongly suggests that natural teleology is not merely a holdover from an outmoded metaphysics or a creationist picture, but something deeply bound up with the nature of living things. As to the second thought above, it's true that most biologists and philosophers who grant the existence of natural teleology understand it ultimately in terms of

---

[29] Foot (1978)—especially "Moral Arguments," "Moral Beliefs," "Goodness and Choice," and "Virtues and Vices"—and (1994), (1995).

[30] See, for example, Bernard Williams' dismissal of what he calls Aristotle's "metaphysical teleology" in Williams (1985, pp. 43–4, 48); and Alasdair MacIntyre's rejection of Aristotle's "biologically teleological account" of the virtues, on the grounds that it is based on a no longer tenable "metaphysical biology," in MacIntyre (1984, pp. 162–3, 196–7).

evolutionary biology, which (as explained later) would make it unfit to play the sort of role Foot wishes it to play in ethics. It cannot, however, simply be assumed without argument that this is the only way to understand natural teleology, since Foot's starting point is precisely the suggestion of an alternative, which makes no such appeal to evolutionary history. There is thus real work to be done in order to show exactly (i) why an appeal to evolutionary history is necessary for understanding present teleological facts about evolved organisms, (ii) how such facts about causal history really come into the account of teleological facts, and (iii) why this precludes welfare-based accounts such as Foot's.

In my view, this has not as yet been carried out satisfactorily, and again, that is a large part of the motivation behind the work in the next few chapters. In chapter two, I'll take up the first project, providing an argument that draws on a revealing parallel with the teleology found in connection with artificial functional systems. In chapters three, four and five I'll go on to develop and defend what I've called an historically-informed, system-oriented view of biological teleology, based on the argument of chapter two and a detailed examination of the principles of natural selection. This will answer the second question more satisfactorily, I believe, than do other historically-informed views. Once this has been carried out, the answer to the third question readily emerges. As I argue in chapter four, section six, given the nature of the principles according to which the special relations definitive of natural teleology are established and shaped, we must conclude (i) that organismic welfare or species-typical flourishing is neither generally nor ultimately the end toward which the various aspects of an organism's natural proper functioning are geared, and (ii) that what *does* play this role is so different from anything like organismic flourishing that the sort of normativity it grounds, even in the sphere of human character and action, couldn't possibly have anything to do with ethical normativity. My attack on welfare-based views of natural teleology, however, goes significantly beyond this. After all, if the only argument against welfare-based views relied entirely in this way on my own positive, evolutionarily informed account, it would be vulnerable to dismissal by someone like Foot, who again will likely just reject my very starting point in chapter two—either rejecting my positive account altogether or claiming that it leaves room for some sort of complementary welfare-based account of the very same subject matter, i.e. objective functional teleological discourse in connection with living things as such.[31] I have therefore devoted chapter six to independent arguments against welfare-based views of natural teleology, which do not rely on my own positive view. Even if my account is rejected, or is regarded as only part of the story, the considerations about organisms that I shall raise in arguing for it are sufficient to pose what I take to be decisive problems for welfare-based approaches. This will be brought out both at an abstract level and in connection with some interesting biological examples that illustrate the problems especially clearly.

4. TELEOLOGY AND HUMAN NATURE

While the natural teleology associated with human psychology and behavior cannot, if I'm right, be of any relevance to ethics, it will still be relevant to an understanding of human nature, telling us something about what we are as members of a particular evolved, natural species. This is the topic of the final two chapters. In chapter nine, I'll examine the ways in which my account of natural teleology plausibly applies to human life, exploring the partial extension of natural function ascriptions to certain kinds of behavior, psychological capacities and dispositions, and even artifacts (including such things as laws and customs). Despite the significant presence of natural teleology in human life, however, I shall also argue that one of the main things that sets us apart from other organisms is our ability—rooted in our sophisticated reflective capacities—to *transcend* our proper biological functioning in systematic and agent-directed ways. Unlike other organisms, we are not destined to live primarily in the unreflective service of the genes that have built and conditioned us, but can instead take over much of the task of designing our lives for ourselves, making them in an important sense truly *our own*.

In chapter ten, I argue that this advanced mental capacity is the foundation of another side of human nature that is frequently

---

[31] Foot has said things in personal communication that suggest she might take the latter position, holding that my account may capture the sort of thing *biologists* are talking about when they use teleological concepts, but that an ahistorical welfare-based account is still tenable with regard to the teleological facts the rest of us allegedly recognize on the basis of intuitions drawn from familiar examples described roughly at the level of nature documentaries—what Michael Thompson (1995) refers to as "Natural Historical Judgments."

overlooked by sociobiologists and evolutionary psychologists who tend to treat human nature as if it were exhausted by those aspects of it that are attributable to the influences of our evolutionary history (as filtered through cultures that are themselves shaped by influences traceable to our evolutionary background). Human nature is in fact complex, consisting of both elements, which means, among other things, that explanations of many aspects of human nature and culture will likewise be complex, appealing to both factors: our evolutionary heritage and the influence of independent, critical reflection. This is particularly true with regard to morality, which again many popular accounts attempt to treat as a set of empirical phenomena entirely explicable in terms of our evolutionary heritage as filtered through various cultural contingencies, rather than recognizing that at least some of the thought and behavior in question may have its source in independent motivation stemming from intelligent evaluative reflection on our lives.

Finally, after examining such explanatory issues, I'll briefly return at the end of chapter ten to the issue of whether there are in fact objective moral standards to which we can legitimately be held accountable. This, again, was Foot's question, which she tried to answer positively by appeal to natural teleology in human life. I have rejected that approach, but that does not mean that I reject the attempt to give a positive answer to the question. In fact, I think that Foot is quite right in insisting (as she has all along) that ethics has largely to do with the promotion of human good and the satisfaction of human needs, and that moral virtues are to be understood largely in these terms. My claim is just that we can't look to *nature* in the way she does—appealing to the teleology bound up with *life*—in the hope of finding a deeper objective grounding for the connection between ethical normativity and human flourishing. But then where should we look in seeking to understand our accountability to moral standards? I will suggest, though only in a very preliminary, forward-looking way, that the source of our accountability to standards that are quite independent of those of natural teleology lies precisely in the second, reason-based aspect of human nature. My primary goal, however, as far as ethics is concerned, is to put teleological naturalism in ethics to rest, thus clearing the way for neo-Aristotelians to return their attention to the business of developing a viable neo-Aristotelianism in ethics, free of this persistent but erroneous appeal to nature in ethical theory.

CHAPTER II

Functional Systems: Artificial and Organic

1. MACHINES AND ORGANISMS

We speak of teleological functions in roughly three kinds of context: (i) in connection with complex systems that can non-arbitrarily be spoken of as *working* in a certain way, with parts and features that play certain roles in that working; (ii) where we can speak of a type of object as *having a certain regular use*; and (iii) where we can speak of a type of object or activity as *being made* or *being done* in a certain way.[1]

In such contexts, certain "how?" and "what?" questions that are characteristic of functional discourse have application. Where we can speak of a functional system, in the above sense, we can ask: "what does it do?" (with reference either to the whole system or to a certain part or feature) and "how does it work?" (again with reference either to the whole system or to a given sub-system). Similarly, where we can speak of an object with a regular use, we can ask: "what is it used for?" and "how is it used?" Where we can speak of technique, we can ask: "what does it accomplish?" (with reference either to the whole

[1] These are rough characterizations (the first of which will be elaborated upon at length below and in chapter four). We might clarify the intended sense of the italicized expressions by adding that in each case they allow for the application of certain evaluative concepts to particulars. So, for example, where we can speak in the relevant sense of a type of working system , we can also speak of a particular system of that type as working *properly* or *improperly*, *well* or *badly*; where we can speak in the relevant sense of a type of object as having a regular use, or as being made in a certain way, we can also speak of particular objects of that type as being used or made properly or improperly, and so on.

technique or to a certain aspect of it), or "what does a practitioner of such a technique, *as such*, do?"; and moving in the other direction, we can ask of the thing made or done in a certain way: "how is it made/done?" Such questions are central to functional discourse because the notions of (a) *what* a system (or one of its parts or features) does, i.e. its *work*, as referred to when we speak of how it works, or of (b) *what* an object is used for (or what it does when it is so used), i.e. its *use*, as when we speak of how it is used, or of (c) *what* is accomplished by a technique, or by some aspect of it, about which we can ask how it is done—these notions are all notions of *function* in the teleological sense.

Consider first the case of something with a regular use. If we are presented with an unfamiliar tool, for example, a natural first question to ask about it, *as* a tool, is what it is used for. Suppose it is a protractor, so that the answer is that it is used for measuring angles. This notion of what it is used for, i.e. its regular use, is a functional notion: A protractor is a measuring device; that is its *purpose* or *function*, in relation to which we can intelligibly ask how it is used—i.e. how one measures angles with it. Similar points can be made in the case of techniques. If we come across someone who is obviously some sort of craftsman employing some sort of technique, though we don't know anything more than that, a natural question to ask is what he is doing. Suppose he is making shoes. We might then inquire as to the *purpose* of the various things he is doing in the course of making the shoes, which are constitutive of the technique—e.g. stretching a piece of leather in a certain way. This notion of purpose is a functional notion: This type of action plays a functional role in the making of shoes; it has a certain *point* in the process, being done—not simply on this occasion by this individual, but in the more general sense of being included as part of the technique—*for the sake of* certain ends, ultimately for the production of shoes. In a slightly different sort of case, suppose we come upon someone playing the violin, and want to know how it is done. Part of the explanation will include angling the bow in certain ways. Of this we might then ask what exactly it accomplishes in the playing of the violin: Is it done *for the sake of* tone, or of control, or of something else? Again, this is a functional notion: Angling the bow in certain ways has an important function in the playing of the violin—it plays a certain role, has a certain point.

There is obviously much more that would have to be said in developing a detailed account of such cases.[2] My primary concern, however, is not with these kinds of cases, but with cases of the first kind mentioned above: systems that can non-arbitrarily be spoken of as *working* in a certain way, in a robust sense that allows for talk of a given system's (or sub-system's) working well or poorly, or functioning properly or malfunctioning. This is because I wish to argue that organisms are to be understood precisely as such functional systems, and that when this is elucidated, we can discover the source of the teleological structures manifested by living things through examining an important parallel with the source of teleological structures of artificial functional systems.

The basic points in the context of functional systems are similar to those in the other contexts. When confronted with an unfamiliar machine, for example, a natural first question to ask about it, *as* a machine, is *what* it does; once we have an answer to that, we may go on to ask *how* it works—how it manages to carry off what it does in the relevant sense (or what it is supposed to do given the type of machine it is). The latter question will be answered in terms of *what* various parts do at the next level down in the hierarchy of levels of organization, which once again invites the question of *how they* work, and so on, until at some point these questions have no further application (i.e.

---

[2] There are some interesting remarks on functions pertaining to techniques in Thompson (1995), and in Broadie (1987). With regard to functions pertaining to *use*, one challenge is to distinguish between cases where something's being used gives it a function and cases where it does not, even where the use is a regular one. For example, while stones that are *used for* paving a road may indeed be said to have the function of preventing erosion, it is not the case that stars regularly *used for* navigation by ship captains have as their *function* the providing of navigational information, or that natural rivers regularly *used for* transportation purposes thereby have this as their function. It seems plausible that the difference has something to do with the amount of control exercised over the thing used, with attributions of function requiring a significant measure, though we should not expect the line to be very clear. Thus, it is natural to say, for example, that the water used by a boat engine for cooling—where it is taken in and circulated, say—has the function of cooling it (at least while it is being so used), whereas it is much less natural to say that the water in a natural body of water used by boats for transportation thereby has as its function holding up boats or providing transportation routes, even temporarily (though it will be different if it is put there for that purpose, in which case both the body of water and the water itself might be said to have functions). The former sort of case, involving *both* the notion of a working system and the notion of something's being used in a certain way (i.e. by the system), is discussed in chapter five in connection with organisms.

when the only questions and explanations left concern chemistry or physics as such).[3]

Consider, for example, a gasoline engine. It obviously does many things, including making noise and polluting the air—just as a protractor may be used for many things, including killing a roach. But the sense of "what it does" that is relevant here is the one directly related to the notion of its *work*, in the sense employed in the question "how does it (do its) work?" When we ask of a gasoline engine, "how does it work?", we are asking not how it makes noise, for example, but roughly how it converts the chemical energy stored in gasoline into rotational mechanical power; for this is what its *working* consists in, at least at the most general level—*what it does* in the sense of its *function*, and what it tends to do more or less successfully when functioning/working properly. (How successfully it does this will obviously depend on how well designed it is).

The question of how the gasoline engine works will then be answered in terms of what various parts (such as the pushrod valvetrain, valves, spark plugs, and so on) do, which when taken together and organized in the right way make the whole engine (do its) work. What the pushrod valvetrain *does* in this sense—its function—is to open and close the valves at the proper times during the cycle; that is the role it plays in the working or functioning of the engine. If we then ask how the *pushrod valvetrain* works, the answer will be in terms of the functioning of *its* parts: As the camshaft rotates, the cam lobes push the valve lifters up, which in turn lifts the push rod, which pushes on the rocker arm, which pivots in such a way as to push on the valve stem, opening the valve, and so on. Each lower-level part's role in the functioning of this higher-level part or sub-system (e.g. the pushrod's

---

3 Cf. Dawkins (1987, pp. 11–12) on such explanation in the context of certain hierarchically organized systems. There is a related discussion in Cummins (1975), but while he calls his topic "functional analysis", he is concerned with something much broader, namely explanation of how a system manages to have certain higher level effects in terms of lower level properties, where the higher level effects can be *any* "interesting" effects it happens to have, which need not have anything to do with *working* in the relevant sense. (See chapter eight.) Indeed, as Kitcher (1993, p. 390) points out, on Cummins' view there could even be "functional analyses" associated with what anyone would normally recognize as a case of *malfunctioning*—e.g. where the complex higher level effect in question is the formation of tumors, the lower level property being the possession of a certain mutant DNA sequence. I am interested here in the more restricted case of explanation of how a system *works*, i.e. how it does *what it does* in the *functional* sense, as discussed below.

being lifted by the valve-lifter and shifting the rocker arm) is *its* function, both within the sub-system and hence within the engine as a whole.[4]

It is important to notice that when we speak in this way of roles things play in the working of a system, we mean something more than just *de facto* contributions they may happen to make to the system's working. Just as when we speak functionally of *working* for a system we are interested in something more than just what the system happens to do, so too when we speak functionally of roles things play in that working, as when describing how such a system works, we are interested in something more than just the ways in which various things might happen to promote that working. That is, what we are interested in here are *proper functional* contributions to the working of the system, and these must be distinguishable from merely incidental contributions that things might happen to make toward effects involved in the working of the system. As pointed out in chapter one, this is reflected in the fact that when we are speaking of functional roles—in the sense of proper functions—of parts and features within working systems, it is equally appropriate to employ teleological expressions such as "in order to/that" or "for the sake of" in describing these items and their contributions; and if X occurs *in order that* Y occur, or *for the sake of* Y, the implication is that X's contributing to Y's occurrence is somehow not merely incidental.

For example, it may be said that the pushrod valvetrain is present in a gasoline engine *for the sake of* effecting the proper opening and shutting of the valves, or that the crankshaft gear is half the size of the camshaft gear *in order that* the camshaft make one revolution for every two revolutions of the crankshaft, which in turn brings it about that . . . etc., ultimately ensuring proper valve timing. This is simply another way of stating that opening and shutting the valves in a certain

---

4 The same points apply to the engine itself considered as a functional part of a larger system, such as an automobile. That is, we may speak of the engine's function or work (what it does) in the context of the larger system, which will naturally fall under a new description going beyond that of its more immediate function. This function will be the contribution that the performance of its more immediate function makes to the working of the larger system—what it *does*, or *is used for*, within that system, in relation to the system's working. So, for example, the function of the gasoline engine in the automobile is to power it, providing the force that enables it to move, by converting fuel energy into the mechanical energy of the rotating crankshaft; this in turn will then come into the account of *how an automobile works*.

way is the *function* of the pushrod valvetrain, that ensuring a certain rotational differential is the *function* of the size differential between the two gears, and so on—in the restricted and interesting sense of "function" identified in chapter one. The teleological language, however, helps to make it clear that these facts require that the effects in question be non-incidental. The opening and shutting of the valves in a certain way, for example, is not just something the pushrod valvetrain happens to do, which happens to contribute to the working of the engine: It is *that for the sake of which* the pushrod valvetrain exists as part of an engine, which implies that the effect is not merely incidental. Causing the valves to open and close in a certain way is the *role the pushrod valvetrain non-incidentally plays in the working of the engine*—something that comes into an account of how a gasoline engine works, or of what goes on in a properly functioning gasoline engine.

Now a natural objection is likely to arise at this point, as soon as we ask what it is that makes these things true of machines and their parts. For surely it is at least largely to facts about *design* that we will appeal in such cases.[5] A gasoline engine's work consists (at the most

[5] It would be hasty, however, to suggest that we can give a simple, general account of artificial function in terms of the intentions of designers. That the matter is significantly more complicated has been shown, I think, by Andrew Hsu in an unpublished paper ("On the Natural History of Artifacts"), and I have already noted that there are artificial functions connected with things that are not designed at all, but only used in a certain way—such as paving stones. Hsu argues that even some cases of what I am calling working systems would resist treatment in terms of the intentions of designers, and in *any* case there is plausibly more involved than just intentions. With regard to the first point, a complex system might come together accidentally and then be copied and given a certain regular use, so that we might still speak of the devices as having a function even though no one *designed* them. There will also be more realistic cases where things originally designed for one purpose come to be regularly used for another, which again loosens the connection between design and function (or perhaps calls for a distinction between two uses of "function"). With regard to the second point above, even where such complications arising from the factor of *use* do not arise, it would be wrong to think that a designer's intending a machine to ϕ is sufficient to make this its function or work. As Hsu points out, someone might design a system with the (hopeless) intention of creating a perpetual motion machine, but it is not plausibly the case that the system actually *has* the function of exhibiting perpetual motion, thus *malfunctioning* every time it runs down; the exhibiting of perpetual motion is merely the function it was intended (but failed) to have. (He gives other convincing examples that don't turn on physical impossibility, such as the first failed attempts at creating a semiconductor amplifier, which did not thereby produce the first transistor. Clearly intention alone is not sufficient.) But these interesting points needn't concern us here: I am not attempting to give anything like a general account of artificial function, which would indeed require a careful look at the roles played both by

general level) in converting fuel energy into constant rotational power—rather than in polluting the air or making noise, which are mere side-effects of its working—because that's what the gasoline engine was designed to do; it is the fact of design that provides for a non-arbitrary determination of *working* for such a system. Similarly, what makes it the case that the aforementioned effects of the pushrod valvetrain constitute *non-incidental* contributions to the working of the engine is the fact that the pushrod valvetrain was designed to have those effects, as part of the design of the engine. It was designed to play such a role in the working of the engine, and this is what makes its doing so non-incidental. But if that's right, then it might be objected that a system-oriented approach to biological function, treating organisms as working systems analogous to complex machines, cannot possibly succeed—unless of course we settle for an account according to which functional discourse in biology is merely metaphorical, treating organisms *as if* they were designed by an intelligent creator.[6] Organisms, after all, are not literally designed, as machines are (at least if neo-Darwinian evolutionary theory is roughly correct, as I'm assuming), so that the crucial factor that explains teleological structure in artificial functional systems is glaringly absent in the case of organisms.

The answer to this objection is simple. Facts about design are indeed relevant to the teleological structures of machines, especially in such straightforward cases. But notice that such facts have come in only indirectly, by way of explaining—in the case of machines—how there can come into being a type of system that can non-arbitrarily be said to *work* in a certain way, with parts and features that play *non-incidental roles* in that working. It is the latter concepts, rather than the concept of design itself, to which we directly appealed in explicating the idea of a functional system; the notion of design came in only as the explanation

---

use and by design in various kinds of case. My point is only that in at least a very wide range of cases of genuinely functional machines, we can discover what *working* consists in for it by discovering what it was designed to do and how it was designed to do it. That is sufficient for the rough contrast I wish to set up with organisms, which are my central concern.

[6] Cf. O'Grady (1986, p. 1013), who maintains that the teleological approach to living things treats an organism "*as if* it were designed by an intelligent agent." On his view, evidently, functional teleological claims in biology are literally false, though they are useful heuristic devices; the only way they could be literally true would be if biological things were something like God's artifacts.

of how, in the context of machines, these concepts come to apply to such systems. The question we need to ask, then, is whether intelligent design is the *only* way of bringing about such relations within a system, or whether something else can equally play the same role, serving to provide a non-arbitrary determination of a system's work, and to set up the sorts of relations that allow for a distinction between incidental effects and non-incidental effects relevant to the system's working. If, as I shall argue, the latter is the case, then the absence of intelligent design is no obstacle to treating living organisms as working systems in the same robust sense in which machines are working systems—where an organism's *work* (i.e. what working consists in for it) is non-arbitrarily determined, and it possesses parts and features that play non-incidental roles in that working. As long as there is some other process that does for living systems what designers do for machines, in the relevant respects, we will be able to apply to organisms no less than to machines questions such as "how does an F work?" (where "F" stands in for a species term), "what role does such and such a part play?" or "what is it for?" (e.g. "what role does the liver play?", "what are teeth for?"), and "does this particular F work well or badly?" And if so, this looks like a natural way to approach functional teleological discourse in biology.

The idea of treating an organism as a working system, analogous to a machine, might initially seem a bit odd. We do not, after all, commonly speak of how whole organisms work—*how a cat works*, for instance. But in fact the idea is already implicit in familiar talk about the parts and subsystems of organisms. For example, there is nothing unusual in speaking of *how the feline eye works*. This notion goes with functional talk of *what* the parts and features of the eye *do* in the functional sense, i.e. what role they play in the working of the eye, just as talk of *how* the pushrod valvetrain in an engine *works* goes with functional talk of *what* its various parts and features *do*. Or equivalently, if such talk about the eye is legitimate, we can equally employ teleological connectives, just as in the case of the machine, saying for example that tears are secreted by the lacrimal gland and diffused between the eye and eyelids *in order to* keep the cornea moist, so that it is not scratched by the eyelid. This may be understood simply as another way of saying that this is the function of the secretion of tears. The notion of how the eye works likewise goes with the notion of *what* the *eye* itself *does* in the functional sense, just as the notion of how the pushrod valvetrain works goes with the notion of *what it does* in the functional sense; for *what* the thing *does* in this sense is what defines what *working* is for it in the first place—i.e. that about which we can then go on to ask *how* it does it. But in the case of the pushrod valvetrain, *what it does*—i.e. its *work*—is precisely the role it plays in the *working of the engine*. Thus, if we can legitimately follow the parallel, familiar talk of *how the eye works*, which is bound up with the notion of *what the eye does* or its *work*, is equally bound up with the notion of the *working of the organism* of which it is a part.

The question is again just whether such familiar talk about organs or sub-systems is indeed legitimate in the absence of intelligent design, with something else serving to set up the necessary conditions for it. If it is, and if the above structural parallel with machines holds, then the *function* of an organ within the organism would be the *non-incidental contribution it makes to the working of the organism*—just as the function of the pushrod valvetrain is the non-incidental contribution it makes to the working of the engine. Again, that is the line I shall be pursuing in the next several chapters, with special attention to the interesting and challenging question of how ultimately to understand the notion of *working* for a living organism—something that, as we shall see, is more complicated and less intuitively obvious than it might at first seem to be.

2. CAUSAL HISTORY AND PRESENT-TELEOLOGICAL STRUCTURE

Living organisms are products of natural selection rather than artifacts of a cosmic designer—or so I am assuming.[7] They cannot, therefore, be said literally to be designed to do anything, as machines are. Nor can such a thing be said about their various parts and features, though it is common to speak metaphorically about how, for example, the bill of the sickle-billed hummingbird is specially "designed" for sucking the nectar from Heliconia flowers, the curved shape of which makes it impossible for a straight bill to reach the nectar. (The non-metaphorical

---

[7] I do not, however, mean to be committing myself to an atheistic position, denying the possibility of some kind of divine influence at a very general level, compatible with the theory of natural selection. I leave that matter open. For further discussion of this issue, see chapter five, section fourteen.

way to put the point is that as a result of natural selection the bill is specially *adapted* for sucking nectar from Heliconia flowers.) Yet the example of the eye shows how natural it is to apply the functional teleological language of working systems to biological entities, even while we explicitly deny that they have literally been designed. One possibility, again, is that all this functional discourse in biology—talk of how eyes work, of the functions of rods and cones, and so on—is really just metaphorical too, like talk of design. But let's begin with the assumption that this is not the case, taking functional talk in biology at face value, as evidently literal and largely true discourse about living things, and see how far we can go toward accounting for it.

If organisms are genuine functional systems despite their not having been designed, we must look to something other than facts about design to flesh out our understanding of their functional nature.[8] We may, of course, be able to understand a certain amount about such particular matters as how eyes work—and so about such things as the local functions of the parts of the eye—simply by examining them here and now. But there will so far still be a great deal of unclarity about more general functional relations. For example, it is far from obvious how exactly the hierarchy of functions and ends is structured within an organism (and possibly leading beyond it), and particularly what exactly the ultimate end(s) is (are) toward which lower level functioning is geared, as the lower level functioning of the sub-systems of a gasoline engine are geared ultimately toward the production of controlled rotational power. Is lower level functioning in an organism to be understood as being ultimately for the sake of its individual survival? Its reproduction? A conjunction of the two? (And if so, how are they related as ends, and why?) The survival and reproduction of its close relatives as well? (And if so, is this always part of the ultimate end, or only sometimes, and why? And how do these compare with the ends of personal survival and reproduction? Does this vary from species to species, and why?) The survival of a broader group? Species survival? Individual *flourishing*? Group flourishing? Or is it perhaps something else altogether, to which an organism's survival and reproduction (and often that of others too) is instrumental?

---

[8] An exception would be cases of artificial breeding of plants and animals by humans, where there *are* facts about design. This, however, is a very different matter from that of natural biological function, and must be treated separately.

As long as such matters are unclear, we are in a position with regard to organisms similar to the position we are in when confronted with an unfamiliar machine: We accept, let us suppose, that it is a genuine functional system—something that can legitimately be spoken of as *working* in a certain way, and so on—but we don't yet know the details of its overall functional structure; we may understand certain local functional facts, such as the immediate function of a valve, but we don't yet have a firm grasp of how exactly the hierarchy of functions and ends is structured, and what exactly the ultimate end(s) is (are) toward which lower level functioning is geared—what the system's *work* ultimately amounts to. Now we have said that in the case of machines, the natural place to look for a non-arbitrary specification of these functional facts is to facts about design—facts about what it was designed to do and how it was designed to work. But what exactly is it about facts about design that makes them so obviously relevant to present functional facts about machines?

The answer, I suggest, may be put as follows:

> D: Facts about design matter to a machine's teleological profile because (i) it is the process of design that is ultimately causally responsible for the compresence of at least many of the system's parts and features, and for their organization into a single coherent system wherein they combine interactively to bring about certain effects, and (ii) the nature of this causal influence is such that the facts about design are also responsible for its being the case that *it is no accident* that the system possesses just such parts and features and just such an organization among them that they interactively bring about such effects.

Consider again the example of an engine. It is the process of design that is ultimately causally responsible for the compresence of such things as the camshaft, valve lifters, push rods, etc., and their organization into a single coherent system. And because of the nature of that causal influence—i.e. the fact that the various parts were brought together and organized in a certain way *with a view toward* their bringing about certain effects—it has also established special *non-accidental relations* between the system's parts and features and those effects. What I mean by this is that it is thereby *no accident*, for example, that the pushrod valvetrain possesses just such parts and features, and just such an organization among them, that they interactively bring about the opening and closing of the valves with a certain timing; and this allows us to speak of this effect of the pushrod valvetrain as being in an important sense *non-incidental*, as opposed to its making a certain rhythmic noise, for example.[9] My claim, then, is that it is because the process of design was ultimately causally responsible for all of this that it makes sense to look back to it in seeking to clarify our teleological understanding of the engine. For surely whatever is ultimately causally responsible for the non-accidental compresence and organization of a set of parts and features into a coherent system that produces certain special effects—without which relations we would not have a genuine functional system at all—is surely equally responsible for determining the functional facts pertaining to the system. Whatever ultimately set up the details of these crucial relations must equally be responsible for determining how exactly the hierarchy of functions and ends is structured, what working ultimately consists in for the system, which effects constitute functional contributions and which are mere side-effects, and so on. It would be highly implausible, after all, to suppose that in general it is one thing that ultimately causes a functional system to have its complex and non-accidental organization of parts, features and activities—without which it would not even exhibit many of its higher level parts, features and activities, and would not be a true functional system at all—and quite another thing that determines the specification of the system's proper functioning.

So if we want to clarify our understanding of a machine's *working* at various levels, and of how various functions and ends are related

---

[9] There is perhaps a derivative sense in which it might also be said to be "no accident" that the pushrod valvetrain is structured in such a way that it makes a certain rhythmic noise; for it is no accident (in my sense) that it is structured in such a way as to open and close the valves with a certain timing pattern, and its making such a noise is a natural concomitant of its doing this. I am, of course, concerned with *non*-derivative non-accidentalness, where the effect in question is relevant to whatever it is that plays the causal role identified above—in this case the process of design—rather than in derivative non-accidentalness, where the effect is etiologically irrelevant, being just an incidental concomitant of the first kind of effect. It is in relation to this non-derivative structure of non-accidental relations that we can speak of certain effects as being interestingly non-incidental. (I am not, however, suggesting that every function within a functional system must have a consequence-etiology. My departure from that view is discussed in chapter five.)

within its hierarchical functional structure, we need to look at the non-accidental relations established by the process of design among the system's parts and features, on the one hand, and certain of their effects, on the other. That is in fact precisely what we do when we look to facts about what various parts and features were designed to do and what was behind the designer's organizing the parts and features in the way she has (or her directing manufacturers to do so, etc.) in order to arrive at a better understanding of the machine's functional structure—of the functions and ends non-incidentally served by the parts, features and activities *so organized.* In discovering, for example, that various parts and features of the gasoline engine were designed and organized as they are with the intention ultimately of creating a machine that would produce controlled rotational power from the combustion of gasoline—this intention being what is ultimately responsible for such organization—we discover the nature of the ultimate end toward the realization of which the lower level functions and ends are orchestrated in the gasoline engine.[10]

Now there are obvious and important differences between the process of intelligent design, through which a type of machine is given

---

[10] As mentioned in a previous footnote, there will be special cases that cannot be treated in this way. For example, a machine may be designed to do one thing but gradually come to be *used* regularly to do another, so that there is some pull toward saying that its function has changed. Here its present function would not be illuminated by looking at the ultimate source of the organization of its parts, features and activities, since it was originally designed to do something other than its present function. But there are two points to notice here. First, while the highest level of a machine's function may plausibly change with a change in use—e.g. from turning the axle in a car to turning a paddlewheel in a boat—it is much harder to see how much could plausibly change about the nature of its working at lower levels through such changes in the use to which the machine is put. It would not be hard, of course, for such lower level working simply to be rendered *irrelevant* to the new use—for example, if engines came regularly to be used as wrecking balls. But short of this, it is hard to see how it could be changed very much. The situation with working systems is thus rather different from that with simpler artifacts like tools; there is much less susceptibility with the former to major changes in their entire functional nature due to changes in how they are used, at least if the new use still involves their use *as* working systems. Second, even to the extent that design and function may diverge in machines, this will not affect the point I am ultimately concerned to make. In the biological case, there will not be the possibility of such a tension arising through a change in how organisms are *used,* since biological function is not to *any* extent a matter of how organisms are used. If there is a kind of function associated with the uses to which organisms are put by us, it is an entirely different matter, closer to artificial function than to biological function. I will therefore limit myself to fairly straightforward cases of machines for the purpose of the parallel with organisms.

a certain organization of parts, features and activities, and the process of cumulative natural selection, through which a type of organism is given a certain organization of parts, features and activities over evolutionary time. For example, natural selection is not itself a teleological process at all—and certainly not a psychological teleological process, as is the process of design and creation; natural selection does not take place *for the sake of* any end, psychological or otherwise. But despite such important differences, there is much that the process of design and the process of cumulative natural selection have in common.

In particular—and this is the crucial point—the two processes have in common exactly the feature that has been suggested to lie at the heart of the relevance of design to teleology in the case of machines: Each is *what is ultimately causally responsible for its being no accident that certain parts, features and activities are compresent and systematically organized in such a way that they interactively bring about a certain special subset of the effects they do, allowing us to distinguish non-incidental effects or contributions from incidental ones.*[11] In other words, the processes constituting the natural selection history behind the evolution of a given type of organism play the same *causal role*, with respect to the selection and organization of organic parts, features and activities at various levels, that is played by the process of intelligent design in the case of machines. (It might also be noticed that as with machines, but even more so in the case of organisms, many higher level parts and features (e.g. the eye) are themselves the *result* of the organization among lower level parts and features (e.g. the retina, lens, muscles, etc.), so that there is an intimate relation between the two things mentioned above: Often what is responsible for the *organization* among parts and features at lower levels is *ipso facto* responsible for the very *presence* of certain parts and features at higher levels.)[12]

---

[11] I will illustrate this just below for the biological case, and will return in chapter four, section seven to explicate this idea more fully—explaining why, for example, it does not plausibly capture too much (e.g. even inorganic natural systems.)

[12] Notice that we are not here interested in the proximate causes of the assembly of traits in a particular case, i.e. the causes involved in the ontogeny of a particular individual. We are no more concerned with this in biological cases than we are in the movements of the fingers of assembly-line workers in the case of a machine. What is relevant here are the *ultimate* causes behind the assembly of traits into a certain organism-type, i.e. the causes involved in the evolutionary development of a given species. This natural selection history is what is ultimately causally responsible for the presence and coherent

We may give a rough and preliminary illustration of this causal parallel by considering not yet a whole organism, but a sub-system such as the eye, keeping in mind the parallel with the sub-system of the engine discussed earlier. The claim is that it is the processes of natural selection throughout the evolution of the eye that are causally responsible for the evolved eye's non-accidentally having just such a combination of parts and features, organized in just such a way that they cooperatively make possible visual perception. Consider how incremental developments accumulated in the evolution of the eye. This was not just random—though of course each small change was initially the result of a random mutation. What happened at each evolutionary stage was that a certain incremental change was incorporated through natural selection into the standard construction of the eye by virtue of its making some positive contribution to sight. It was, after all, on the basis of such a contribution that the mutant gene coding for such a change in the eye out-propagated rival alleles at that genetic locus and came to be standardly included in the genomes of the organisms in question; and so likewise the phenotypic modification caused by that gene came to be part of the standard construction of the eye.[13] It is through the accumulation of such incremental modifications through natural selection that the eye's present organization and coherence as a functional system is to be explained.

The result, then, is that it is *no accident* that a present, evolved eye has the various parts and features it does, organized as they are, such that they cooperatively bring about visual perception (or rather, those aspects of visual perception for which the eye is responsible). And so, for example, the ciliary muscle's focusing effect, which is an adaptation, is *non-incidental*; it is a proper, functional contribution to the eye's working, where this is all made out in terms of the non-accidental relations between traits and certain of their effects, set up by the eye's natural selection history. Thus, even though the eye was not

organization of the traits of a given type of organism, and for establishing certain non-accidental relations between certain traits and certain effects.

[13] The biological terms will be clarified in the discussion of natural selection in the next chapter. Basically, *alleles* (or allelomorphs) within a gene pool are the various genes of different types that occupy homologous loci of homologous chromosomes throughout the population or species. The *phenotypic effects* (or *expressions*) of a gene are the effects its presence has—in the context of its environment, including the rest of the genome—on the traits of the organism (i.e. its morphological, physiological or behavioral characteristics).

literally designed to do anything, there is nothing stopping us from speaking—as we already naturally do—of the eye as a genuine working system (or sub-system), with parts and features that play various non-incidental roles in that working; and we can in principle confirm or clarify the nature of the eye's proper work (what Aristotle meant by its *ergon*) and the nature of the functional roles involved in it by considering the natural selection history behind its construction, and the special connections it sets up.

It is largely the purpose of the next two chapters to explore what I've briefly illustrated in the case of the eye more generally, showing what principles are ultimately involved and what this implies about the teleological structures of entire organisms—in particular about what plays the role of an organism's ultimate end within the framework of biological teleology. For the moment, what I want to stress is just that if the natural selection history behind a given type of organism does indeed play the same causal role as that played by the process of design in the case of machines, then it makes sense to look to the details of natural selection to learn more about the structure of functions and ends presently served by organic parts, features and activities.

In summary, the overall argument rests on a general principle that I take to apply both to artificial systems and to natural ones:

> P: With regard to a given type of functional system S, whatever it is that is ultimately causally responsible for the compresence and organization of its parts and features—in such a way as to make it the case that it is *no accident* that the system possesses *such* parts and features and *such* an organization among them that they interactively bring about a certain special subset of the *effects* they do, thus allowing us to distinguish non-incidental effects or contributions from incidental ones—is thereby also responsible for determining the specification of working for S; that is, it equally determines the nature of the functions and ends served by these parts, features and activities so organized, and the way in which these functions and ends are related at various levels, i.e. how exactly they fit into the hierarchical teleological structure of S.[14]

---

[14] There will, however, be exceptions with respect to particular functions both in the case of machines and in the case of organisms—in the former as a result of changes in the way

If P is correct, and it is true that natural selection plays the causal role in question with regard to organisms, then it makes sense to look to facts about natural selection history in seeking fully to understand what it is that constitutes *working* for a living organism, including the specification of the ultimate end(s) toward which lower level functioning is ultimately geared. We thus have a promising direction for inquiry into the questions about biological teleology raised earlier in this section.

The strength of this argument obviously depends on the plausibility of principle P. I do not claim to have said enough to convince all potential opponents of the truth of this principle, and I recognize that some may find the whole approach to organic functioning *via* the comparison with machines unattractive. But I think this approach and this principle have been sufficiently motivated to justify an investigation into the sort of account of biological function to which it points, and this is what I shall undertake in the next three chapters. What I hope to show is that this approach is indeed fruitful—that the account I shall develop, which naturally emerges from the above argument together with a careful consideration of the principles of natural selection, is a powerful and attractive one that covers in an intuitive and non-arbitrary way the whole spectrum of cases found in biology, and meets the major challenges facing any account of biological teleology.

---

the machine is used (as noted earlier), and in the latter as a result of various possible evolutionary changes, such as those resulting in a trait's vestigiality. This issue will be dealt with in detail when I present the account of biological function in chapters four and five, and will be set aside for the moment; my concern here is only to bring out the general importance of the ultimate cause of organization in likewise shaping the general functional structure of the system in question.

CHAPTER III

Natural Selection, Genes and Organismic Welfare

1. A CENTRAL QUESTION ABOUT BIOLOGICAL TELEOLOGY

If we pursue the system-oriented approach to biological teleology described in the previous chapter, a central question we face in attempting to provide an account of biological teleology is this: Given that organisms are products of natural selection, what is it that we can non-arbitrarily point to as constituting *working* or *proper functioning* for organisms, in relation to which the biological functions of organic parts and features are to be understood? What is it that organisms have, as it were, been "designed" by natural selection ultimately to do, this being their *work* in the functional sense?

It should be clear from the argument of the last chapter that the use of the design metaphor here is innocuous, and is not intended to be doing any philosophical work. Its use is justified as a simple convenience on the grounds that the natural selection history behind a given organism-type shares with the process of design and creation the important causal role discussed earlier, and therefore plausibly shares the role of shaping the teleological structure of the systems in question; it is solely on the basis of these shared roles that the design metaphor is here introduced, and so long as this is clear, there need be nothing misleading about it. The fact that natural selection is not itself a teleological process, like the process of design and creation, is simply beside the present point, since the metaphor was never intended to link the two cases in *those* respects; those parts of the metaphor are irrelevant, and do not enter into the argument given in chapter two.

The metaphor could thus easily be dropped if it proves distracting or arouses suspicion. I find it convenient, however, and will continue to employ it as a reminder of the important role that *is* shared by natural selection and the process of design and creation; the use of quotation marks, indicating merely metaphorical usage, is a reminder of what they do *not* share. If at any point one wishes to eliminate the metaphor, to be sure that we are not being misled, it can easily be converted into literal terms as follows. To say that a given type of organism has been "designed" by natural selection to ϕ is simply to point to the fact that it is the natural selection history behind that type of organism that was ultimately causally responsible for the compresence and organization of parts, features and activities into a coherent working system that includes ϕ-ing as part of its working—ϕ-ing being among the effects for which it is the case that it is *no accident* that the system is arranged in such a way as to bring them about. Something parallel can be said regarding a machine that has been designed to ϕ, with the relevant causal role being played by the designer, which is why the metaphor is apt, and need not involve any smuggling of psychological elements into talk of biological function.[1]

Let us then take up this question: What is it that natural selection has "designed" organisms to do? Or more precisely: *What is it that the processes constituting the natural selection history responsible for putting together a given organism-type have "designed" such organisms to do*? On the face of it, the answer probably seems obvious: While the details vary greatly from species to species, organisms have been "designed" by natural selection to survive and to reproduce; survival and reproduction, it will be said, are the formal biological ends

---

[1] The term "design" is also sometimes used in a deliberately extended sense, as for example, by Plantinga (1993, p. 13) and Kitcher (1993, p. 380), where talk of design or design-plan is not supposed to be taken to imply the existence of an intelligent, designing agent (though Plantinga eventually argues for the existence of such an agent behind biological design-plans). I prefer to stick with the ordinary use of the term, indicating metaphorical or extended uses with quotation marks and making the point and the limits of the metaphor explicit, as above. It should be stressed that while my use of "design" is metaphorical in the context of biology, my uses of the central teleological terms and expressions—"working", "function", "end", "in order to", and so on—are *not* meant anywhere to be metaphorical. (The quotation marks in the last sentence, of course, simply indicate mention vs. use.)

in terms of which *working* for living organisms is ultimately to be understood.

Now I certainly grant that the notion of working for organisms will pertain to their survival and reproduction; and it will therefore plausibly have something to do with their welfare—i.e. with meeting various needs they might be said to have in connection with survival and reproduction—at least in a broad range of cases. But the question we need to ask is how exactly survival and reproduction are to be understood insofar as they are truly ends to which functions are to be related. This will shed light on the question whether the meeting of the various needs of organisms—the promotion of *their good*—can properly be regarded as a *general* and *ultimate* end, or whether, as I shall argue, it comes into only a certain range of cases, and even there only as a *proximate* end, *instrumentally* related to something else altogether.

We may begin to get a sense of the difficulties lurking behind the deceptively simple notion of "survival and reproduction" by considering certain questions that must arise as soon as we consider the actual biological world in any detail. Is the end of *reproduction*, to which various functions are related, simply that of producing viable offspring at a rate sufficient to keep the species going? Or might it be something more complicated, like maximizing personal reproduction in relation to other members of the same sex within the population, so as to *out*-reproduce them? If the latter, why should that be? And can the meeting of such an end really be construed as promoting the good of organisms, increasing their well-being? Regarding the end of *survival*, does the survival in question pertain only to the survival of the individual organism and its offspring, or can it also pertain to the survival of other kin, for example—so that some functional traits might promote the survival of siblings and parents, perhaps even at the expense of personal survival (as with honeybee stings, for example)? And if the latter, why exactly should that be—and why does the survival of close kin seem to have such a special place as an end in comparison with that of others in the population? Finally, how are the ends of survival and reproduction related to each other: Are they on a par, or is one subordinated to the other, so that there could be functional traits that contributed to increased reproductive output even at the expense of the end of personal survival *without thereby being dysfunctional*? And again, if the latter, why is it that the end of reproduction should be given theoretical precedence in this way?

If the argument in chapter two is sound, then an examination of the process of natural selection should be helpful in answering these questions. The natural selection history behind the assembly of various traits into a coherent organism-type ought to shed light on what *working* ultimately amounts to for such an organism—not just its proximate and immediate functioning at lower levels considered in relative isolation (which is already well accounted for in physiology textbooks, and is not of any general philosophical interest), but its overall functioning as a coherent, hierarchically organized system, geared toward the realization of certain high-level and possibly quite remote ends. I will therefore devote this chapter to the topic of natural selection, with the aim of shedding light on the facts that will prove to be relevant to the issue of biological function. In particular, I will seek to make clear that the principles of natural selection, which govern the "selection" of traits to be incorporated into the characteristic "designs" of organisms of various species, do not generally or ultimately have to do with the welfare or need-satisfaction of organisms.

Note that the argument concerning the nature of biological function will not be completed until chapter four, combining the argument from chapter two with the arguments of the rest of this chapter; only then will we be in a position to answer the questions raised above. Since my account of biological function will not be an etiological account, i.e. it will not analyze the present functions of particular traits *directly* in light of their natural selection histories, it is important to be careful not to draw conclusions about what I wish to say concerning biological function from the discussion of natural selection taken in isolation. (Again, the distinction between my account and etiological accounts will become clear in chapters four, five, seven and eight.) It is also important not to be misled by any metaphorical expressions used in the following description of natural selection (such as talk of adaptations "benefitting" genes, or of "common projects" advanced by co-adapted sets of genes). When I return to the topic of functional teleology in chapter four, the account will concern functions and ends *literally* understood, and the argument will not have relied in any way on such metaphors.

2. THE PERSPECTIVE OF GENIC SELECTIONISM

The process by which complex adaptations are put together over evolutionary time, integrated into coherent systems constituting living organisms, seems to be primarily a matter of intra-specific microevolution by natural selection—i.e. evolution due to cumulative natural selection operating within populations of members of a given species. There are no doubt relatively simple *macro*evolutionary trends (such as the increase in average horse size over the last 30 million years) that are the result *not* of cumulative microevolution, but of a kind of higher-level natural selection among closely related species following speciation events; such differential extinction of species may perhaps also account for such phenomena as the predominance of sexually reproducing species over asexual species. And there are of course many other factors that contribute to evolution, such as genetic drift and various cataclysmic events. My primary interest, however, is with complex adaptations and their organization and integration into coherent organisms, in connection with which we commonly apply functional teleological concepts in relation to the characteristic lives of organisms. Thus, since it is microevolution through cumulative natural selection that is plausibly both necessary and (for the most part) sufficient to account naturalistically for this, it will be my focus here.[2]

The process of natural selection, even in the context of microevolution, can be described in various ways for various ranges of cases. In many cases, we may appeal to the effects of an organism's biological traits ultimately on its reproductive success, whether directly or through contributions to its survival, *via* all sorts of mechanisms at various levels; in other complicated cases, we may have to appeal instead to a trait's effects on the reproductive success of kin; and there are still other cases of natural selection that cannot be described in

[2] See Dawkins (1982, pp. 104–8) and (1987, 265–9) for a discussion of intra-specific microevolution as the "significant force in the evolution of the complex machinery of life," in contrast with the comparatively simple changes potentially resulting from "species selection". The argument for why species selection (as opposed to microevolution) cannot generally be responsible for the evolution of complex adaptations is basically just that there have not plausibly been anywhere near enough speciation events to account for the evolution of such adaptations (e.g. the eye) in terms of random mutations and differential extinction at the species level. It is necessary instead to appeal to microevolution within populations, where there can be heritable mutations and differential reproduction with every generation—allowing for enough incremental changes to make for a tenable evolutionary story.

terms of contributions to the reproductive success of organisms at all, and must be described at a lower level. (Examples of each will be given below.) There is nothing wrong with having such a plurality of descriptions for various ranges of cases. But if we are interested in getting clear about the essence of natural selection in the context of microevolution—about what is *ultimately* going on in the whole spectrum of cases the theory is meant to cover—then it makes sense to adopt the most general and unified perspective, from which any case may in principle be described in an accurate and illuminating way.

It has been argued, persuasively in my view, that we have just such a perspective in *genic selectionism* (described in detail in the following section), provided that such things as environmental factors and population structures—such as the sorts of group dynamics often stressed in the name of multilevel selection theory—are properly taken into account and not ignored.[3] The latter qualification is important: The "gene's eye view" of things certainly does not imply looking only at genes and ignoring everything else, any more than a bird's eye view of things implies looking only at birds. It's perfectly compatible with genic selectionism to recognize that the long causal chains mediating a gene's effect on its own propagation in the gene pool relative to rival alleles will involve such factors as the concentrations of different phenotypes in different groups and the rates of dispersal of members of various groups back into the general population, not to mention all sorts of other environmental factors.[4] Similarly, the emphasis on the genetic perspective is certainly not meant to imply any denial of the relevance of factors at the organismic level, or a denial of the need to consider the complex interactions among genes within genomes. Nowhere is it suggested, for example, that genes have the effects they do on embryological development *in isolation*, either from other genes in the organism's genome or from the external environment; nor is there any absurd commitment to a one-to-one correspondence between genes and

---

[3] See, for example, Dawkins (1982), and Kitcher and Sterelny (1988).

[4] See Sober and Wilson (1998). Though this is perfectly compatible with genic selectionism, some of its implications tend to be overlooked by genic selectionists, with the resultant overlooking of the possibility of traits evolving due to beneficial group-level effects, despite their being less than optimal from the individual point of view, and hence inherently unstable when considered only *within* groups. Consideration of this possibility, in part, will lead (in section thirteen of chapter five) to an interesting qualification of the account of biological teleology given in chapter four.

complex adaptations.[5] Many of the criticisms of genic selectionism—particularly concerning the unhappy controversy over whether it is organisms or genes that are the true "units of selection"—have been grounded in misunderstanding over these and related points.

Indeed, much of the controversy can fairly easily be seen just to be semantic rather than factual. Gould, for example, insists that the individual organism is the "unit of selection", and criticizes Dawkins for holding the view that it is genes that are the "units of selection".[6] But a simple consideration of what each *means* by "units of selection" reveals that they are talking about two quite different things. Dawkins wishes to use that expression—as a matter of stipulation—to refer to the replicating entities that can potentially increase their frequency, in the form of copies, over the generations as a result of natural selection—as everyone agrees happens to genes within a gene pool over evolutionary time. (This will be described in detail below.) Gould, by contrast, uses the expression to refer to the entities that must be taken into account in understanding the effects at any given time that are relevant to natural selection, and comes to the unsurprising conclusion that it is whole bodies, rather than isolated genes, that must be considered here; this is what he means when he says metaphorically that "selection views bodies", and Mayr says much the same thing in his discussion of "targets of selection".[7] Thus the stage is set for misunderstanding, which is exactly what happens. Gould proceeds as if Dawkins were suggesting that genes are the "units of selection" in *Gould's* sense (or the "targets of selection" in *Mayr's* sense). But Dawkins certainly never claims that the effects of genes on embryological development occur atomistically, independently of the context of the rest of the genome; nor does he think that the further evolutionarily relevant effects of these traits can be understood in isolation from the rest of the phenotype and the external environment. As Dawkins puts it: "*Of course* genes are not directly visible to selection. *Obviously* they are selected by virtue of their phenotypic effects, and certainly they can only be said to *have* phenotypic effects in concert with hundreds of other genes". There is no disagreement over

---

5 Dawkins (1982, pp. 116–17, 92–3) makes this clear in his response to criticisms by Gould (1980).

6 Gould (1980).

7 Mayr (1988, p. 101).

the facts here, but only a shift of focus to the perspective of genes over evolutionary time.[8] And far from denying the importance of organisms, this perspective on selection actually explains more precisely the crucial role that organisms *do* play in the process, and why it is, for example, that other organisms—such as close kin—should also sometimes be expected to play important roles (as I'll bring out below).

The claim, then, is just that the genic selectionist perspective provides the most general framework within which ultimately to understand the various processes of natural selection and their evolutionary significance; it is not a rival to theories that appeal to "individual selection", "kin selection", or "group selection", but just the most comprehensive and revealing framework within which to describe what is *ultimately* going on in all of these phenomena. And it should be no surprise that this perspective is privileged in this way, given the special role that genes play in neo-Darwinian evolutionary theory, as described below. I shall therefore examine natural selection from a genic selectionist perspective, bringing out the general and unified view of natural selection that emerges from this fundamental perspective. I will not bother here to rehearse the standard criticisms and defenses of genic selectionism. In my view, the case for its legitimacy—once the view has been properly understood—is compelling enough, and has commanded sufficient respect in biology, that I can justifiably make the philosophical points I wish to advance largely conditional upon it. The discussion of genic selectionism that follows is therefore not meant to be a defense of it, but primarily a careful description of it, which will help in bringing out the implications of the fact that organisms are products of natural selection for our understanding of their teleological nature.

3. THE GENETICALLY ORIENTED PRINCIPLES OF NATURAL SELECTION

From the perspective of genic selectionism, natural selection may be defined as the process occurring over multiple generations whereby: (i) given a certain amount of genetic variation within a population of organisms, (ii) genes of a certain type increase their frequency in the gene pool relative to other alleles, (iii) by exerting phenotypic effects in

---

[8] This point is emphasized by Kitcher and Sterelny (1988).

the organisms in which they find themselves, such that (iv) these effects ultimately serve to promote the propagation of genes of this type over the generations more effectively than the phenotypic effects exerted by the other alleles (in the organisms in which these other alleles find themselves) promote these other alleles' propagation. As these more "successful" genes thus *out*-propagate the other alleles and increase their frequency in the gene pool relative to the other alleles, this naturally results in (v) the concomitant spread of these traits (i.e. the phenotypic effects exerted by the more successful genes) over the others throughout the population.[9]

For natural selection to get going, there must be genetic variation in the sense that there must be some variation among the members of the population with regard to which alleles they carry at a given locus; this genetic variation must in turn result in some variation in the phenotypic traits manifested by various individuals, and finally this variation in traits must somehow or other tend to result ultimately in the differential propagation of the respective genes. Suppose, for example, that L is a genetic locus affecting some aspect T of the phenotype of

---

[9] *Alleles* (or allelomorphs) within a gene pool are the various genes of different types that occupy homologous loci of homologous chromosomes throughout the population or species. For example, the various alternative genes that can be found at a particular position on chromosome 7 in various human beings are alleles, constituting an allelic set of genes. Note that by "*genes*" I do not necessarily mean only cistrons (segments of chromosomal material that code for a single protein), but also, following Dawkins, linkage groups of cistrons, so long as the complex genetic unit is still small enough potentially to undergo differential replication relative to allelic complexes over many generations, thus potentially figuring in this way into the process of natural selection. (See Dawkins, 1976, pp. 30 f., who takes this definition from G.C. Williams.) Hence, there is no clear distinction between genes and gene complexes as I am using the terms. The *phenotypic effect* or *expression* of a gene is the effect its presence has—in the context of its environment, including the rest of the genome as well as the external environment—on the higher level traits of the organism (i.e. on its morphological, physiological or behavioral characteristics). Different alleles often have importantly different phenotypic expressions from one another—a fact that is crucial to the possibility of natural selection. Finally, to say that genes of a certain type *out-propagate* rival alleles at a given locus, or increase their relative frequency in the gene pool, is to say that they manage to get copies of themselves into relatively more members of subsequent generations than the other alleles do, so that the percentage of individuals carrying such a gene—and hence likewise bearing the trait that is the gene's phenotypic expression—increases relative to the percentage carrying other alleles and bearing those traits. (By contrast, we are not interested here in the sense in which genes of one type might be said to out-propagate genes of another type simply by being replicated more times *within* the bodies in which they are found, e.g. by tending to make their possessors fat. That kind of differential replication is not relevant as such to natural selection.)

organisms belonging to a certain population, and that A1 and A2 are allelic gene-types found at L; some members of the population carry copies of A1 at L, others carry copies of A2 at L. Suppose further that the presence of A1 at L tends—in the context of the other genes with which it is typically combined in a genome, and in the context of the standard environment—to make a certain difference T1 to the organism's phenotype, and likewise with A2 and T2; that is, the aspect of an organism's phenotype that is affected by L, namely T, tends to be modified in the direction of T1 if the organism contains A1, or alternatively in the direction of T2 if it contains A2. This may be put more succinctly by saying that A1 "codes for" phenotypic effect T1 while A2 "codes for" T2, or that A1 is a "gene for" T1 while A2 is a "gene for" T2—though it is important to be clear that the "for" here is *not* a functional teleological one (as in "for the sake of"), but just a causal one.[10] Now the differences that A1 and A2 make to the phenotype may be very small, but if they are significant enough that they can affect the alleles' respective propagation over the generations, then natural selection can occur. So, for example, even if T1 is a relatively minor variation in the phenotype of the organisms carrying copies of A1, there will be natural selection in favor of A1 *if* T1 proves ultimately to make a more effective contribution to the propagation of A1 over the generations than T2 makes to the propagation of A2—for example, by more effectively promoting its possessor's chances of attracting mates.

---

[10] Cf. Dawkins (1982, pp. 22f.) for a discussion of this notion of "genes [coding] *for*" a given trait. See also Kitcher and Sterelny (1988, p. 348 f.), who go on to provide a more technical and precise reconstruction of these notions, taking careful account of environmental sensitivity (involving both the complementary part of the genome and the external environment) and fleshing out the notion of a standard environment. The rough formulation above is sufficient for my illustrative purposes, however. Certainly if one were concerned to articulate a workable model for purposes of research, one would have to take careful account of the fact, for example, that genes sometimes code for very different traits in different genetic environments—e.g. coding for one trait when in the body of a male, and for another when in the body of a female—and work out the statistics to see exactly what particular hypotheses about the selection of certain genes are viable and which are not. Similarly with the fact that rival alleles will often be combined within a single organism (assuming we are dealing with diploid organisms), which adds complexity to the story, or the fact that most genes have more than just one effect on the organism's phenotype. These complications do not affect the basic points I shall be making, which involve general claims given certain statistical assumptions, and will thus be set aside.

That is, there will be natural selection in favor of A1 if the modification of this type of organism's phenotype by T1 ultimately does more to get copies of A1 passed on to subsequent generations than the modification by T2 does to get copies of A2 passed on to subsequent generations, so that in this sense A1 *out*-propagates A2, increasing its relative representation in the gene pool (meaning that a relatively greater percentage of organisms in the population carry copies of A1). T1 may help A1 to accomplish this in any of a variety of ways, though in a wide range of cases this will involve somehow contributing to the reproductive output of the organism in which it is manifested: If organisms carrying copies of A1 and hence exhibiting T1 tend to have more (viable) offspring than organisms carrying copies of A2 and exhibiting T2, then A1 will be passed on to the next generation, in the form of copies, with relatively greater frequency than A2, so that its representation in the gene pool will increase over the generations.[11]

The result at the phenotypic level is, of course, the familiar evolution of living things. As genes of a given type increase their relative frequency in the population *via* a certain type of effect they tend to exert on organisms, i.e. the trait that is the typical phenotypic expression of this type of gene, this trait likewise spreads throughout the population, increasing its incidence among members of the population relative to the traits that are the phenotypic expressions of the less successful alleles; eventually it may spread to such an extent as

---

[11] If trait T1 (which again, need be only a relatively minor *difference* made to the phenotype due to the presence of A1, against the background of the rest of the genome) tends to make the organism bearing it more reproductively successful, then of course *in any given case* it is not just A1, but also the rest of the genes in the genome of the organism, that will tend to be replicated into more offspring than they otherwise would. But due to the significant genetic variation among the genomes of individuals within a population (both at a time and over generations, due to recombination—at least in sexually reproducing species), it is only A1 that is *consistently* present in organisms with T1 (at least until A1 begins to become a standard part of the species-typical genome), thus *consistently* enjoying T1's replication-enhancing effects. This is why it is said that T1 promotes the propagation of A1 in particular, over evolutionary time, helping it to out-propagate rival alleles. (Cf. Dawkins 1987, p. 122.) Note also that it is not necessary that T1 have a positive effect on genetic replication (i.e. through increased reproduction) in every single case, or even that A1 give rise to T1 in every single case. Sometimes copies of A1 will be grouped with other genes that will result in early death or sterility, or that will change the phenotypic expression of A1 to something other than T1. What matters is just that *on average* copies of A1 will give rise to T1, which in turn will tend to increase the reproductive success of the organism more effectively than T2 does, resulting in A1's out-propagating A2. For a helpful analogy here, see Dawkins (1976, pp. 40 f.).

to become characteristic of the species, thus coming to be a species-typical adaptation. Thus, in sum: The traits we find as adaptations in current organisms are the ones that have been favored by natural selection, and traits are favored by natural selection for one fundamental reason: *They are ultimately more effective in promoting the propagation of the genes whose phenotypic expressions they are than the traits produced by the various other historical alleles were in promoting the alleles' propagation.*[12]

Consider a simple example to illustrate the point. The coloration pattern we find on the flower petals of a given species of plant, which is so marvelously effective at attracting pollinating insects, was incorporated into the "design" of current specimens for one fundamental reason: The genes contributing to this coloration pattern in this species of plant better promoted their own propagation—on average, at each stage of evolutionary development at which they were

---

[12] Three points should be noted here. First, as already stressed, it goes without saying that it is only in the context of their given natural environment that genes of a given type exert certain phenotypic effects and that these effects influence the propagation of these genes in certain ways; and the environment here obviously includes other genes in the gene pool, with which the genes in question are combined in the genomes of various individuals. If, for convenience, I do not always mention the importance of environmental factors, this should not be taken to imply the erroneous view that genes act in isolation, or are selected simply because of their intrinsic qualities. Some critics have wrongly (and rather surprisingly) attributed such a view to Dawkins. For a lucid defense against this and related criticisms, see Dawkins (1982) and Kitcher and Sterelny (1988). Secondly, it should be noted that while it is genes that are the primary "units of selection" in Dawkins' sense, i.e. the replicating items "selected" over evolutionary time over rival alleles, I have departed slightly from Dawkins' terminology in speaking likewise of the causally relevant phenotypic expressions of such genes as being "selected". I do not see any objection to speaking derivatively of these phenotypic traits as "selected" for inclusion within the characteristic phenotype (or range of phenotypes) of the species, since they are, after all, the effects that were causally relevant to the genes' "selection". I don't think Dawkins should object to this, and in fact he sometimes speaks this way himself (1987, p. 60). Finally, the above claim in the text should not be taken to imply either that every adaptation is *optimal*, or even that every adaptation is better disposed with respect to propagating the genes responsible for it than all of its historical rivals, all else being equal. As Kitcher (1993, p. 388) points out, natural selection works only on the actual variation provided by mutation and recombination, and there could very well be possible variations for more optimal traits that simply never arise, leaving a trait that is still far from optimal to be selected. Also, a generally less optimal trait may be selected in part because of some *chance* occurrence that eliminates a more optimal trait (e.g. a peculiarly selective disease happens to wipe out most of the organisms bearing the more optimal trait). Thus, we cannot assume that every adaptation is, all else being equal, better at propagating the genes whose phenotypic expressions they are than all of their historical rivals. These points should be kept in mind throughout.

present, and in the context of the rest of the genes in the gene pool, with which they were combined in various particular genomes—than did the various other alleles, such as genes tending to produce different colors. How did these genes' phenotypic contributions enable them to out-propagate rival alleles? In this case, it was because, all else being equal, specimens containing one of these genes, and hence exhibiting its usual phenotypic contribution to the coloration pattern (i.e. the contribution it tended to make in concert with the rest of the genes making up typical genomes), proved on average to be more effective at attracting pollinating insects than specimens containing one of the other alleles, which exhibited the allele's less effective phenotypic contribution to the coloration pattern. The former specimens, being more effective at attracting pollinators, all else being equal, enjoyed statistically *greater reproductive success*, thus passing on their genes—including the genes contributing to this effective coloration pattern—to relatively more offspring than did the latter specimens. The result is that with each generation the genes in question increased in frequency relative to rival alleles, as did the incidence of the sort of coloration pattern we now regard as a standard adaptation in the species.[13]

---

[13] The matter is somewhat more complicated in cases of frequency-dependent selection, where the effectiveness of a trait depends to some degree on its frequency in the population relative to alternative traits. Here, rather than one trait's eventually becoming uniform, the result of natural selection may be a stable balance of two or more alterative traits in the population (balanced polymorphism), or a balanced mixture of alternative traits in each individual in the population, or something in between. (A good example is the nest-digging vs. nest-adopting behavior of the solitary digger wasp, as described by Dawkins, 1982, pp. 122 f.) This complication does not alter the fundamental point, however, which is that the presence and balance of traits in a population, to the extent that this is due to natural selection, is to be explained in terms of their relative success in getting the responsible genes propagated in their natural environment—whether or not this success is frequency-dependent. Frequency-dependence is, after all, just a special case of dependence on environmental conditions, which is always a factor. Cf. also Griffiths (1993, p. 415). It should also be noted again that for the purposes of this discussion I am ignoring various genetic complications, such as the fact that in sexually reproducing species the genetic loci at which alleles "compete" are diploid (with the exception of e.g. hymenopteran males, which are produced asexually and are haploid), so that copies of competing alleles may often share bodies, and occasionally a gene that is successful when possessed heterozygously proves to be unsuccessful when possessed homozygously (as with the sickle cell gene). I take contemporary biologists at their word when they claim to have shown that such complications do not undermine the general theory. (See Grafen (1984), who cites, for example, the work of Maynard Smith in this connection.) In general, I should stress that the purpose of the examples offered in the text is to illustrate the essence of natural selection from the perspective of genic

Most cases of natural selection are like this one in that the genes in question out-replicate rival alleles over the generations, thus increasing their frequency in the gene pool, by exerting phenotypic effects that tend to make a more effective contribution to the eventual reproductive success *of the organisms bearing those phenotypic traits*. This might be a contribution to the organism's survival fitness, which thus indirectly contributes to its reproductive fitness (e.g. by increasing its chances of living long enough to reproduce at all, or by increasing its probable number of offspring through increasing its probable reproductively-active life span, etc.); or, as in the case considered above, it might be a direct contribution to the organism's reproductive fitness, having nothing to do with its survival. But there are also cases of natural selection, falling under the heading of "kin selection", where the genes in question increase their frequency in a less direct way. The same fundamental principles outlined above still apply, but the particular ways in which differential propagation is effected are more complicated than in the simple example given.

Generally, what happens here is that, as in other cases, genes of a given type exert phenotypic effects in the organisms in which they reside, but instead of these effects' promoting the genes' propagation by making some contribution to the reproductive success of the organisms exhibiting these phenotypic effects, they do so by making some contribution to the reproductive success of relatively close *kin* (as e.g. by doing something to help kin to survive). The reason why this works is that close kin are closely related genetically—full siblings in sexually reproducing species, for example, typically sharing 50% of their non-species-standard genes, which is far greater than what they share with non-kin. This means that any recent mutation, A1, whose copies are present in a given organism, O, has a relatively high chance of having copies present in this organism's close kin, K, as well—so long as the mutation traces back to a common ancestor (a common parent, grandparent, etc.); and that in turn means that if the phenotypic expression of A1 happens to be a "kin-altruistic" trait, such that the copies of A1 in O have the effect of tending to make O aid K in some way that is ultimately significant to K's reproductive success, then there is a good chance that in doing so the copies of A1 in O will be

selectionism, and *not* to provide fully sophisticated and articulated *models* of the sort that working biologists would need in order to do research.

promoting the replication of copies of A1 in K (and vice versa). More precisely, copies of A1 that are active in the relevant cells of O will exert phenotypic effects in O that in turn have an effect on K, such as ultimately to promote the replication of copies of A1 in K's gametes (and likewise if O and K are reversed). The effect of this will thus be to promote the propagation of A1 in the gene pool—just as in more familiar cases where the copies exerting the effects and the copies enjoying increased replication into the next generation are all in the same body, the former in the body's various somatic cells, the latter in its gametes. There is, therefore, the same general feedback whereby a gene of a given type promotes its own propagation through the phenotypic effects it exerts, but in this case the feedback loop goes from copies in the somatic cells of one organism, through their phenotypic expression in that organism, to effects in *another* organism, and then finally to the copies of the gene in the gametes of this latter organism, which tend to get replicated into the next generation more successfully than they otherwise would.

It is easy to see how a gene like A1 could out-propagate an allele that did not have this effect. Consider, for example, two possible alleles, A1 and A2, in a population of bees. A1 is present, due to recent mutation, in one colony (consisting of an already-mated queen, some reproductive sons and daughters, and a number of sterile daughter "workers"), while A2 is present, also due to recent mutation, in another. Suppose these alleles are such that they are active when in the bodies of sterile workers but "silent" when in the bodies of reproductives. The difference between the two is that A2 tends to make its worker-possessor refrain from stinging when it would normally be done to protect the hive, while A1 tends to make its worker-possessor especially effective at stinging when necessary. Now as Dawkins and others have pointed out, A2 would help workers carrying copies of it to *out-survive* their sisters who carry A1, but this has no significance in terms of natural selection. The workers are all sterile in any case, so their enhanced personal survival is in itself of no consequence to the spread of the types of genes they carry. On the other hand, A1, while it proves relatively detrimental to the survival of the workers who carry copies of it, has an effect that does indeed promote its propagation. In making the workers who possess copies of it protect the hive more effectively—and in particular, protect their *reproductive* siblings and queen, which is what really matters—they are contributing to the probable reproductive success of these reproductives; since these are close kin, they likely contain copies of A1 themselves (though these genes are not expressed in their phenotypes), with the result that these genes are thus passed on to future generations all the more effectively. That is, there will tend to be greater reproductive success among the reproductives of colonies where A1 is found than among the reproductives of the colonies where A2 is found, leading to the founding of more new colonies of the A1 type; A1 will thus spread throughout the population more effectively than A2, with the concomitant spread of the aggressive, self-sacrificing disposition among worker bees.[14]

It is worth pausing at this point, without going into the details, to note the distinction (which will come up again in later chapters) between the phenomenon of *kin altruism* resulting from kin selection—as just discussed—and superficially similar phenomena involving *reciprocal altruism*, manifested in cases of symbiosis or other "cooperative" relationships among non-kin. Cases of kin altruism involve behaviors that promote the reproductive success of kin even where this confers no advantage at all, in terms of reproductive output, on the individual itself. Again, such traits can evolve because of the close genetic ties among close kin, so that the behavior promotes the propagation of the genes responsible for such behavior even though it doesn't promote the reproductive success of the organism exhibiting it, as in the case of bee stings. By contrast, cases of superficially "altruistic" behavior in evolved *symbiotic* relationships among members of different species—e.g. ants' protection of aphids—must ultimately be understood to promote each organism's own reproductive success better than alternative "selfish" behaviors would. Symbiosis is just a result of ordinary natural selection occurring in an environment that includes other living things and their behavior. In such an environment, there can easily be selection pressures that favor certain forms of cooperation in non-zero-sum, iterated prisoner's dilemma type contexts—i.e. where the genes for each cooperative behavior would tend to be favored over rival alleles in their respective gene pools (provided there are also "enforcement mechanisms" to prevent exploitation), because organisms of each type who manifest such

[14] For a discussion of this and related types of case, see Dawkins (1976, ch. 10) and (1982, p. 85).

behavior in an environment that includes the other generally do better than conspecifics who display more selfish behavior and miss out on the benefits of the cooperation.[15] Similarly for intra-specific cooperation among non-kin (or cooperation among kin that is not attributable to kin selection): Such behaviors can evolve and remain stable so long as the circumstances are such that genes producing them tend to be favored by natural selection over rival alleles. That is, such behavioral traits can constitute "evolutionarily stable strategies" if, in the context of the whole environment, *which includes other members of the population and their behavior*, these traits tend best to promote the reproductive success of the organisms that exhibit them, providing advantages not available to non-cooperators—and continue to do so even when the trait becomes widespread.[16]

In summary, the general point that I wish to make about natural selection, following Dawkins, is that in whatever specific form it takes, *natural selection is driven ultimately by the effects genes have, through their phenotypic expressions in bodies, on their own propagation over generations*. Alleles—i.e. various alternative genes that can be present at a given locus in members of a given population—tend to have different effects, against the background of the rest of the typical genome, on the phenotypes of the organisms in question. Some of these traits come to be incorporated into the standard phenotypic "design" of a species of organism by virtue of the fact that in the context of the rest of the genome and its phenotype (as well as the external environment) these traits promote the propagation of the genes that code for them more effectively than other traits promote the propagation of the other alleles that code for those traits; this is how species evolve through natural selection—how the various traits we find in organisms have come to be "selected" and bundled together into coherent organic systems, while countless other traits have been "rejected" along the way. I have already mentioned some very different general ways in which a gene can promote its own propagation through its phenotypic expression, and of course at a more particular level there is virtually no limit to the intricate ways in which this is accomplished.

---

[15] Symbiosis is further discussed in chapter five, section thirteen.

[16] For a lucid discussion of both kin altruism and evolutionarily stable strategies (or mixes of strategies) involving "reciprocal altruism"—and particularly the game-theoretic approach emphasized in the work of Axelrod and Hamilton—see Dawkins (1976, ch. 10) and (1989, ch. 12). See also chapter ten, section one.

I shall consider a number of other examples below, both to illustrate the wide variety of phenomena captured by the genic selectionist framework, and to provide a useful stock of examples to have in mind later, when we return to thinking explicitly about biological teleology. Before going into them, however, it is worth pausing to consider how the above view of natural selection bears on one of the central concepts brought into play in chapter one: namely, that of organismic welfare. This is not yet to ask explicitly what role welfare plays in biological teleology, which will have to wait until the account is given in the next chapter.[17] But inasmuch as natural selection history has been argued to be relevant to biological teleology, and one of the central questions I have raised is whether a welfare-oriented account of biological teleology is tenable, it makes sense to consider in our discussion of natural selection how organismic welfare enters into it—and how it does not. This will make it much easier later to see why, as I shall claim, it cannot play the role of general or ultimate end within an organism's teleological structure.

4. ORGANISMIC NEEDS AND WELFARE

From what has already been said about natural selection, it should already be clear that there is simply no reason to suppose that the traits that prove best at getting the genes that code for them propagated over the generations—and hence wind up being incorporated into the "designs" of organisms—will necessarily be the traits that best promote the *good* of organisms, securing the satisfaction of *their needs*. Indeed, we should expect this not to be the case wherever the most effective replicating "strategies" of genes diverge from what is most beneficial to organisms.[18] But in order to make this point satisfactorily it is necessary to give some attention to the concepts of *need* (particularly species-typical needs, which is what is relevant here, as opposed to idiosyncratic needs in special circumstances) and *welfare*.

---

[17] The discussion of the role of welfare in biological teleology, according to my account, is in section six of chapter four. Independent arguments against welfare-based views of biological teleology are given in chapter six.

[18] Again, I owe much here to the work of Dawkins (1976, 1982, 1987), who has argued that traits are naturally selected and become "adaptations" *not* generally or ultimately because they are good for the species, or good for groups of organisms, or even good for individual organisms, but because of their consequences for the propagation of the genes whose phenotypic expressions they are, as explained above.

We attribute species-typical biological needs to plants and animals in connection with both their survival and their reproduction. (I will set aside the human case for now.) By this I do not mean that we attribute to them the need to survive and the need to reproduce; such judgments on the face of it would lack clear sense.[19] The idea is rather that we attribute various needs to them in connection with their species-typical modes of survival and reproduction. Various things are said to be needed by organisms of a given kind insofar as they play important causal roles in the characteristic patterns of survival and reproduction of such organisms. Often when we say that Fs need X, this is clearly equivalent to saying that any F (*as* an F) typically needs X, and the need is straightforwardly related to the individual's survival or reproduction. This is the case, for example, with the claim that human beings need vitamin D for the absorption and metabolism of calcium; it would equally be the case with a claim that some nutrient is needed in certain amounts for the proper formation of certain reproductive organs or for the production of hormones governing sex drive.

In other cases, however, the situation is more complicated. Honeybees need stings, we say, for the protection of the hive; but in light of the fact that the act of stinging is typically fatal to the stinging bee, it may seem less clear that we can say of any particular worker bee that *she* needs her sting.[20] The difficulty here is that the sting does not aid either the survival or the reproduction of its possessor (who is sterile in any case), so that something seems odd about the claim that the individual needs the sting; after all, wouldn't *she* ultimately be better off without it? On the other hand, we have no trouble in saying that the individual needs her sting for protecting the hive: For she can't protect effectively without the sting, and protecting the hive is plausibly part of her proper functioning (an example of a behavior function), the function of the sting being to aid in that activity.[21] Similarly, we would not hesitate to say in relation to this that a worker honeybee needs such

---

[19] Though they could be given sense by relating the need for survival to reproduction or the need for reproduction to survival, as in: "Fs need to survive for at least three months in order to reach maturity and reproduce," or "Fs need to reproduce (at such and such a rate, etc.) if the species is to survive."

[20] Foot makes use of this example, which is taken from Geach (1977, p. 17).

[21] I will say more about function in relation to this case later, once an account of function has been given. For the moment I shall simply assume that these are plausible function attributions, the point being that if they are correct, then so are the attributions of need.

and such nutrients in her diet, on the grounds that they are crucial to the manufacture of the sting. If it is then asked whether the individual needs the hive protected even at the cost of her own life, this should probably be dismissed as a misguided question. The point is that given the nature and organization of honeybee life, the protection of the hive is important to the survival of the members of the colony generally and crucial to the reproduction of the queen; the protection of the hive is thus something that is needed in relation to characteristic honeybee survival and reproduction. That would seem to be enough to ground the general claim that the stings are needed, i.e. *they* need stings. And if it is further true that the stings have a crucial *function* in aiding the worker's protection of the hive—however exactly this fact may or may not be related to the general fact about need (which will come out later)—then we can say that *she* needs the sting in order to protect the hive.

We must proceed carefully here, however, especially when we begin to consider the relation of the notion of *need* to the notion of organisms' *welfare* or *good*. There appear actually to be two different applications of the concept of need in the above discussion. When we say generally that honeybees need stings for their protection, the idea is indeed that the stings are important to their welfare, though in the complex, general way noted above: *They* (generally) need stings if they are to flourish under normal conditions; without them honeybees would not *get on* as well as they characteristically do. But the notion of need that is employed in the claim that the individual needs her sting for protecting the hive is different; it is so far simply relative to the individual's proper functioning: She needs her sting if she is to carry out the full range of her proper functioning, which includes protecting the hive. (Again, I am just assuming for the moment that once we have an account of biological function—such as I will provide in chapter four—these plausible functional claims will indeed turn out to be true.) It is therefore so far an open question whether *this* notion of need is generally connected to welfare. It would be if functions were generally related to welfare, so that things needed in connection with functions would thus be needed for the promotion of welfare, whether that of the individual or that of the group (as would be the case here). But that is precisely what is at issue at this point, and cannot just be assumed to be the case. It is possible that functions are to be understood in relation to something other than the promotion of welfare, and if that is indeed the case, then attributions of need that are made in relation to function—as in the case of the worker bee's needing her sting, or the nutrients necessary to manufacture it—are not necessarily related to welfare.

If it sounds implausible that there could be function-related needs that don't connect up with the promotion of welfare (i.e. needs whose satisfaction doesn't confer any benefit), recall that we speak of such needs all the time in relation to artifacts. A gasoline engine *needs* oil, or *needs* a carburetor, for example: It *needs* these things in order to function properly in certain respects; but there is no such thing as the *welfare* of gasoline engines. It is true that certain things are *good for* engines and others *bad for* them, but this again is to be understood simply in relation to their proper functioning. Something is bad for an engine because it in some way makes it less well disposed with respect to carrying out its proper function, not because it makes it less *well off*. Engines do not flourish or languish, so the things they need in this function-relative sense are obviously not things that contribute to their flourishing. Thus, we already have a familiar use of the concept of need that is related to function and is not essentially tied to the notion of welfare, and there is no obvious reason why that same concept of need should not be found also in connection with functional systems that *do* flourish or languish, such as honeybees; that is, it is possible that *both* a non-welfare-related notion of need (connected to function) *and* a welfare-related notion of need have application to living things.

The point, then, is that when we apply this function-relative notion of need—i.e. saying that something is needed for proper functioning—we cannot just assume that it ultimately connects up with the *welfare* or *good* of the organisms in question (or of any other organisms); that will depend on whether biological function essentially connects up with the welfare of organisms, which is precisely what I wish to question (and what I shall eventually deny). Thus, the attribution of need in claims such as "the individual worker bee needs her sting" cannot be assumed to connect up with honeybee welfare; even if in this case there happens to be a plausible connection to honeybee welfare through the protection of the hive, which bees may be said to need in the welfare-related sense, the possibility remains that there is no essential connection here, and that in some cases an item may be said to be needed by an individual in relation to its proper functioning (on the model of the engine and the oil), even though the item does not tend to promote any organism's welfare.[22] What this means is that at this stage we must at least be aware of the possibility that not all needs in the context of biology are to be understood in terms of welfare, whether that of the individual or that of the group or species.

I bring up these complications only to clarify what it is that I wish to deny—not yet in connection with functional teleology, but first just in connection with natural selection. My claim, again, was that there is no reason to expect that adaptations will generally promote the *good* of organisms, or satisfy *their needs*. We can now clarify this by noting that the needs spoken of here are what I have been calling *welfare-related* needs—that is, needs that are understood in connection to the welfare of organisms, such that the things needed in some way promote their welfare, and they *benefit* from the satisfaction of these needs. Again, some might think that all needs in biology, including all function-related needs, *are* welfare-related; I shall eventually argue that this is not the case, but in any event it is *welfare-related* needs that I am interested in at the moment, the claim in question being that adaptations cannot be expected generally to be geared toward the satisfaction of the welfare-related needs of organisms, or (equivalently) the promotion of their good.

I think we already have a fairly good intuitive grasp of what sorts of things generally count as welfare-related needs, but it is worth making a few observations about this. Welfare-based needs are attributed both generally (honeybees need to protect their hives from other animals, and for this they—in some general sense—need stings) and individually (human beings need such and such amounts of vitamin D). They are also attributed both in relation to survival and in relation to reproduction; and when we speak of survival and reproduction here, what is meant is not just barely warding off death and managing to leave progeny, but the manner of survival and reproduction characteristic of the species, at least insofar as this ties into welfare.[23]

---

[22] Examples will be given below.

[23] That is, in my terms, it appears that not only needs pertaining to survival, but also at least some needs pertaining to reproduction count as *welfare-related* needs. (Cf. Foot, 1978b, p. 38.) The above qualification "at least insofar as this [characteristic survival and reproduction] ties into welfare" is necessary because there are some aspects of reproduction that are characteristic of a species but that are not plausibly related to welfare, and so do not ground welfare-based needs, as I go on to argue below in connection with e.g. birds of paradise and elephant seals.

Thus, for example, a certain nutrient may be needed by plants of a certain kind—in the welfare-related sense—either because it is necessary for proper metabolism or because it is necessary for the proper development of reproductive organs; in either case, we may speak of the nutrient as being *beneficial* to such plants. In both cases we can certainly attribute the need in a general way, saying that *they* need this nutrient (as bees need stings); without it, they could not get on as well as they do. In the former case, it seems clear enough that this need can also be attributed directly to individuals: An individual plant is benefitted by the presence of this nutrient. This is plausibly true also in the latter case, but it is less clear. The presence of the nutrient is necessary to the individual's overall health, which includes its reproductive capacities, and thus seems connected to its welfare; but there is room to question how tight the connection is between reproductive-health and individual welfare.

Consider, for example, a hypothetical species of plant that typically diverts a significant portion of its resources to reproduction, leaving an individual more vulnerable to various diseases or environmental changes than it would be if it didn't reproduce. An individual that for some reason did not divert these resources to reproduction (perhaps because its reproductive system never developed properly) would plausibly be defective or malfunctioning in certain respects, but it is at least not obvious that its overall welfare would be decreased; it might, after all, tend to weather various diseases better and to live longer than the others. The matter is still less clear in the case of animals. A cat that is sterilized, for example, is certainly harmed inasmuch as its reproductive functioning is impaired; similarly if its mammary glands are damaged. But it is not obvious that this sort of harm necessarily amounts to a decrease in individual welfare; in fact, castrated male cats evidently tend to live longer than intact ones, due to the absence of the male sex hormone, and this seems at least to count in favor of their being *better* off.[24] If it seems that they may nonetheless be *worse* off, this might just be because we suspect that the damage leaves the animal frustrated, depressed or uncomfortable, and these things can plausibly outweigh the advantages of increased longevity. But the issue is whether there is decreased welfare *simply by virtue of diminished reproductive functioning*, and it is this that is not clear. If a castrated

---

[24] See J.Q. Wilson (1993, p. 168).

male cat is not frustrated, depressed or uncomfortable, and lives longer than he otherwise would have, it is hardly obvious that we should regard him as nonetheless worse off—or even that the diminished reproductive functioning should count significantly toward making him so.

These are difficult issues, and I admit to being unclear about the precise relations between reproduction and welfare. Fortunately, however, my argument does not require a complete account of such relations. I will concede for the sake of argument that there are welfare-related needs in connection with both survival and reproduction, and that in at least some cases such needs may be attributed even in relation to individual reproduction. All that I have to deny for my purposes here is that *everything* that plays a role in the sort of reproduction that is characteristic of a species can thereby be said to promote the welfare of such organisms, and hence to be needed in the welfare-related sense. Consider, for example, a trait that is perhaps not even essential to successful reproduction, but serves only to increase personal reproductive output, and at the cost of detracting dramatically from the organism's own survival prospects or of increasing its risk of serious injury. If it turns out that this is all part of such an organism's proper functioning, once that notion is understood, then we might attribute *function-related* needs here (e.g. needs for nutrients necessary for the development of the trait in question). But why should we think that the satisfaction of such function-related needs must automatically amount to benefitting or promoting the *good* of the organism (or of any other organism for that matter)—that is, that these needs are *welfare-related* needs?

As I have already emphasized, the traits that are "selected" for incorporation into the "designs" of organisms and become characteristic adaptations—such as the reproduction-enhancing trait posited above—are simply those which prove best at getting the genes that code for them propagated over the generations; and there is no necessary connection between this ability and promoting the good of organisms—unless, of course, one just *defines* an organism's good in such (ultimately genetic) terms. But what reason is there for thinking that the notion of an organism's welfare or well-being is reducible to such genetic considerations in this way, especially in the case of sentient animals? Such a stipulation would be a radical departure from intuitive notions of organismic welfare or well-being, even when we grant as part of the latter that there are *some* welfare-related needs associated with reproduction. In the absence, then, of some good motivation for such a move, we should retain our ordinary, if somewhat fuzzy, conception of organismic welfare, and recognize that given the nature of natural selection, there is at any rate no reason to suppose that adaptations will generally promote it.

What misleads us into thinking otherwise is no doubt the tendency to focus on a certain broad range of cases where the traits that successfully promote gene-propagation equally promote welfare. For example, a trait might contribute to the propagation of the genes for that trait by serving to camouflage the organism against predators. In helping to protect the organism from attack, the trait thereby increases the organism's chances of surviving, and hence also of reproducing, thus doing more ultimately to propagate the genes for the trait than alternative traits lacking this effect do to propagate their genes, all else being equal. Here, the trait ultimately promotes the propagation of the genes for the trait by way of a proximate effect that does indeed benefit the organism, namely concealing it from predators. Concealment from predators is among the welfare-related *needs* of such organisms, and this need is met through the camouflage provided by the trait.

Similarly in the case of flower petal coloration already discussed in relation to natural selection: This adaptation has the proximate effect of attracting pollinators, which is how it ultimately contributes to the propagation of the genes that code for it; and this effect does indeed satisfy one of the needs of such plants: They *need* to attract some kind of pollinator in order to reproduce, and in this species of plant that need is met through a certain petal coloration that attracts a suitable insect. Again, this seems to be not only a matter of purely function-related individual need (the individual needs to attract a pollinator if it is to carry out the full range of its functioning, assuming that reproduction is part of its proper functioning) but also plausibly a matter of welfare-related need, perhaps individually and in any case certainly in a more general way. Such coloration is beneficial to such plants, making a crucial contribution to their flourishing; without such coloration, they could not get on nearly as well as they do, and might even die out altogether.

What we must recognize, however, is that even in these cases—where the means by which the trait promotes the propagation of the genes that code for it are perfectly aligned with the satisfaction of organisms' welfare-related needs—the relevant factor in the trait's coming to be included in the integrated repertoire of traits through natural selection is *not its benefit as such* to organisms, but simply its genetic consequences. The benefit is relevant, if at all, only insofar as it happens to be the case that the *means* by which the trait promotes the propagation of the relevant genes (i.e. the more proximate effects it has that promote the propagation of its genes) happen to satisfy certain welfare-related needs of the organism—something which is not always the case, as we shall see presently. Thus, even where naturally selected traits do straightforwardly serve the welfare-related needs of organisms, they are "selected" for inclusion in the phenotypes of such organisms simply because of their gene-propagating abilities. If the two had diverged, so that the traits had the same gene-propagating abilities though they did not serve the needs of organisms—neither the needs of the organisms bearing the traits nor the needs of any other organism on earth—they would just as surely have been "selected" for inclusion in the "design" of the species.[25]

The biological world is full of examples of such divergence—many of which are far from being fringe cases—and this is exactly what we should expect if the biological world is indeed a Darwinian one, where the force that shapes species is not a concern for the well-being of organisms, but simply allelic competition. Let me turn now to a consideration of some of these cases to help illustrate and reinforce the point.

5. ADAPTATIONS, GENES AND WELFARE: SOME ILLUMINATING CASES

Consider, for example, cases of what Darwin called "sexual selection". The peculiarly long tail of male birds of paradise, for example, is an encumbrance that almost certainly decreases personal survival-fitness. But it also happens to be very attractive to females, who prefer longer-tailed males as mates.[26] And this is sufficient, with a few additional

[25] I will often for convenience speak simply of the needs of organisms, as above, meaning welfare-related needs, except when otherwise indicated.

[26] We need not be concerned here with the question of *why* exactly female birds of paradise are preferentially attracted by the longer tails. In some species, such "preference" for a given trait amounts simply to the fact that the trait in question provides a more powerful "sexual sign stimulus", serving more effectively to "release" patterns of sexual behavior in females. This is evidently the case, for example, with the oversized

assumptions, to explain the evolution of progressively longer tails, at least up to a point. The extremely long tail has become incorporated into the organisms' "design" for the simple reason that tails up to that length prove so effective at attracting mates that they tend to contribute more to net reproductive output than they subtract from it due to making the possessor less likely to survive long enough to reproduce; males with longer-than-average tails have tended to enjoy greater reproductive output than the others, and that greater reproductive output naturally leads to the spread of the relevant genes, with the process continuing until the average tail length reaches the point where further increase ceases to have an overall reproductive advantage (because the costs in terms of encumbrance or wasted energy now outweigh the advantage of being more attractive to females). Note that it is not essential to the selection of such a trait that it be absolutely necessary to the male's ability to attract a mate and reproduce. It may well be that the longer tail length simply enables him to attract *more* mates than he would otherwise attract, thus spreading his genes—including the gene for longer tail length—more effectively.[27]

Now it is hard to see how such a trait can literally be said to *benefit* its possessor simply because it tends to *increase his reproductive output*—especially since it tends at the same time to threaten his very survival. As I have already made clear, I do not wish to deny that an organism can be benefitted or harmed with respect to its reproductive functioning as well as with respect to its personal survival. But it is implausible to insist that something that merely tends to increase probable reproductive output, so that an organism tends to have more offspring in the long run than it otherwise would have, or more offspring than "rival" males in the population with shorter tails, can

---

claw of the fiddler crab. (See Gould, 1982, p. 41.) In other cases, something more complicated may be going on, whereby an initial female preference for slightly longer-than-average tails (which may just be arbitrary) itself co-evolves with male tail length in a runaway positive feedback loop. (See next footnote.) For my purposes, I will just take it as a given that the females do prefer the longer-tailed males as mates, and consider the consequences for the evolution of male tail length.

[27] The above is a greatly simplified account of a complicated process that, according to Dawkins (1987, ch. 8), plausibly involves a linkage disequilibrium among genes for preferring tails of a certain length and genes for the corresponding tails lengths (both of which tend to be present in both males and females, though only the former are expressed in females, and only the latter are expressed in males), and resulting positive feedback whereby female preference for longer tails and male production of longer tails advance together in a way that can explain the evolution of such dramatic traits.

automatically be said literally to be beneficial to the organism—especially when it is at the expense of personal survival-fitness. It certainly seems wrong to say that attracting more mates than would otherwise be attracted is something an individual *needs*—in the sense of welfare-related needs, i.e. needs the satisfaction of which promotes genuine welfare—simply on the grounds that it contributes to his tendency to have more offspring than shorter-tailed "rivals" in the population. Likewise, it seems wrong to say that the added tail length is something the individual *needs*, in the sense of welfare-related needs, simply on the grounds that it enables him to attract relatively more or healthier mates. And it is no more plausible to suppose that the long tails are somehow *needed* in a more general way by birds of paradise—that it is something "they" need, in the way that bees need stings, connecting up ultimately with their good. The good of organisms, whether considered individually or generally, simply doesn't have anything to do with such traits. They become part of the "design" of the organism (in this case, male birds of paradise) simply because of their genetic effects in comparison with the genetic effects of traits produced by alleles, in spite of the fact that they do not serve the needs of organisms.[28]

It is easy to find other examples to underscore the point that adaptations often cannot be construed as contributing to the good of organisms. Male elephant seals, for example, do not go about the business of reproduction in the way that one would expect if they had been designed by a benevolent creator interested primarily in the welfare of elephant seals. They are found not merely pursuing mates in a way necessary to keep the species going (meeting what we might call the survival and reproductive needs of the species), but expending a great deal of energy fighting with one another in an attempt to win *exclusive* control of a harem—hell-bent, as it were, on *out*-reproducing their peers, even at the risk of significant personal injury. Nor do females settle for reproducing with just anyone. Subordinate males who try to mate with them are typically rebuffed in favor of the dominant

---

[28] If it turns out that the added tail length is nonetheless to be regarded as a *functional* trait, as it is on the view I will present in chapter four, then it may be said that the individual *needs* the extra tail length in relation to its proper functioning. But this attribution of need is so far merely an attribution of function-related need, on the order of an engine's need for a carburetor, and does not imply that the satisfaction of such a need promotes the *good* of this or of any other organism.

male.[29] It may also happen that resources are distributed unevenly to offspring, with a son receiving significantly more than any one daughter—not simply because he requires more to survive, being larger, but because the extra nourishment puts him in a better position someday to become an alpha male himself and dramatically *out*-reproduce his peers.[30]

All of these behaviors are adaptations which make perfectly good evolutionary sense. It is not difficult to explain why genes tending to produce such behaviors (i.e. fighting, rebuffing, and uneven resource-allocation), once they arose through random mutation, would have out-propagated rival alleles, thus coming to dominate in the gene pool, resulting in such characteristic traits. But it is hard to see how such traits could be construed in terms of meeting the *needs* or promoting the *welfare* of elephant seals, either individually or in a more general way. Can we really suppose that animals fighting desperately with their peers simply in order to out-reproduce them are thereby acting "for their good", or for any organism's good for that matter, making themselves or others better off? On the contrary, it seems that Elephant seals could at least in principle *get on* just as well without these traits—perhaps even better, expending less energy fighting, avoiding the inevitable injuries, and so on. (Though it may be true that in the long run the species would deteriorate, as weaker genes were passed on with greater frequency.) That is simply beside the point, however, since natural selection does not bother to consult facts about the welfare of organisms in "selecting" traits for inclusion in their modes of life.

Similar observations can be made in a variety of other cases. In what has been called the "Bruce effect", for example, a male mouse secretes a chemical that can make females who are pregnant by another male abort, freeing up resources for the immediate and optimal production of his own offspring.[31] It is easy to see why a gene for such a trait would tend to spread in the population over rival alleles, leading to the incorporation of this trait into the phenotypes of male mice throughout the population. Males possessing the gene, and hence the trait, would have a reproductive advantage in at least certain situations over males who did not; and this means that all else being equal, this

---

[29] Cf. Dawkins (1976, pp. 169, 173).
[30] Cf. Dawkins (1976, p. 156).
[31] Dawkins (1976, p. 159).

gene would be passed on to subsequent generations with greater frequency than rival alleles, until eventually the trait becomes largely characteristic of male mice. None of this, of course, has anything to do with the good of mice or the meeting of mouse needs, nor should we be surprised that it doesn't. All that matters for the trait's becoming part of the male mouse's phenotype is that it is especially effective at propagating the genes that code for it, which it does by contributing to increased reproductive output—however that may happen to affect mouse welfare.

The same point holds with respect to the more familiar phenomenon of male "philandering". Male snow geese, for example, remain to guard their mates when the latter are fertile; but when the females are no longer laying eggs, the males take the opportunity to go to neighboring nests in search of fertile females to impregnate.[32] Again, such behavior tends to maximize the male snow goose's reproductive output, and this is why it has come to be part of their behavioral repertoire: Genes for such behavior naturally out-propagated alleles for "faithful" (and therefore less reproductively bountiful) behavior. But I can see no reason at all for regarding such behavior as in any way thereby promoting the *good* of snow geese—whether of the philandering males themselves or of any others. Indeed, to speak of benefit and need here, simply in connection with increased reproductive output relative to other males in the population, would stretch the notions beyond all plausibility.

So far I have considered adaptations that contribute in some way to an organism's reproductive output, though not in a way that plausibly relates to the satisfaction of organisms' needs or the promotion of their welfare. But there are also cases of adaptations that do not even contribute to increasing net reproductive output—either of an individual or of a colony. I'll mention two. The first involves a battle between the queen and sterile female workers in a colony of ants over

---

[32] Diamond (1992, p. 93). Another, rather extreme, example of male philandering among birds involves superb fairy-wrens. In one study, it was found that fully three quarters of the two hundred nestlings tested were fathered by non-resident males. Females were also found, as might be expected, to show selective receptivity to invading males, cooperating with those displaying recognizable indications of greater fitness, while rebuffing others. See Mulder (1994, p. 61).

the sex-ratio among the queen's reproductive offspring.[33] In this struggle, sterile female workers will contrive to manipulate this sex-ratio in such a way as to bias it 3 to 1 in favor of females—directly thwarting the queen's own efforts to maintain an equal sex-ratio among her reproductive offspring. The behavior involved here might involve anything from selective feeding/starving of female and male larvae to possible pheromonal manipulation of the queen's reproductive system (influencing the ratio of eggs she allows to be fertilized to those that are left unfertilized so as to develop into males). Such behavior is an adaptation in the sense defined earlier, existing ultimately for the same reasons as any other naturally selected trait: Genes tending to make worker ants exhibit female-biased sex-ratio manipulation behavior are more effective at propagating themselves than are alleles for alternative traits.

The explanation for this turns on the genetic peculiarities of the Hymenoptera (a group of insects that includes bees, ants and wasps). Among Hymenoptera, female siblings turn out to be three times more closely related to each other genetically than they are to male siblings, though the queen is equally related genetically to both female and male offspring.[34] This affects the ways in which genes promoting various

---

33 See Dawkins (1982, pp. 74–8) and (1976, pp. 189 f.). This is evidently still to some extent a matter of speculation, and it is uncertain to what extent the purported manipulation actually occurs, but this is beside the present point: It is precisely the sort of thing that might be expected to evolve by natural selection, whether it has actually been confirmed or not, and thus can be useful in helping to illustrate the nature of adaptations.

34 See Dawkins (1976, pp. 188 f.). This ratio of genetic relatedness pertains to the genes over and above the high baseline percentage of genes shared by all members of the species; if the entire genome were considered, the difference would of course be much smaller. The reason why Hymenopteran sisters are more closely related to each other than they are to their brothers is to be found in the haplodiploid genetic system characteristic of the Hymenoptera. Males develop from unfertilized eggs and are hence fatherless and haploid, inheriting all of their genes from their mothers (though each has only half of his mother's genes), whereas females develop from fertilized eggs and are therefore diploid, inheriting half of their genes from their mothers and half of their genes from their fathers. Consider, then, a given genetic locus L in two daughters, D1 and D2. In each daughter, there will be a pair of genes at L, one on each of a pair of chromosomes; one gene will have been inherited from the mother, the other from the father. Now the gene in D1 that was inherited from the father has a 100% chance of being present also in D2, since the father, being haploid, gives *all* of his genes to each offspring (putting the complete set in each sperm), and D2 has his singular contribution to L as well. The gene in D1 that was inherited from the mother, however, has only a 50% chance of being present in D2, since the mother is diploid, and gives only half of her genes to each offspring (putting only half in each egg, through meiosis)—with different combinations each time; that is, there is

kinds of sex-ratio manipulation behavior would propagate through the population over generations. In particular, genes in female workers that tend to make them behave in some way that leads to a female-biased sex-ratio among their reproductive siblings would be favored by natural selection. For example, genes in the bodies of female workers that made them give preference in the allocation of food resources to their reproductive sisters over their reproductive brothers would enjoy an advantage, since there is a far greater chance that genes of the same type are present also in the bodies of sisters than there is that they are present in the bodies of brothers. These genes enjoy an advantage in terms of propagation because in making their possessor favor other highly likely possessors of the same (types of) genes, who will therefore go on to survive and reproduce on average more effectively than their rivals who were not favored, genes of this type will naturally tend to out-propagate alleles for other behaviors, all else being equal.[35]

Now this case is interesting because the trait in question obviously does not in any way contribute to the workers' own reproductive success (which is zero in any case), and does not even necessarily contribute to the numerical reproductive success of their queen. The result of the manipulative behavior is simply to change the sex-ratio among the reproductive offspring of the queen, and this does not have anything to do with the *good* of ants—either individually or more generally. It cannot be said that the manipulation in any way benefits the workers who perform it; how is their good possibly served by changing the sex-ratio among their reproductive siblings? Nor can it be said that such behavior is somehow important to the flourishing of ants generally, as if they somehow couldn't get on as well without it. In the case of differential feeding of larvae, it may be true that female larvae tend to benefit; but it is equally true that male larvae are harmed as they

---

only a 50% chance that the gene the mother contributed to D1 at L was also among the half of her genome she contributed to D2. Thus, on average, there is a 75% chance that any given gene in D1 will also be present in D2; sisters share *all* of the genes inherited from their fathers, and *half* of the genes inherited from their mothers, meaning that they share 75% of their genes. By contrast, a female shares only 25% of her genes with a brother. This is because *none* of the genes she inherited from her father will be present in her brother, unless for independent reasons (since he doesn't inherit anything from her father); and as far as the 50% of the genes she inherited from her mother are concerned, there is only a 50% chance in each case that they are also present in her brother, since she and her brother each contain only half of the mother's genes, randomly distributed.

[35] At strange as this case may seem, it is really just a special and more restricted case of the familiar phenomenon of "kin selection".

are starved by the diversion of resources to their sisters, and the tradeoff here has nothing to do with a net benefit for anyone or for the species, but is simply a result of genetic factors. And in any case, if the manipulative behavior is of the other sort, involving some kind of pheromonal influence on the queen, tending to make her produce more females than males, no organism is benefitted at all, and it is perfectly clear that this naturally selected behavioral trait was selected and exists as part of the repertoire of worker ants not because it serves the good of organisms, but simply because of its ultimate genetic consequences.

Finally, it is worth mentioning a case of an adaptation that not only does not promote the survival or reproduction of organisms, and does not plausibly promote their good in any way, but that usually harms them significantly. It might be objected that this sort of case distorts the sense of "adaptation" as it is commonly understood, but this only points to the need for another term that does not reflect an unwarranted preoccupation with the organismic level of biology in thinking about natural selection. We have already seen that what is essential to adaptations in the sense of naturally selected traits is their ultimate consequences for the propagation of the genes that code for them, and not their consequences for the welfare, the survival, or even necessarily the net reproductive output of organisms, whether individually or even "inclusively" (i.e. including kin). The following case illustrates the point even more dramatically. It involves a "renegade gene"—a gene that promotes its own propagation in a way that is independent of the propagation of the rest of the genes in the genome of the organism, generally tending to hinder it.

Most active genes promote their own propagation by being such that their copies make contributions to the reproductive success either of the organism in which they reside, or of their kin that are likely to carry other copies of the same genes; either way, all of the genes represented in the genome tend to "benefit" equally in terms of propagation, since they all have an equal chance of being copied into the gametes or of having copies present in the kin who are aided, which would then be copied and passed on through their gametes, etc. (Note that the use of the term "benefit" here is, of course, purely metaphorical. Genes are not the sorts of things that can literally be benefitted. I use the term here, as Dawkins does, only as a convenient shorthand, and trust that it will not be misleading. To say that something "benefits" a gene is just to say that it promotes its propagation.) Renegade genes, however, are different. For example, in a phenomenon known as meiotic drive, *segregation distorters* propagate themselves not by making a contribution to the organism's reproductive success, which would "benefit" the whole genome, but simply by manipulating meiosis in such a way as to increase their chances (which would normally be 50% in any given case) of winding up in gametes. Such a gene tends to spread very quickly through natural selection even if it has generally deleterious effects on the welfare of the organism, and even if it detracts from the propagation of the rest of the genes in the genome—so long as it doesn't prevent the organism from reproducing at all. This is because it has such an enormous advantage over rival alleles at the meiotic level. The organisms in which the alleles reside might reproduce with greater frequency, but due to normal meiosis the alleles will be present in only 50% of those offspring, whereas the segregation distorter could at least theoretically rig things so that it is present in every offspring produced by the organisms in which it resides.[36]

As I have already mentioned, there are important differences between this case and the others that have been considered. In particular, the phenotypic trait in question—whatever the mechanism is that distorts meiosis in favor of the genes that produce the mechanism—has the effect it has *not* by making a "cooperative" contribution to the rest of the organism's phenotype so as to help to bring about a result that helps all of the contributing genes to propagate (such as increased reproductive success), but simply by doing something that helps its own coding genes to propagate even at the expense of the rest of the genome. I will argue in chapters four and five that this makes an important difference as far as biological functionality in concerned: Although proponents of the etiological view are forced to maintain that such a trait has a biological function in the same sense that the heart does, I avoid this implausible result. But for the moment, the point is just that this sort of trait is as much a product of natural selection as any other, even though it does not promote the *good* of any organism whatever, and is the last thing that any organism *needs*.

The conclusion I wish to draw from what has been said in this chapter, along the lines of the thesis advocated by Dawkins, is therefore the following. Natural selection is a process that is governed ultimately

---

[36] See Dawkins (1989, ch. 13), (1982, ch. 8). An example is the *t allele* in mice.

by the genetic consequences of various phenotypic traits in their natural environments. That is, the traits that are "selected"—from among the various phenotypic traits that actually arise through genetic mutation and recombination—for inclusion in the characteristic phenotypes of organisms are ultimately those that *on average* prove best (against the background of the rest of the phenotype of the organism in each case) at promoting the replication of genes of the types that (in the context of the environment, including the rest of the genes in the gene pool, with which the genes in question are combined in the genomes of members of the population) cause those traits to be manifested. The traits we find as adaptations in current organisms are therefore simply the ones that have managed to propagate their coding genes more effectively than traits produced by historical alleles managed to propagate those alleles. Many of these "selected" traits also promote the good of organisms or satisfy their needs in various respects, but many do not.

We should not, then, expect organisms to exhibit the sets of traits they might exhibit were they biological artifacts of a benevolent supernatural designer, who gave them traits geared ultimately toward promoting their welfare or satisfying their needs, whether individually or as a species. What we should expect instead is to find traits that are geared in every case toward the propagation of their coding genes, doing the sorts of things—through various causal pathways—that would have helped them to propagate their genes more effectively than alternative traits propagated alleles in the bodies of other members of the species at various stages of evolutionary history; and all of this will be the case regardless of how such traits may or may not promote the welfare of the organisms that are the living conglomerations of such genes and traits. Naturally selected traits need not have "biological value" for their possessor, or for the group or species to which their possessor belongs; what matters to their having been naturally selected and integrated into the characteristic phenotypes of various types of organisms is what might be called their "propagation value" (in the context of their total environment) to the genes that give rise to them. Thus, we might say that living things should be expected to behave *not* as the artifacts of a benevolent designer would behave (assuming here that such a designer would be interested primarily in the flourishing of his organisms), but more like the artifacts of a rather odd designer who, due to some bizarre fascination with genes, designs organic systems primarily with maximally efficient gene propagation in mind. (See chapter six, section one for a discussion of such an imaginary example.) Recall some of the examples considered above, involving fierce battles to out-reproduce peers, induced abortion, selective starvation of young, etc.—all primarily in the service of genes. The actual biological world everywhere bears the stamp of gene-driven natural selection, whether this is striking (as in these rather dramatic examples) or not (as in relatively mundane cases of physiology). We should not forget that it therefore looks much different from what it would look like if it were not the product of evolution by natural selection, and were instead something designed with more familiar ends in view.

CHAPTER IV

An Account of Biological Teleology

1. ORGANISMS AS INTEGRATED SYSTEMS: CO-ADAPTED GENOMES AND PHENOTYPIC STRUCTURES

I argued in chapter two that what makes it plausible to suppose that organisms are indeed genuinely teleological entities is that they are at least for the most part coherent, hierarchically organized systems that can intelligibly be spoken of as *working* in certain ways, with various interacting parts and features making non-incidental contributions to higher-level aspects of that working, just as with complex machines. (I will say more about the general conditions for the application of these concepts in section seven below.) The difficulty was to know how exactly to understand *what working ultimately is* for organisms—not just proximate functioning at lower levels (such as the proximate functioning of cells and organs), which is relatively unproblematic, but functioning at the higher levels of the hierarchy, toward which lower level functioning is geared; what was needed was a better understanding of high-level ends, such as survival, reproduction, and perhaps a still more ultimate end, that would clarify the overall functional structure of living things.

We are now a step closer to such an account, having examined the general principles governing the natural selection of traits for incorporation into the characteristic phenotypes of organisms of various types. So far in the discussion of natural selection, however, I have considered organisms only as collections of genes and traits such that each trait happens to be good at propagating its coding genes when in the company of the rest. In principle, that would be compatible with the various genes' each somehow employing different and independent methods of propagating themselves which merely happen to be most effective when in the context of the others—in which case there would not be anything that could plausibly count as a single, coherent system, or as the high-level functioning of the organism to which various traits make functional contributions. But again, a living organism is something far more complex and unified than such a mere collection, and this is crucial to the emergence of genuine teleological relations. Thus, before we will be in a position to develop an account of biological function, we must say a little more about organisms as well-orchestrated systems of naturally selected traits.

It is helpful to begin by imagining the sort of process that plausibly led to the emergence of life in the first place, at least if we accept a thoroughgoing naturalism rooted in the theory of natural selection.[1] Suppose there are various replicating molecules floating freely in solution. The ones that will over time increase their frequency relative to the others are obviously the ones with properties that, in their environment, ultimately make them better at replicating themselves than others are. But now suppose also that some replicators of different types are capable of combining with each other into larger units; and suppose further that in at least some cases the properties of the members combine in such a way that the net effect is to promote the replication of the "team"—in fact, to promote it so effectively that members of the "cooperating" team are thus better able to replicate themselves better than they are on their own, and better than other replicators are on their own. These replicators will naturally increase in frequency relative to others, and they will tend increasingly to be grouped together. Over time, copying errors ("mutations") will occur, with consequent variation among teams of replicators and the combined effects they exert; and once again some will prove more effective at replicating than others, and the members of such teams, so grouped, will increase their relative frequency. In suitable conditions, this will naturally lead to teams of replicators that combine to produce

[1] This is, of course, highly speculative, but it is plausible enough for my purposes here. See Dawkins (1976, ch. 2), (1982, ch. 14), (1987, chs. 5–7), and (1989, pp. 257 f.), of which what follows is a rough summary.

increasingly complex and effective sets of effects which help the team to replicate. Eventually this may even come to involve the synthesis of new molecular compounds, which somehow contribute to replication.

Finally, while this is obviously oversimplified, the thought is that this sort of process could potentially lead to the emergence of simple cells—the cell being a complex structure consisting of the combined effects exerted by a compatible team of genetic replicators, geared toward promoting their replication. Over time, this could lead—and, according to the theory of natural selection, has indeed led—to still more complex forms of organization among the effects exerted by teams of replicators, as some eukaryotic cells begin to hang together after splitting, forming many-celled structures (i.e. multicellular clones) that prove still more effective at promoting the replication of the genes that build them.[2]

If this picture is even roughly on target, then the genetic replicators that build living cells effect their replication over generations *not* through a variety of independent methods that simply happen to work well in the context of the others in the group, but by each member's making some contribution to a "common project"—the construction and operation of cells, which eventually leads to the replication of the members of the "cooperating" team of replicators, continuing indefinitely down the generations. Again, this sort of "cooperative" organization would just be a result of primitive processes of natural selection, and would continue to be reinforced by the sort of natural selection that goes on within the gene pools of established species:

> Each gene is selected for its capacity to 'cooperate' successfully with the population of other genes that it is likely to meet in bodies . . . [For] one of the most important aspects of the 'climate' in which a gene is favored or disfavored is the other genes that are already numerous in the population—the other genes with which it is likely to have to share bodies. Since the same will obviously be true of these

---

[2] As Dawkins puts it: "Genes working through the organs and behavior patterns of many-celled bodies can achieve methods of ensuring their own propagation that are not available to single cells working on their own. Many-celled bodies make it possible for genes to manipulate the world, using 'tools' built on a scale that is orders of magnitude larger than the scale of single cells." (1987, p. 177)

> 'other' genes themselves, we have a picture of teams of genes . . . evolving towards cooperative solutions to problems.[3]

So, for example:

> Genes for making teeth suitable for chewing meat tend to be favoured in a 'climate' dominated by genes making guts suitable for digesting meat. Conversely, genes for making plant-grinding teeth tend to be favoured in a 'climate' dominated by genes that make guts suitable for digesting plants. And vice versa in both cases. Teams of 'meat-eating genes' tend to evolve together, and teams of 'plant-eating genes' tend to evolve together. Indeed, there is a sense in which most of the working genes in a body can be said to cooperate with each other as a team, because over evolutionary time they (i.e. ancestral copies of themselves) have each been part of the environment in which natural selection has worked on the others.[4]

The point, then, is that natural selection leads to the co-evolution of teams of compatible genes—that is, *co-adapted genotypes*—such that each member gene makes some "useful" contribution to the team's collective phenotype—"usefulness" here being measured ultimately in terms of promoting the replication of the members of the team. And this applies not only to particular subsets of genes within the genomes of organisms of a given type, but to whole genomes—or at any rate, to most of the active genes found in a genome. That is, since the environment in which genes have been selected includes in a very direct way the other genes in the gene pool, with which these genes are combined in the cell nuclei of members of the population, the result is that individual genomes tend to be for the most part mutually compatible teams of "cooperating" genes. Any one gene might have a very small effect, for example coding for a certain enzyme that is crucial to a certain chemical reaction. But the phenotypic effects of the various co-adapted genes in the genome will add up to constitute a coherent, well-orchestrated system of traits, causally geared ultimately toward the replication of the member genes—or, as we noticed earlier,

---

[3] Dawkins (1987, pp. 170–1). Cf. also Dawkins (1982, p. 239f.).

[4] Dawkins (1987, p. 172).

geared partly or sometimes even wholly toward the replication of copies of those genes in the bodies of kin.[5]

A living organism, then, is at least for the most part such a combination of mutually compatible genes along with their phenotypic effects, woven together into a coherent and unified system that tends ultimately to promote the inter-generational replication of genes of the types that built it—whether of copies in the organism's own germ-line cells (gametes and their mitotic ancestor cells), or of copies in the gametes of kin; this will be brought about through the building and maintenance of these systems and either their own reproduction or their promotion of the reproduction of kin, or both. (Actually, the qualification regarding kin will have to be extended to certain other cases as well, as discussed in section two below, though I'll set that aside for the moment.) The genes in the *co-adapted genome* "cooperate with one another . . . because they share the same outlet—sperm or egg—into the future".[6] That is, their evolved method of propagation is to "cooperate" with each other in the construction and conditioning of certain organic systems—i.e. organisms—that either reproduce themselves or contribute to the reproduction of other organisms that are closely related genetically, or both. (As already noted in chapter three, this is not to deny the relevance of higher levels of organization—the importance of group dynamics, for example. Organisms employ a wide variety of methods in achieving even individual reproduction, often including along the way such things as cooperative relations with other organisms—even members of other species, in cases of symbiosis—in the context of certain group structures. The point is that these things come in as proximate links in the endlessly varied causal chains leading from genes to the reproduction of organisms carrying copies of them, and hence ultimately to the propagation of those same genes down the generations.) It is this gearing of selected traits toward such common

---

[5] In sexually reproducing organisms, it is of course not the case that the entire team or genome is replicated with each offspring, as in simpler cases. Due to meiosis, only half of the genome is represented in each gamete, most likely with a different selection in each case. But since meiosis is a random process, any member of the genome has a 50% chance of being represented in any given gamete. Hence, anything that promotes the reproduction of the organism, for example, promotes the propagation of each of its genes equally. That is, they all equally stand to "gain" in terms of propagation from the same thing, namely, the reproductive success of the organism. (And similarly with the reproductive success of its close kin, for the reasons explained earlier.)

[6] Dawkins (1989, p. 245).

ultimate effects that gives organisms their coherence and unity of structure and operation, making them such clear candidates for individual functional systems.[7]

The above claim was qualified with "for the most part" because there are, of course, exceptions to this coherence and unity, showing how much more complicated the notion of an individual organism is than it might at first appear to be. As I have already mentioned, some genes, such as segregation distorters, do not promote their propagation by making a contribution to any such "common project" as getting sperm or eggs advanced into the next generation; their phenotypic expressions—the mechanisms that distort meiosis in their favor—do not make a cooperative contribution to the reproductive success of the organism (or of its kin with closely related genomes), but have effects that independently promote the propagation of the segregation distorter gene, often at the *expense* of overall reproductive output. So whereas the genes in the co-adapted genome each make phenotypic contributions arranged in a coherent hierarchy leading to a single small range of high-level effects (i.e. the reproductive success of the individual or of its kin), giving the organism its striking unity and coherence of operation, the segregation distorter operates outside of this framework, pulling the organism in a contrary direction suited only to the differential propagation of that particular gene. An organism undergoing meiotic drive will thus naturally appear to some extent divided against itself, working largely toward maximizing the reproductive success of itself or of its kin (whatever exactly that involves in its given environment), while at the same time doing things that pull it in another direction favoring only the replication of the segregation distorter, and thus behaving less as a unity than we should ordinarily expect.

---

[7] I will use the expression "*co-adapted genome*" to refer to the whole set of co-evolved, "cooperating" genes in an organism's genome, which together give rise (within their environment) to a coherent and unified set of phenotypic traits. It does not, therefore, include "renegade" genes, such as segregation distorters: A segregation distorter is part of the genome of the organism in which it is found, but not part of the co-adapted genome in my sense, for reasons given earlier. I will also speak of the *co-adapted genome* characteristic of a given *type* of organism, such as the male elephant seal, by which I mean really a certain *range* of genomes, given that there will be a certain amount of natural genetic variation among individuals. For convenience, I will generally ignore such variation where I am concerned with what is common to and characteristic of the type, and speak simply of "the" *co-adapted genome of a given type of organism*.

Another factor that sometimes detracts from the unity of organisms is their manifestation of what Dawkins calls the "extended phenotypic effects" of genes in the bodies of other organisms—even of organisms belonging to different species.[8] The basic idea is simple. Genes exert phenotypic effects at various levels in the bodies of the organisms in which they reside. At a very proximate level, these effects involve influences in the manufacture of various proteins, which leads, at a slightly less proximate level, to changes in overall cell shape and behavior, and hence to differences in the construction of tissue; this in turn results, at a still more remote level, in differences in the gross biological traits of the organism. Now the causal chain whereby phenotypic traits feed back to promote the propagation of their coding genes obviously does not end here; it extends in various ways beyond the mere manifestation of traits in the organism, which in themselves, with no further effects, would hardly do the trick. And Dawkins' suggestion is simply that in at least some cases it makes sense to include these further effects as just more remote aspects of the phenotypic expressions of the genes in question—i.e. as extended phenotypic effects. This is especially tempting where the effects are actually biological traits of other organisms.

A good illustration of this is giantism in flour beetle larvae, caused by a parasite called Nosema.[9] The parasite infests the larvae and manufactures a chemical analogue of the flour beetle's "juvenile hormone", which latter normally maintains the larvae stage until the time is ripe for adulthood. The chemical is manufactured by the parasite in great quantities even after the point where the larvae would normally cease to produce it and turn into adults, thus causing the larvae simply to grow into giant larvae, providing a superior environment for more Nosema parasites. The manufacture of the chemical analogue is the proximate phenotypic effect of certain genes in the Nosema; but the further effect of that chemical in the host beetle larvae, which is equally part of the causal story behind the natural selection of these genes in Nosema, is also a phenotypic effect of the Nosema genes—an "extended" or "remote" phenotypic effect. (In both cases, of course, environmental factors play a role.) We might, therefore, even say that in a sense giantism in flour beetle larvae turns out to be an "adaptation"

[8] See Dawkins (1982, chs. 11–14), (1989, ch. 13).
[9] Dawkins (1989, p. 242).

not of beetles but of Nosema, though it happens to reside in the beetle's body.

In any case, the important fact is that the giantism is a phenotypic effect that, like any other naturally selected phenotypic effect, is geared toward promoting the propagation of the genes that produce it; and what is interesting here is that those genes happen to lie in the body of *another* organism. Thus, we would certainly be in a mess if we looked at giantism in the beetle larvae with the expectation that it should somehow fit in with the rest of the *beetle's* way of life, making some contribution to the "common project" of its reproductive success. Like the mechanism that produces meiotic drive, this trait lies outside of the cooperative framework of the beetle's co-adapted genome and phenotype, and again makes it look at least slightly less like a single coherent entity, and more like a set of traits that are largely, but not wholly, orchestrated into a single working system. Generally speaking, then, to the extent that an organism's traits are extended phenotypic expressions of the genes contained in other organisms, or expressions of "renegade" genes in their own bodies, they will appear more like a collection of traits at "cross purposes", i.e. without a single unifying principle behind their operation, and we will be less inclined to regard them as constituting a single functional system.[10]

My claim that organisms are coherent and unified systems of well-orchestrated traits and activities, and hence natural candidates for functional systems, must therefore be qualified. And the qualification is not quite as straightforward as it might appear to be. So far, it may seem that we can simply restrict the claim to apply only to the traits that are phenotypic expressions of the organism's *own* genes which belong to the co-adapted set of genes that "cooperate" in the realization of a common result; this would rule out extended phenotypic traits and traits caused by "renegade" genes, and leave us with traits that form coherent

---

[10] There are many other more familiar examples of traits that are extended phenotypic results of genes in other organisms, which need not involve internal parasitism. Baby cuckoos, for example, have various ways of manipulating host parents into feeding them. Not only is the cuckoo's manipulation behavior a phenotypic expression of its genes, promoting their propagation, but so too, in an extended sense, is the host's "sucker" nurturing behavior toward the cuckoo: It is an extended phenotypic effect of the cuckoo's genes, whereby they promote their own propagation. The host is thus pulled in one direction by the genes in its own co-adapted genome (which would never countenance wasting food on cuckoos), and in another direction by the genes in the manipulative cuckoo. Cf. Dawkins (1989, p. 248).

and well-orchestrated systems. But this move is complicated by the fact that sometimes genes from *other* organisms actually contribute to the same "cooperative project" advanced by the genes in an organism's own co-adapted genome, thus *contributing* to the unity of the organism's functioning; and in some cases it may not even be clear whether certain genes should be regarded as the organism's own or as foreign genes.

For example, some parasites—such as certain bacteria in wood-boring ambrosia beetles—propagate their genes by way of the host's sperm or eggs. Natural selection has thus arranged things here such that both the parasite's genes and the host's genes stand to "gain" in terms of propagation from the same thing, namely the host's reproductive success; and in such a situation, the parasite's genes can be expected to work in unison with the genes in the host's genome, making some cooperative contribution to the "common project" of the host's reproductive success. So here what are ostensibly foreign genes will actually tend to contribute to organic *unity*. Indeed, in such a case, "over evolutionary time [the parasite] will cease to be a parasite, will cooperate with the host, and may eventually merge into the host's tissues and become unrecognizable as a parasite at all"—at which point the distinction between an organism's own genes and foreign genes is obviously blurred.[11] Along these lines, it has even been argued that all eukaryotic cells reflect such a merger that took place roughly two billion years ago, when various kinds of bacteria united together, the descendents of which now include such things as mitochondria and chloroplasts (which, in line with this hypothesis, both have their own DNA).[12]

As long as we recognize these complications, however, they need not stand in the way of a reasonably clear formulation of the general claim that has so far been made about organisms. Despite the sorts of exceptions we have seen, organisms are still for the most part unified systems of well-orchestrated traits, resulting from the natural selection of groups of genes that tend to "cooperate" because they share a common means of propagation, to which they each make some phenotypic contribution. To the extent that this is so, an organism can

---

[11] Dawkins (1989, p. 243).

[12] The argument is given by Margulis (1981), and is discussed in Dawkins (1987, p. 176).

be regarded as a coherent and unified system, with a variety of traits orchestrated in such a way as eventually to promote the reproduction of the organism or of its kin—this in turn being geared ultimately toward the replication of genes of the types represented in the organism's co-adapted genome. With this picture of the form imposed on organic systems by natural selection, we are now ready to move on to develop an account of biological function and to consider its implications.

2. THE TELEOLOGICAL STRUCTURE OF ORGANISMS

At the beginning of chapter three, I asked how exactly the notion of *working* for living organisms is to be understood, beyond the obvious fact that it has to do with survival and reproduction. How is the hierarchy of functions and ends generally structured in an organism, and what exactly is (are) the ultimate biological end(s), the realization of which constitutes the organism's *work* at the highest and most general level—comparable to the engine's converting fuel to constant rotational power? Combining the arguments up to this point, we can now answer this question.

The conclusion of the argument of chapter two was that what we need to investigate in seeking to clarify these matters are the ultimate causal principles behind the assembly and organization of biological traits into coherently integrated organic systems, i.e. organisms. For whatever is ultimately responsible for this selection and organization of traits—without which we would not have coherent entities that could be considered functional systems at all—is plausibly thereby likewise responsible for determining the teleological structure that belongs to such sets of traits *so organized*. More specifically, we ought to be able to answer the above questions about organisms by learning more about the non-accidental relations that are established between organic traits and certain of their effects by whatever it is that is ultimately causally responsible for the selection and organization of traits into coherent organism-types to begin with.

Now the assumption I have been making is that it is natural selection that is ultimately responsible for the compresence and organization of biological traits characteristic of a given type of organism, and for establishing the non-accidental relations that exist between the traits and a certain subset of their effects. The way this works for an organ has already been illustrated with the example of the evolution of the eye given in the second section of chapter two. And we can now extend the same treatment to an organism as a whole. After all, sub-systems of organisms, such as organs, do not evolve independently, as entities in their own right with their own self-contained ends. Rather, the forces of cumulative natural selection put together coherent subsystems like the eye over evolutionary time as part of the more general process of putting together and shaping the entire organism (- type) as a coherent system consisting of such sub-systems, non-accidentally organized in just such a way as cooperatively to bring about certain evolutionarily relevant effects at the level of the whole organism. That is, the various incremental developments in the parts and features of a given type of organism in the course of its evolutionary history, which have led to its present extremely complex organization into a coherent functional system, occurred as a result of contributions that certain changes at each stage made to certain evolutionarily relevant effects at the level of the whole organism; for it was on the basis of such effects that mutant genes coding for such changes in various aspects of the phenotype out-propagated rival alleles and came to be standardly included in the genomes of the organisms in question, with the corresponding phenotypic change thus coming to be part of the standard construction of the organism—and so on, cumulatively, through evolutionary history.

In light of the examination of natural selection in chapter three, we can now say more about those "evolutionarily relevant effects at the level of the whole organism", on the basis of which traits come naturally to be incorporated into the phenotypes of organisms. Phenotypic traits are incorporated into the standard "designs" of organism-types through natural selection because they ultimately do more to *propagate their coding genes* than the alternative traits produced by rival alleles do to propagate those alleles. But as illustrated in section one, the way this generally happens is through the trait's fitting into the previously established coherent network of traits in such a way as to make a positive, cooperative contribution to certain complex organismic effects that tend ultimately to promote (on average) the replication of all the cooperating members of the co-adapted set of genes underlying that network of traits.[13]

---

[13] Traits produced by "outlaw genes", such as the segregation distorter described in chapter three, are of course exceptions. But they are precisely the sorts of traits that do

What makes an evolved organism such a marvelously coherent system is the fact that it is constructed (for the most part) by a coalition of genes that have cumulatively been selected together by virtue of their ability to "cooperate" with each other in constructing phenotypes that succeed in *getting the member genes passed on to the next generation.* The integration of the various phenotypic traits of an organism into a coherent system is a result of the fact that the members of the coalition of genes responsible for those phenotypic traits share this common means of getting themselves copied into the next generation: They work together to build and direct organisms that do things that help them all to propagate down the generations.[14] These things, of course, include all kinds of biological activities at various levels—from cellular metabolism, to the beating of hearts, to the seeking of mates, to the helping of close relatives (in cases where kin-selection has been a factor), to cooperating with symbiotic partners, and so on. But what all these elements of survival and reproduction have in common is that they ultimately promote the inter-generational replication of copies of the genes belonging to the organism's co-adapted genome, as described in section one.[15]

We thus arrive at an answer to the question of what it is that formally constitutes *working* for an organism in the relevant sense—what it is that, as a result of the shaping of the organism (-type) by the processes of natural selection, all the various adaptational parts and features of the organism non-accidentally cooperate ultimately in bringing about. Though it will be qualified in various respects throughout this chapter and the next, it may be put roughly as follows:

---

not fit into the coherent functioning of the organism, and are not generally taken to have functions within the organism. They are discussed further in chapter seven. I will, however, consider a case shortly below in which a trait produced by what might be called a "quasi-outlaw" gene might actually expand the scope of the organism's ultimate biological end.

[14] See Dawkins (1982, p. 239 f.), (1987, pp. 170–2).

[15] I use the term "copy" here in a broad sense, such that one gene can be said to be a copy of another if it is a token of the same type, even if the one was not copied from the other. Thus, in this sense, genes in the body of one organism might code for certain behaviors that promote the replication of (germ-line) copies of those genes in siblings, this being how they promoted their propagation and spread (i.e. through kin selection involving siblings).

> O: An organism is a complex system of integrated parts and features that non-accidentally *works* ultimately to promote the *inter-generational replication of certain germ-line 'copies' of the genes belonging to its co-adapted genome*; for that is what the coalition of genes making up the co-adapted genome have ultimately built the organism, as a single coherent system, to do—as a result of the co-evolution of that coalition of genes.[16]

Again, the working of organisms obviously involves all kinds of biological activities at various levels, but the point is that all these things—including such things as socially-oriented adaptations—are properly understood as playing *non-incidental roles in the working of the organism* by virtue of making some contribution to the ultimate effect identified above, which is likewise what is ultimately responsible for their incorporation into the coherent repertoire of the organism in the first place. To put it another way, it is precisely with regard to this ultimate effect that we can say, as before, that it is *no accident* that the various organic parts and features are compresent and organized in such a way as to promote it, by way of giving rise to these various biological activities.

The thesis as stated so far requires an important clarification. I have said that the ultimate biological *end* for an organism—the realization of which is the *work* of the organism at the highest and most general level—is the inter-generational replication of certain germ-line copies of genes of the types represented in the organism's co-adapted genome (i.e. genes of the types that built the organism insofar as it is a coherent whole). But *which* copies are relevant here? Recall the reason why gene replication was argued to be an end in the first place: It was only because of the special part played by certain kinds of genetic replication in the natural selection history behind the formation of a given type of working system that these kinds of genetic replication were argued to constitute the ultimate biological end for such a system. Thus, it is only these sorts of genetic replication, which *actually* figured

[16] Recall that by "inter-generational gene-replication" I mean the replication of genes down the germ line, into the bodies of new individuals—as opposed e.g. to the replication of genes in somatic cells during growth, which is of course one low-level biological end among others, but not the ultimate biological end for the organism. And as before, the term 'copy' is used broadly, such that genes of the same type in the body of a sibling count as copies of an organism's genes (even though they are not descended from them).

into the natural selection history, that figure into the ultimate biological end for that type of organism.

In the case of an organism-type whose selection history does not involve any sort of kin selection, the organism will generally have been "designed" ultimately to work simply toward the inter-generational replication of copies (of the genes of the types represented in its co-adapted genome) *residing in its own germ-line cells*.[17] In the case of an organism-type whose selection history involves some degree of kin selection, the end will be broader: Such an organism will have been "designed" ultimately to work toward the intergenerational replication of gene-copies residing *both* in its own germ-line cells and in the germ-line cells of relevant kin (or in some cases, such as sterile worker bees, only the latter). Which kin are "relevant", of course, again depends on the nature of the actual selection history: If the type of organism in question has evolved adaptations for altruism toward siblings but not for any altruism toward second cousins (which latter would be less likely to evolve, as the genetic relatedness is significantly more distant), then while it can be said that the organism has been "designed" in part to work toward the replication of gene-copies residing in the germ-line cells of siblings, it cannot be said that it has been "designed" to work toward the replication of gene-copies residing in the germ-line cells of second cousins; to put the point literally, the promotion of the replication of gene-copies in the bodies of second cousins simply does not count as an end for such an organism, just as the promotion of the replication of gene-copies in the bodies of siblings does not come into the end for organisms whose selection history does not contain any kin selection at all.

There is, however, a complication that should be mentioned here, which requires a bit of a digression. As noted in chapter three, there are other ways for genes to promote their propagation than by the two routes mentioned above—i.e. promoting the reproductive success of their possessors or promoting the reproductive success of their possessor's close kin. That is, organisms can be shaped over evolutionary time by selective processes other than those of "individual selection" and "kin selection", even though these are generally taken to account for the vast majority of naturally selected traits. Various "outlaw genes", such as the segregation distorter discussed in chapter

[17] Though there are possible exceptions, as discussed just below.

three, spread *not* by making some contribution to an effect that equally promotes on average the spread of all the genes in the organism's co-adapted genome (i.e. the reproduction of the organism or of its kin), but by promoting their own propagation independently. Another example of such an "outlaw gene" would be one that somehow makes its possessor tend to act altruistically toward others who are likely to carry copies of it *regardless of whether or not they are kin*, i.e. simply on the basis of some specific trigger associated with their being carriers of *that* gene (rather than on the basis of some trigger associated just with their being kin, which would tend to serve not only this gene, but all those in the organism's co-adapted genome, since they all have an equal chance of being shared by kin). This raises the possibility of a type of altruistic trait that is distinct from both kin altruism and standard reciprocal altruism.

A particularly fanciful, hypothetical form of such a phenomenon is something that has been called the "green beard effect", named after a hypothetical case in which a gene tends, through incidental pleiotropy, to give rise both to a green beard and to the disposition to act altruistically toward individuals with green beards; such a gene could obviously spread, just as in the case of kin altruism, thus resulting in organisms that are disposed in part to be genuinely altruistic toward certain non-kin.[18] Again, however, an important difference is that unlike with kin altruism, a gene for green beard altruism would be an "outlaw gene", promoting its own propagation at the expense of that of the rest of the members of the organism's genome. There would thus be selection pressure favoring the spread of a potential modifier gene that inhibited the expression of green beard altruism: Such a modifier would tend to spread over other alleles, since carriers of the modifier would avoid the loss of resources through green beard altruism (and the organism's green beard altruism consistently promotes only the spread of the green beard/altruism genes, not the others in the genome, so the modifier gene would lose nothing in cutting off such altruism).[19] Along

---

[18] Dawkins (1976, p. 96), (1982, pp. 145 f.).

[19] Note that a modifier gene that neutralizes the expression of the green beard/altruism gene *entirely* would not in fact tend to be successful, as its possessor will miss out on the altruism that would have been accorded *to it by others* as a result of *its* having a green beard. The successful modifier would be one that inhibits the expression of green beard altruism, while allowing the green beard itself to be expressed. See Dawkins (1982, pp. 148–9).

with the improbability of the sort of incidental pleiotropy needed to get such an example going, this inherent instability is one important reason why this sort of altruism is relatively unlikely to evolve and to be maintained—as opposed to standard kin altruism or reciprocal altruism, where there is no such pressure in favor of modifier genes that would undermine the altruism.

There are, however, also less fanciful and more interesting possibilities. As Dawkins points out, the problem of the improbability of the relevant sort of incidental pleiotropy could be eliminated by having the trigger be *not* something incidental like a green beard, but the exhibiting of altruistic behavior itself:

> Imagine a gene programming the behavioural rule: 'If you see another individual performing an altruistic act, remember the incident, and if the opportunity arises behave altruistically towards that individual in the future'. This might be called the 'altruism-recognition effect'. [Such a gene] might spread because it was, in effect, recognizing copies of itself. It is, in fact, a kind of green-beard gene. Instead of using an incidental pleiotropic recognition character such as a green beard, it uses a non-incidental one: the behaviour pattern of [altruism] itself.[20]

Such a case could arise only among fairly intelligent social animals, capable of the relevant types of recognition and memory, so its potential applicability is rather limited; but it is worth considering nonetheless.[21] What we have here is behavior that could regularly involve genuine individual sacrifice (unlike with standard reciprocal altruism), but would not have to be directed at kin (unlike kin altruism) and is at least in some sense an "outlaw trait" (unlike either kin altruism or standard reciprocal altruism). This last matter is actually rather tricky. The gene in question appears to be an "outlaw" gene, like the green beard/altruism gene and *unlike* genes for kin altruism or standard reciprocal altruism, insofar as it *directly* promotes its own propagation at the *expense* of the rest of the genome. On the other hand, it is not obvious that there would be selection pressures favoring inhibiting

[20] Dawkins (1982, pp. 154–5).

[21] I am indebted to Robert Adams for drawing my attention to the possibility of this sort of case.

modifiers in this case; in fact, it seems doubtful that there would be. This is because the rest of the genes in the genome of a given organism would benefit *indirectly* from the presence of the gene in question, since the altruism to which it gives rise acts as an attractor of the altruistic attention of *others* who carry the gene for rewarding altruism with altruism, thus benefitting the whole genome as the organism receives this reward. Therefore, since the whole genome—including potential inhibiting modifiers that might arise—stands to gain from the altruism-*attracting* feature of the gene, *and* since there is here no possibility of preserving this feature while inhibiting the altruistic behavior (as there was in the green beard case, where there was a distinction between the indicator, i.e. the green beard, and the altruistic behavior, so that the former could remain while the latter is inhibited), there would not after all be selection pressure favoring a modifier that would inhibit the expression of the gene.

What we seem to have in such a case, then, is a gene that behaves like an "outlaw gene" in its direct effects, but does not ultimately qualify as one, inasmuch as it does not give rise to selection pressures favoring genes that would inhibit its expression in altruistic behavior. It thus seems possible for such altruism to evolve, at least in certain higher animals, as a third alternative to kin altruism and standard reciprocal altruism. Moreover, wherever there is intraspecific reciprocal altruism on something like the "Tit for Tat" game-theoretic model, the kind of selection at issue in "altruism-recognition" cases would seem to be occurring *in addition to* the ordinary kind of selection. That is, the genes in question would tend to spread over competing alleles within the gene pool not only because copies promote their own replication (i.e. by making their own possessors interact with others in such a way that their possessors benefit overall, as in iterated prisoner's dilemma situations), but also partly because copies of the same gene are thereby also promoting *each other's* replication.[22]

---

[22] To see that these two processes are distinct and do not always or necessarily go together, consider (*inter*specific) symbiotic relations, which evolve through the first process (i.e. each partner does better for itself than it would do without the relationship—and hence out-reproduces its conspecific peers), but not through the second. The second process is not involved because the traits in question—e.g. the ants' protecting behavior and the aphids' nutrient production—are governed by genes belonging to entirely different gene pools, the relevant competition being among rival alleles within each gene pool. In the shared environment that includes both ants and aphids, genes for the ants' behavior out-propagate rival alleles within the ants' gene pool (because in the company

We must therefore qualify the claim that each individual has been "designed" simply to promote the inter-generational replication of copies of its genes residing in its own germ-line cells and/or copies in those of kin (if kin selection was involved as part of its natural selection history)—recognizing in addition that in certain cases an organism may have been "designed" partly to promote the replication of copies of the relevant *altruism gene* in non-kin. This is consistent with the general point above, i.e. that which copies of genes are relevant to the ultimate biological end(s) of any organism depends on the particular natural selection history of the type of organism in question; what we are recognizing here is simply that there are other potential forms of selection that might figure into that natural selection history, and they complicate the picture a bit. (I shall discuss a further qualification involving the possibility of a certain kind of higher-level selection for environmentally sustainable traits in chapter five, section thirteen; again, this will in some cases complicate the description of an organism's ultimate end.)

It is important to keep these points in mind throughout the following discussion, where I will generally speak for convenience only of "inter-generational gene replication" as the ultimate biological end for any given organism. What is meant is not just *any* sort of inter-generational gene replication, but the sort that is systematically promoted by an organism of a given type as a result of the particular selection history that has shaped that type of organism into the sort of working system it is. The fact that other members of the species—or for that matter, members of other species—may carry germ-line copies of many of the same genes possessed by a given organism does not *by itself* do anything to make the replication of those copies part of the ultimate end of the organism, or to imply that it can in any way be faulted as being defective or dysfunctional for failing to promote their replication. Again, we must always consider the natural selection history of the type of organism in question to determine what exactly it is that such organisms have been "designed" ultimately to do—i.e. which gene-copies it is whose replication is relevant. And since for the

---

of aphids these ants do better than ants without that behavior), while genes for the aphids' food production out-propagate rival alleles within the aphids' gene pool (because in the company of ants, aphids with that trait do better than aphids without it), and the symbiotic relation evolves. Symbiosis is further discussed in chapter five, section thirteen.

vast majority of cases what is involved is standard "individual selection" and/or "kin selection" (rather than the sorts of selection involved with "outlaw" or "quasi-outlaw" genes), I shall speak for convenience of organisms generally having as their ultimate biological end the inter-generational replication of copies of their genes residing either in their own germ-line cells or in those of kin, or both. But the above qualification should be kept in mind, as the situation is probably not that simple.

My claim, then, is that for any organism, all lower level biological functioning is coordinated and geared ultimately toward the end of inter-generational gene replication—by way of promoting the reproductive success of the organism or that of its kin, or both (depending on the type of organism in question), whether directly or by way of promoting survival—just as the functioning of the spark plugs and carburetor is coordinated and geared ultimately toward the production of controlled rotational power.[23] The functional concepts apply here just as they do to complex machines. The *ultimate end* of a functional system is the accomplishing of its *work* at the highest level, whether this consists in converting gasoline into controlled rotational power, or in effecting inter-generational gene-replication; and *subordinate ends* are aspects of its work at lower levels that must be accomplished if the ultimate end is to be achieved, e.g. the oxidation of the fuel, in the case of an engine, or the acquiring and metabolizing of food, in the case of organisms. Various *functions* serve various ends, and various ends are orchestrated in such a way as to promote the satisfaction of the ultimate end(s). The point, then, is that the ultimate end for a living organism, as a biological functional system, is to be understood in terms of inter-generational gene replication. From the first primitive cells to contemporary multicellular bodies of great complexity, natural selection has put together what are ultimately, functionally speaking, organic *gene replicating systems*—complex systems of complementary genes and their well-integrated phenotypic expressions that at least for the most part work as coherent units "designed" primarily to bring about the replication of genes, as

[23] Of course, one disanalogy between the two cases is that in the case of the engine the realization of the ultimate end is a steady state simultaneous with the lower level functioning of the system, whereas in the case of the organism it is a more remote effect. This does not substantially affect the point, I think, but if it proves distracting I provide a closer parallel below.

specified above. They accomplish this end not simply and directly, but by a variety of means, which of course include all the familiar aspects of survival and reproduction that fill biology textbooks.

This picture of organisms, at least *qua* biological-functional systems (which is of course all that is intended), is bound to strike us initially as somewhat grotesque. It is certainly a long way from our ordinary conception of living things. But the suggestion is precisely that our ordinary conception is in need of radical revision in light of what is now known about the relations between genes and organisms, particularly from the perspective of the evolutionary processes that have made organisms into the kinds of systems they are. Our ordinary conception of living things is shaped largely by the fact that we invariably learn about a variety of relatively accessible traits and activities pertaining to the survival and reproduction of plants and animals long before we ever hear about replicating bits of DNA. From the beginning, then, organisms seem in some sense to be ends in themselves, fighting simply for their own survival and reproduction, so that when we do hear about genes, the natural tendency is to wonder *what good they are to organisms* in their struggle for survival and reproduction. Indeed, as Dawkins remarks, "[even] some biologists go so far as to see DNA as a device used by organisms to reproduce themselves, just as an eye is a device used by organisms to see!"[24]

This way of conceiving the relation between genes and organisms, however, has things backwards. It might have been correct if genes had been designed by a benevolent supernatural designer as devices to further the good of organisms in various ways. But if the theory of natural selection is correct, then as Dawkins has argued at length and I have tried to make clear as well, there is an important sense in which genes have explanatory priority over organisms, even to such an extent as to change our conception of what organisms are *qua* functional entities. Organisms have been put together as the sorts of coherent working systems they are according to principles that have ultimately to do with gene propagation—and not ultimately with welfare promotion, whether at the level of individuals, groups, species, or ecosystems; at each stage in the evolution of a given type of organism, traits have been "selected" for inclusion in the standard "design" of the system simply on the basis of how well they contribute to the propagation of their

---

[24] Dawkins (1989, p. 237).

coding genes when they are part of such a system (in an environment which of course includes other organisms, groups, and the rest of the ecosystem). Organism-types are thus assembled, by natural selection over evolutionary time, to be at least for the most part coherent systems geared ultimately toward the replication of certain germ-line copies of genes of the types represented in their genomes, as discussed above. And again, for the reasons given earlier and summarized above, this means that they must be thought of from a *functional* point of view (as well as from the evolutionary point of view) as complex gene replicating systems, refined through generations of natural selection, and not ultimately as ends in themselves whose good is properly served by genes.[25]

3. AN ACCOUNT OF BIOLOGICAL FUNCTION

It is now time to step back for a moment from the above conclusions about biological teleology and to clarify the general concepts involved, in order to formulate a reasonably precise account of biological function. The key concept, which is also central to an account of the functional teleology associated with complex machines, is that of a working system—a system that can non-arbitrarily be spoken of as working in a certain way, with various parts and features that make functional contributions to that working. What is the relevant sense of *working* here? Based on the discussion so far, we may explicate this roughly as follows:

> For a system of a given type to *work* in a certain characteristic way is for it to possess parts and features that enter into certain causal interactions characteristic of such a system, tending to bring about certain effects characteristic of such a system, where it is *no accident* that the system possesses such parts and features and such an

---

[25] As Dawkins rather provocatively puts it: "An organism is a tool of DNA, rather than the other way around" (1982, p. 158). It is important, however, not to misunderstand this claim as implying that genes cannot be said to have functions in organisms—though Hampe and Morgan (1988) interpret Dawkins as making the latter claim as well. I will take this up below. I also discuss in section four why this "revolutionary" view of organisms does not fly in the face of the very plausible claim that Harvey, for example, knew about the function of the heart long before anyone knew anything about genes or natural selection.

> organization among them that they tend to enter into such causal interactions resulting in such effects.[26]
>
> The system's *working* is thus to be understood as the carrying out of these activities at the various levels involved.

According to such a conception of a working system, it is thus a necessary condition that the system possess a *non-accidental unity* of the sort described here, along with possessing the relevant sort of *structure of causal relations among its parts, features and activities.*[27] The former notion is also relevant to distinguishing between incidental and non-incidental effects of a system's parts, features and activities, which in turn is crucial to an explication of what it is for something to have a proper function within a system:

> For a certain type of part or feature or activity of a given type of working system (or some other type of item that may be used by such a system) to have a *proper function* within such a system is for it to play some *non-incidental role* in the working of such a system, this being its function (as opposed to various other "mere effects" it happens to have).
>
> For something to play a *non-incidental role* in the working of such a system is for it to tend to have some effect contributing to the working of the system at some level (understood as above), where at least one of the following two conditions is satisfied:
>
> (i) it is *no accident* that such an item is present within the system and integrated into it in such a way as to have such an effect, where this is understood in terms of the *etiological relevance* of the effect to the presence and integration of the item within the system;
>
> (ii) such an item tends to be used by the system in such a way that it has such an effect, and the use in question is itself functional.[28]

---

[26] The relevant notion of non-accidentalness is clarified below.
[27] I will discuss examples in section seven that further illustrate the importance of both conditions.
[28] Examples of the latter will be given in chapter five.

> Finally, for a token part or feature of a token working system to have a proper function is for it to be a token of a functional type of part or feature (regardless of whether or not the given token part or feature actually tends to have the relevant effect), its function being that of its type.[29]

Likewise, the notion of an end within a functional system may be explicated as follows:

> For something to be an *end* within a given type of working system is for it to be a type of effect *non-incidentally* promoted by a *proper function* within such a system, as part of the system's *working*. As with functional parts or features, token working systems have the ends associated with their types (whether or not such ends are actually promoted in a given token system).
>
> The *ultimate end(s)* of a working system is the accomplishing of its *work* (i.e. the activities constitutive of working for such a system) at the highest level(s). *Subordinate ends* are lower level ends that tend to promote the realization of the ultimate end(s), as part of such a system's working.

If the above explications of the general concepts we have been using are reasonably clear, then we are in a position to offer what should be a reasonably clear, general account of biological function in these terms. The account will be elaborated upon at length in the following chapter, which will include some qualifications (e.g. in section thirteen) in addition to those already discussed in section two above. But keeping this in mind, the account may be summarized roughly as follows:

> For a given type of entity (broadly construed to include such things as organs and other parts, substances, processes, physical features, behaviors, and products of behavior) to have a *biological function* is for it to play a *non-incidental role* in the *biological working* of a

---

[29] The reason for putting the explication of function primarily in terms of types, and only derivatively in terms of tokens, is discussed in section eight of chapter five.

> given type of organism, this being its biological function; if such an entity plays multiple roles, then it has multiple functions. And for a token entity to have a biological function is for it to be a token of a biologically functional type of entity, its function being that of its type.
>
> For any given type of organism, it was the processes involved in its natural selection history that were ultimately causally responsible for the non-accidental organization of its parts, features, and activities, and for the non-accidental relations that obtain between them and certain of their effects. Thus, according to the above explications of the notions of function, end, and working, the nature of biological working for a given type of organism and the structure of its biological functions and ends will reflect the ordering principles of natural selection.
>
> That is, biological *working* for a given type of organism consists in the adaptive activities carried out at various levels, either directly as a matter of adaptation or indirectly under the influence of adaptations. And the various biological functions and ends are hierarchically structured in such a way that for any organism of that type, the *ultimate biological end*, served by all the organism's subordinate functions and ends, is the inter-generational replication of certain germ-line copies of genes of the types represented in the organism's co-adapted genome; this may also be described as the organism's *work* at the highest, most general level.[30]

My claim, then, is that when we speak of various types of things as having biological functions or playing natural functional roles *in the lives of* various types of organisms, this is how such talk should be understood. The crucial point is that an organism's biological functions and ends are geared ultimately toward the inter-generational replication of certain copies of genes of the types represented in the organism's genome, whether through the reproductive success of the organism, or that of its kin, or some of both—depending on the natural selection

---

[30] As more carefully spelled out in section two above.

history of the species in question.[31] Some ends, such as the facilitated intake of food that might be promoted by a certain aspect of tooth shape, are of course far removed from inter-generational gene-replication. The claim is therefore certainly not that every functional entity serves this end *directly*, but only that every functional entity serves some end associated with the working of the type of organism in question, and that as a result of the crucial role played by natural selection in "designing" organism-types, these ends are geared ultimately toward gene replication; that is, the notion of *working* for organisms must be understood *at the highest and most general level* in terms of accomplishing inter-generational gene replication of the relevant sort.

4. THE "HARVEY OBJECTION"

Nothing I have said should be taken as a denial of familiar facts about the functions of biological traits and the *proximate* ends served. I can say as well as anyone else that the function of the heart is to pump the blood, for example, and that the end served is the blood's circulation. The only difference is that where others might appeal to the survival or reproduction of the organism or to the meeting of its various needs as some kind of ultimate end in its own right, I insist that where such ends come into the picture they must be placed within the wider framework revealed by the natural selection background.

Earlier, in connection with the ahistorical, welfare-based approach discussed in chapter one, I considered the initially plausible thought that it couldn't possibly make a difference to the function of the heart, say, whether organisms with hearts are products of natural selection or whether they are products of divine design or even of sudden cosmic accident: The function of the heart is perfectly clear from its role in the life of current organisms. In one sense, of course, I agree. Our conclusions about the *proximate* functions of particular traits or organs would not plausibly be affected by the theological debate. But in another sense, the above thought is simply mistaken, and we have now seen why. It makes all the difference to a *general* account of biological function—and to a *full* theoretical account even of the function of the

[31] Recall the qualification from section two: This is only a close approximation, and there are possible forms of selection in cases involving higher animals that would complicate the description of the ultimate end(s).

heart in particular—whether organisms are products of natural selection or whether they are products of an intelligent designer who could have given them their traits with an view to *their good*, taken as an end in itself. Not only would the biological world look a lot different if the latter were the case—with male elephant seals living more "democratically", for example, and lions raising the cubs in new prides rather than killing them to make room for the production of their own—but since it is not (as I have been assuming), we cannot understand any biological function *ultimately* in relation to the good of organisms. Thus, the issue of the true causal background *does* matter even to our understanding of the functions of particular organs, such as the heart, no matter how much physiologists might know about it in abstraction from that background.

It should be clear from this, I think, how we should answer what may be called the "Harvey objection" to accounts of biological teleology that make appeal to natural selection. The basic idea is neatly summarized in the following passage from Prior:

> A statement like "the function of the heart is to pump the blood" is accepted by us as true because physiology tells us that it is true . . . Physiology is a lot older than evolutionary biology, and physiologists were making claims about the functions of organs long before there were any claims about their evolution. For example, William Harvey was able to claim that the function of the heart is to pump the blood. So . . . [as to the matter of who has the option on the word "function"] a good case could be put in favor of physiology.[32]

My response is as follows. Biological function has primarily to do with the roles played by certain entities in the working of organisms. Now the claim that this working is properly to be understood ultimately in terms of gene replication certainly does not imply the claim, for example, that "biological function" just *means* "standard contribution to gene replication". If it did, then this would indeed be a problem, since we would have to conclude that pre-Darwinians used the term "biological function" in some altogether different sense, and weren't even talking about the same thing we are talking about. As far as I can see, "biological function" just means roughly "standard role played in

[32] Prior (1985, p. 317).

the working of an organism", and there is thus no good reason to suppose that pre-Darwinians could not think about, investigate, and even learn a great deal about biological functions (and similarly with current skeptics); for there is no obvious reason to suppose that a knowledge of genes and evolution is necessary to the ability to recognize organisms to be working systems, with parts and features that play standard roles in that working, or even to the ability to learn at least some of the facts about this working. The fact that prior to Darwin most people wrongly (so I am assuming) thought of organisms as acquiring their nature as working systems from an intelligent designer does not mean that they were wrong or unjustified in thinking organisms to *be* working systems. Nor does it imply that they could have no knowledge at all of that working. Harvey is indeed a fine example of someone who knew quite a bit about it. It is thus perfectly consistent with my view of biological function that people like Harvey spoke of functions in just the same sense as we do, and discovered certain limited, but by no means trivial, truths about them—such as the fact that the function of the heart is to pump the blood.

What I am claiming is just that we now have a far more complete understanding of what *working* amounts to for living things, so that we can have both a more comprehensive understanding of biological function in general—particularly with regard to the structuring of the remote, highest level ends (the ultimate end in any given case being one that Harvey would not even have imagined)—and a more complete understanding even of particular functions, not necessarily with regard to the proximate ends served, which someone like Harvey may in principle have understood perfectly in certain cases, but with regard to the ultimate end(s) served. It is important not to go too far in granting knowledge of function to pre-Darwinian physiologists (or to post-Darwinian physiologists and ethologists who ignore evolutionary biology). It is one thing to discover the proximate function of the heart or of the eye—which does not require any knowledge of the nature and structuring of remote, high level ends, beyond the fact that survival is among them—and quite another to understand something like the uneven resource distribution among elephant seals, or the manipulation of sex ratio among Hymenoptera; there will also be difficulties in drawing the distinction between functions and accidents in ignorance of evolutionary history (as brought out in detail in chapter six). My conclusion is that while it is possible to know a great deal about biological function quite apart from any considerations of natural selection, there is also much that cannot be known in ignorance of such considerations—including the ultimate end(s) to which biological functions are directed in any given case. The account I have offered reflects this fact, and brings out the true relevance of an understanding of natural selection to an understanding of biological function.

5. THE HIERARCHY OF ENDS

In the first section of chapter three, I raised a series of questions about biological function in relation to survival and reproduction. It is now clear, I think, how they should be answered. Survival and reproduction are, of course, paradigmatic general biological ends, toward which various forms of functioning and various lower level ends are directed in all species of living thing. But neither survival nor reproduction is an *ultimate* end—that for the sake of which everything else ultimately takes place. Nor are the two on a par. Because organism-types receive their teleological structures from natural selection, shaped in accordance with principles that have ultimately to do with genetic propagation, the end of survival naturally takes a subordinate position to the end of reproduction. In the sort of organic gene replicating system that is "designed" by natural selection, survival is relevant only insofar as it is a means to reproduction—something which is evident in the way in which the various traits of the system are in fact orchestrated and geared toward the promotion of reproduction. As we have seen, the ways in which reproductive success is promoted often involve the promotion of survival, but often they do not, and may even tend to detract from it; by contrast, traits will *not* generally be found that tend to promote survival without tending thereby to promote reproduction, since such traits would never have been selected for and included in the organism's repertoire. Since the end of survival is subordinate in this way to the end of reproduction, there is no difficulty in knowing what to say about a case such as the cumbersome but attractive tail of the male bird of paradise. It is not to be regarded as partly functional (insofar as it promotes reproduction) and partly dysfunctional (insofar as it detracts from survival fitness), but is rather to be regarded as a straightforwardly functional trait, playing a role in the working of the male bird of paradise as the gene replicating system that, functionally speaking, it most fundamentally is; the fact that it lessens the individual's survival fitness or tends to shorten his life in no way detracts from its functionality, since survival and longevity are not ultimate ends in themselves whose realization must universally be promoted by functions.

Something similar should be said in the case of organs such as mammary glands, whose function is to nourish offspring. We should not confuse the issue by speaking in the following way, for example:

> From the point of view of the functioning of single organisms, mammary glands are dysfunctional. From the point of view of passing on one's genes, they are extremely important.[33]

On my view, the distinction drawn here is a bogus one. The functioning of single organisms *is* directed ultimately toward gene replication, not just toward personal survival; the promotion of personal survival comes in not as a distinct kind of functioning, but simply as one—admittedly large—aspect of an organism's functioning, directed toward gene replication. Mammary glands, then, are not *dysfunctional* from any point of view, though it is true enough that they do not contribute to personal survival; they are straightforwardly functional, playing a standard role in the working of the organism as a gene replicating system, by promoting the (survival and hence eventual) reproduction of offspring.[34]

Now just as survival is not an ultimate end, but is subordinate to and geared toward reproduction, reproduction is not an ultimate end either, but is subordinate to and geared toward the inter-generational replication of germ-line genes of the types represented in the organism's co-adapted genome. Because reproduction in general is relevant as an end due to the role it plays in this sort of genetic

---

[33] Hull (1982, p. 305). Hull cites four others who evidently share this way of putting things.

[34] As Dawkins (1976, p. 101) points out, parental care is of course just a special case of kin altruism, which I have all along included in the account of function. Thus, in this case, the mammary glands contribute ultimately to the replication of germ-line copies of genes in offspring. The difference between this sort of kin altruism and non-offspring cases is that in this case, what is being promoted is actually the continuation of a lineage of germ-line genes stemming back to the organism's own germ-line genes, whereas in the case of sibling altruism, it is a different lineage (though the genes are of course still generally of the same types).

replication, *whose* reproduction is relevant when we speak of it as a high level end in the teleological structure of organic systems is not as simple a matter as it initially appears to be. For, as we have seen, it is not just personal reproduction that contributes to the replication of genes of the types represented in the organism's co-adapted genome: The reproduction of kin does so as well, and it is therefore equally an end that may be promoted by various functions (e.g. behaviors geared toward aiding kin reproduction). This, of course, also implies the relevance of the survival of kin as an end, at least insofar as its promotion would tend to promote their reproductive success. This explains, then, why the survival and reproduction of kin have a special place with regard to biological function, as compared with that of other members of the population or species, or of organisms of different species. (Though as noted earlier, in the case of cooperative relationships among non-kin—whether within a population or among members of different species involved in symbiosis—the *survival* of the partner(s) may come in as an end to which certain functions are directed (though their reproduction is beside the point, except insofar as it may result in more useful partners). Ants in symbiotic relationships with aphids, for example, care for aphid eggs, feed the young and protect them from various dangers. But again, this enhanced survival on the part of the other organism is relevant as an end only insofar as it is really a *means* toward realizing the personal "benefits" of the relationship, understood ultimately in terms of gene replication; it is not in any way an end in itself. The ants nurture and protect the aphids not ultimately for the sake of aphid flourishing, but simply as a means of guaranteeing a supply of the "honeydew" the aphids produce, which the ants "milk" for their own nutrition.[35])

The natural selection background equally explains the way in which the ends of personal survival and reproduction and the ends of kin survival and reproduction tend to be weighted. I have already mentioned the rather dramatic case of sterile worker bees, for whom the ends of the survival and reproduction of their queen and reproductive siblings clearly trump the end of their own survival; thus we find worker bees functionally disposed to sting an intruder to protect the reproductives of the hive—ultimately to promote the replication of the germ-line genes they carry—even though this results in their own

[35] The example comes from Dawkins (1976, p. 195).

death. Naturally, they won't be expected to sting wantonly—for their death does constitute a significant cost to the reproductives of the hive, since they consequently lose the service they would have enjoyed in the future had the worker remained alive to give it. What we should expect, given the shaping of natural selection, is simply that worker bees will tend to sting in situations where stinging tends to contribute more to the relevant genetic replication (through promoting the survival and reproduction of the queen and/or reproductive siblings) than their consequent death subtracts from it (through depriving the reproductives of future service). In any case, the point is that in this way the reproductive success (and derivatively, the survival) of the reproductives of the hive will take precedence as ends over the worker's own survival—all as a result of the "design" imposed on the organic system by natural selection.

The hymenoptera are, of course, a special case. But we can equally calculate the ways in which various ends will tend to be weighted in more ordinary cases. I will just give one hypothetical case, based on an illustration of the principles behind kin altruism given by Dawkins. Imagine a runt in a litter, struggling with its siblings for a limited amount of food. If we ignore the phenomenon of kin selection, we might expect that it would struggle right to the end to get whatever food it can—all of its functional behavior being directed toward its own survival and eventual reproduction. But if kin selection has been a factor in "designing" organisms of this type, then this is no longer a safe assumption. Dawkins explains the behavior we should expect as follows:

> As soon as a runt becomes so small and weak that his expectation of life is reduced to the point where benefit to him due to parental investment is less than half the benefit which the same investment could potentially confer on the other babies, the runt should die gracefully and willingly. He can 'benefit' his genes most by doing so. That is to say, a gene which gives the instruction 'Body, if you are very much smaller than your litter-mates, give up the struggle and die', could be successful in the gene pool, because it has a 50% chance of being in the body of each brother and sister saved, and its chances of surviving in the body of the runt are very small anyway. There should be a point of no return in the career of a runt. Before he

> reaches this point he should go on struggling. As soon as he reaches it he should give up, and preferably let himself be eaten by his litter-mates or his parents.[36]

The point made in this passage is simply that this is the sort of behavior we might expect to find as a result of natural selection—behavior that on the face of it looks generous, but is in fact just another case of behavior directed toward the relevant sort of genetic replication, just as we should expect. The point I want to make is that this behavior is perfectly functional, according to the account I have offered, and it illustrates nicely the way in which the ends of personal survival/reproduction and kin survival/reproduction may co-exist and be weighted. It is safe to say, I think, that all of this would remain entirely obscure if we simply ignored the natural selection background and focused only on the various contributions we presently observe to be made by various traits to the survival and reproduction of various organisms, as on the ahistorical welfare-based view. The natural selection background provides clear and non-arbitrary principles explaining *whose* survival and reproduction count as ends in relation to a given organism's functions, *how* they are weighted, and *why*.

---

[36] Dawkins (1976, p.140). The sense of the "should" here is, of course, simply that this is what we should expect to find, given the operation of natural selection (assuming that it had sufficient variation to work with, sufficient time, etc.): Genes coding for such behavior are the ones that, once they arise through random mutation, will tend to out-propagate rival alleles, such as genes for struggling right to the end. The significance of the 50% figure is that due to meiosis, any gene (type) in the runt, such as a gene for the conditional behavior in question, has a 50% chance of being found in any given sibling. We are considering a gene that tends to make its possessor sacrifice itself for siblings when the situation is such that this sacrifice will ultimately do more to propagate the gene than efforts to keep the possessor alive and propagate the gene through the possessor's own reproduction. Such a situation obtains when the runt is such a bad risk that the resources it consumes make less than half the contribution to its probable survival and reproduction that the same resources would have made to the survival and reproduction of a sibling. This is because in such a situation the probability that the gene will propagate by encouraging the inefficient use of resources on a body sure to contain it (i.e. the runt) is less than the probability that it will propagate by encouraging the more efficient use of resources on bodies that have a 50% chance of containing it. The self-sacrificing "strategy" is more successful once the various probabilities are factored in. The point, then, is that the gene we have been considering will tend to spread in the population over alleles that make the individual struggle to the very end, eventually resulting in the fixation of the trait in the population (and perhaps eventually in the species).

Because reproduction is a subordinate end geared toward gene replication, all within the context of teleological systems shaped by natural selection, we can also see why the end of reproduction must often be understood in a *competitive* way. A male elephant seal, for example, is a certain type of gene replicating system (which, of course, carries out its highest function—no less than its lower functions—only in the context of its natural environment, which includes female elephant seals). Since its "design" is the product of natural selection, it is no surprise at all that a male elephant seal should be "designed" not simply to reproduce, but to do his very best to *out*-reproduce his peers—e.g. by fighting to control a large harem with which he alone mates, preventing many of his peers from mating at all. For it is genes coding for such competitive, "greedy" behavior that, once they made their appearance, naturally spread through the gene pool over rival alleles for more "democratic" behavior; it is thus such genes and such behavior that were naturally selected for inclusion in the "design" of the male elephant seal, and this conditions the very nature of the end of reproduction in this case. Male elephant seals work to reproduce *not* in the "public-minded" service of keeping the species going, which would not require all the fighting and competition, but in the service of replicating their genes; and because of their particular natural selection background, it is not just "democratic" reproduction that has been selected as the proximate end for them to pursue in working toward their ultimate end, but *competitive* reproduction. Similarly with female elephant seals: They have been designed not simply to reproduce, but to reproduce *selectively*, fighting off all but the dominant males; this is because such behavior best contributes to the replication of their own genes. Thus again, the end of reproduction for female elephant seals is more complex than just that of having offspring; and the ways in which it is complicated are dictated by the natural selection background that has shaped their whole teleological structure.

Similar points can be made in connection to such things as philandering among male snow geese. These behaviors are not just nasty consequences of evolution, on the periphery of the topic of functional teleology, but are right in the center of it. The elephant seal's fighting and the snow goose's philandering are paradigm cases of biological-functional behavior, playing standard roles in the working of each type of gene replicating system. And the same can be said for other rather strange cases mentioned earlier, such as the uneven distribution of resources among elephant seal offspring, the "Bruce effect" in mice, the male lion's tendency to kill existing "step-cubs" when it takes over a new pride, and the manipulation of sex-ratio among reproductive siblings by sterile worker ants. According to the account of biological function given in section three above, all of these count as functional traits. They are both *present* in and *functional* in organisms for the same ultimate reasons that more familiar traits involving such things as nutrition or protection from predators are: They contribute to gene replication of the relevant sort. If it seems counter-intuitive to regard them as equally functional, this is only because we are still caught in the pre-Darwinian mode of thinking of organisms as somehow "designed" primarily to promote their own good (as might have been the case if they were literally designed by a benevolent designer, aiming ultimately at their good), rather than fully appreciating the fact that from a natural functional point of view, given that they are products of natural selection, they must be regarded ultimately as gene replicating systems.

6. THE LIMITED ROLE OF WELFARE

Finally, what does the foregoing imply about the significance of organismic welfare as an end within biological teleology? Given (i) the argument in chapter two for the relevance of natural selection history to biological teleology, together with (ii) the observations in chapter three about the lack of any general connection between naturally selected traits and welfare and (iii) the account in the present chapter of the bearing of natural selection history on an organism's teleological profile, we can conclude that the welfare-based view described in chapter one is false—at least insofar as the preceding arguments have been sound. That is, on the above account, it is plain that the end toward which an organism's natural proper functioning at various levels is *generally* and *ultimately* directed is not the promotion of its welfare, or that of others of its species.

First, biological functions are not *ultimately* related to the welfare of organisms even where they do promote welfare, because the real ultimate end to which biological functions are directed, by virtue of the shaping of biological systems by natural selection, is gene replication of the sort specified earlier; all other ends, including ends the satisfaction of which is *good for* organisms—and I do not deny that there are many ends of this sort—are subordinate to and geared toward the ultimate end of gene replication. The promotion of welfare by any given trait is a genuine natural end *at all* only insofar as it occurs non-incidentally as part of the natural working of the organism; and this in turn is the case only insofar as this welfare-promotion tends further to promote reproductive success and hence genetic propagation. For it was only by virtue of its having done this that the trait came to be incorporated into the organism's coherent working in the first place, in such a way that it presently makes a non-incidental contribution to it. The very thing that makes the welfare-promotion non-incidental also makes it the case that it is only a *proximate* end, relevant only because of its promotion of a certain further effect. One cannot just stop with the welfare promotion and ignore the further effects without which it would never have been relevant in the first place.[37]

Second, biological functions are not *generally* related to the welfare of organisms, and this should come as no surprise. As we've seen, the characteristic traits of a given type of organism came to be incorporated into its makeup and working because (at least for the most part) each somehow helped possessors of the genes that code for it to have more (viable) offspring than conspecifics with alternative traits. Since, as many of the examples we've considered have illustrated (e.g. fighting elephant seals, overgrown bird of paradise tails, philandering snow geese, infanticidal lions, sex ratio manipulation among hymenoptera), this does not align generally with the promotion of organismic welfare, there is every reason to expect that functions will often have nothing to do with the promotion of organismic welfare; that is not generally the end to which they are directed (though again, it's of

---

[37] A qualification is necessary here. It is possible for a trait to promote organismic welfare non-incidentally even though this welfare promotion does not *currently* tend further to promote reproductive success—so long as it originally did. This, however, would just be an example of partial vestigiality—i.e. where a trait still has some of its proximate effects, but no longer has the full range of effects that made it functional in the first place. We might find this odd if we approach it with a pre-Darwinian conception of organisms and natural teleology: How can a trait that still does some *good* be vestigial? But the point I have stressed is that its proper function (i.e. the non-incidental role played in the working of the organism) was never to do such good for its own sake, but to do it as a means of achieving something further, and it is because it no longer does the latter that it is to that extent vestigial. The issue of vestigials is further discussed in chapter five.

course true that many do proximately serve that end—the heart's pumping of blood being an obvious example).

Suppose someone agrees that adaptations that happen to involve the promotion of welfare (such as the heart's pumping of blood) are properly functional, but refuses to recognize other adaptations (such as those mentioned just above) as being properly functional. In that case, some argument would obviously have to be given for such selectivity among traits that are all integrated elements in a single, coherently functioning system, non-accidentally geared toward the promotion of certain ultimate effects—*the very effects* whose promotion was ultimately responsible for each trait's coming to be included in the system. Why, despite all of this, should *welfare-promotion*—something that was relevant to the determination of the system's constitution, organization and non-accidental relations in only a partial way, and even then only insofar as it was relevant to the further effects referred to above—be treated as a *general necessary condition* for natural functionality? (This point will be further discussed in chapter six, section five.) I see no good argument for this, and suggest that the inclination here is just a vestige of a pre-Darwinian conception of organisms, according to which organismic welfare is just assumed to have a special place in biology as a natural end-in-itself.

The account I have offered thus leads to the rejection of welfare-based views of biological function. It likewise implies that *function-related needs* in biology, as discussed in section four of chapter three, do not generally or ultimately relate to the welfare of organisms. Since they are related to functions, they have generally and ultimately to do with gene replication, just as the function-related needs of an engine are related ultimately to the production of rotational power characteristic of that type of engine. Recalling an earlier example, it may be a general truth that honeybees need stings to flourish, this being a matter of *welfare*-related need. But the sense in which the individual worker bee needs her sting, i.e. to carry out her full range of functioning, which includes protecting the hive, is different. This is a matter of *function*-related need, which does not generally or ultimately relate to welfare—though of course it is entirely possible for there often to be overlap, so that things needed in the function-related sense do in fact often promote welfare (as in this case). A worker bee needs her sting if she is to carry out her full range of functioning as a gene replicating system—a functional system geared ultimately toward the replication of germ-line copies of genes in her queen and reproductive siblings, to which end the protection of the hive is related as an instrumental means. Her functioning in this respect may confer various *benefits* on a number of organisms (including even old workers who no longer contribute anything to the eventual reproductive success of the reproductives in the hive, and plants that need the bees to serve as pollinators), satisfying various welfare-related needs of theirs. But that is altogether beside the point. It is not such benefit or welfare-need satisfaction as such that grounds the facts about function and function related needs.[38] Again, this is even more clear in cases where things needed in the function-related sense don't happen to confer any plausible benefits on organisms, as with male elephant seals' need for certain traits that help in fighting for dominance: These traits are needed if they are to function fully as the sorts of gene replicating systems they are, but they don't improve the lives of elephant seals or contribute to their flourishing.

Thus, if the account of biological function that I have offered is correct, then biological teleology is structured *not* in the way that Foot maintains, as discussed in chapter one, but in a very different way that would undermine her metaethical project of construing ethical normativity as a species of natural teleological normativity. If biological functions were related ultimately to the promotion of the *good* (or *welfare*-need satisfaction) of the organism in question, or of others in its group or species, then there might be at least some plausibility in the idea that this same structure, as applied to higher levels of human life, might ground a certain range of ethical judgments.[39] But we have seen that this is not the case, at least if the

---

[38] The example of functions associated with bees in a hive is further discussed in chapter five, section thirteen.

[39] Though this would commit us to regarding morality as having generally and ultimately to do with *human* good alone, which I think is implausible. It is one thing to say that the good*ness* involved in human morality is human goodness, but it is quite another to insist that the welfare that ultimately grounds all the functions pertaining to morality, i.e. all the functional traits of character that are virtues, is *solely human* welfare, any other relevant welfare coming in only derivatively in relation to this. The implausibility of this exclusive focus on *human* welfare (at least as the ultimate source of facts about human virtue) would be especially clear, I think, if we co-existed with non-human descendants of earlier hominids, such as *homo habilis* or *homo erectus*—species that would be far closer to us in intelligence and complexity of life than the actual higher primates.

account I have presented is right. So unless we are to think of morality as aiming ultimately at competitive gene replication of the relevant sort—something which I take to be self-evidently absurd (as would Foot, without question)—there is no hope for the metaethical project of construing ethical discourse as a special case of the sort of evaluative discourse associated with functional contexts in biology generally.[40] As I noted earlier, however, there is more to be said in support of these negative conclusions—both against welfare-based views of biological function and against the attempt to use such a view to do work in ethical theory—which does not rely on the success of my own positive account. That will be the topic of chapter six.

Before moving on to that, however, I shall take some time to fill out the account offered in this chapter. First, in the next section, I'll return to the general idea of a functional system that was spelled out in section three, and explore more fully through a variety of examples just what is generally involved in meeting the conditions laid out, and why they are plausibly met by machines and organisms but not, for example, by non-living natural things. This should lend plausibility to the account I've offered, showing that it does not implausibly wind up treating even such things as crystal colonies or tornadoes as teleological entities. Then I'll go on in chapter five to develop the account of biological teleology in more detail and to examine some of its particular implications.

7. FURTHER GENERAL CONSIDERATIONS REGARDING FUNCTIONAL SYSTEMS

I have said that a functional system, in the relevant sense, is a system that can non-arbitrarily be spoken of as working in a certain way, with various parts and features that make functional (hence non-incidental) contributions to that working. Let's review and consider a little more carefully what exactly this involves, and then look at the factors in connection with some examples.

Clearly such a system involves, to begin with, a certain kind of *unity* among the items identified as the system's parts and features, so that they form some kind of coherent whole and not just a motley

---

[40] On the absurdity of such a genetically-oriented view of ethics, see chapter ten, section two.

collection of objects and properties. There is a loose sense of "system" (common in physics problems, for example) that does not require any such unity, but the notion of a working system, in the sense we are after, is clearly more restricted. But exactly what kind of unity is at issue here? There seem to be two basic necessary conditions for the relevant kind of unity: (i) there must be a certain coherence in the system's organization and consequent behavior, i.e. in the way that the parts and features causally interact to produce further effects, so that there is such a thing as the behavior of the system as a whole, where this is something more than just the sum of the unrelated effects of a collection of independent elements and properties;[41] and (ii) this unity must be something more than just accidental, since a merely accidental collection of objects and properties would not allow for any distinction between incidental and non-incidental effects—crucial to the function/accident distinction—or for any non-arbitrary determination of ends and of the proper relations among them, or for any objective norms of proper functioning.[42]

A working system, then, must non-accidentally possess a certain kind of organization among its parts and features, such that they can intelligibly be spoken of as entering into certain interactive causal relations in the non-incidental production of certain effects; that is, it must be *no accident* that the system possesses at least a certain subset of the parts and features it does, and the organization among them, such that these parts and features so organized interact causally to produce a certain special subset of the effects they bring about. Where such conditions are met, we have the resources to distinguish between functions and accidents, or ends and mere effects, and to give a non-arbitrary specification of the system's *working* at various levels, as spelled out in section three. I believe that the satisfaction of these conditions is not only necessary, but also sufficient for something to

---

[41] This is not to suggest that something counts as a part of a functional system only if it fits in like this. The human appendix, for example, does not participate in coherent interactions with other parts, but it is nonetheless a part of the human body. The point is just that there must be a certain degree of coherent organization and interaction among parts and features in order for there to be a genuine single functional system at all. Given that such a system is established, it can have some parts and features that are irrelevant to this.

[42] These problems, already described in earlier chapters, are discussed further in chapter six, section four, in connection with the hypothetical example of an accidental object that mimics a genuine functional system.

constitute a working system in the relevant sense, to which explicitly teleological concepts and language have application.

It should be kept in mind, however, that there may be wide variation in the ways in which these conditions are satisfied, some being far more robust than others; so it should be no surprise if cases are found where the conditions are technically satisfied but our inclination to apply the concepts of functional systems is still relatively weak. In the case of living things, the relevant cumulative natural selection history of the species in question provides an extremely robust way of meeting both conditions, as it results in hierarchically organized, integrated systems of unsurpassed complexity, each with a rich set of non-accidental relations between its evolved traits and certain of their effects within the evolved structure, as explained earlier. Similarly, in the case of complex machines, the design background provides a comparably robust way of meeting the above conditions. Our intuitions are therefore predictably strong that living things and complex machines constitute genuine working systems. If there are cases involving design but very little complexity of interaction, for example, or some kind of natural selection background that is much less robust than that responsible for putting together complex adaptations, then we should expect that our intuitions will be weaker in the relevant respects, without this counting against the general view.

Having said this, let us move on to the examples, to see how the above observations may be applied in thinking about cases other than organisms and complex machines. Consider first a very simple case: a "system" consisting of nothing but a bowl and a small sphere resting at the bottom; this might be something sitting on a coffee table (with the ball placed in the bowl for no particular reason—maybe even by a pet), or even a natural formation out in the desert. Now it might be pointed out that this "system" exhibits a certain sort of homeostasis. Despite various "disturbances" involving changes in the position of the sphere the "system" tends to return to the same state, with the sphere coming to rest at the same point at the base of the bowl. In many cases where homeostasis is found, it is clearly functional; body temperature homeostasis and blood sugar homeostasis both serve important functions in various organisms, for example. It is intuitively obvious, however, that the homeostasis found in connection with the bowl and sphere is *not* functional, and indeed that nothing about the "system" that contributes to this homeostasis is functional. Why is this so? Why do we not speak of *working* here, and of the sphere's remaining at or returning to the bottom as an *end*?[43]

An initially tempting answer might be that in the case of organic homeostasis some *good* is served, whereas in the case of the bowl and sphere no good comes from the ball's remaining at or returning to the bottom.[44] If the account I have provided of function in biology is at all plausible, however, then such an appeal to good is generally misguided. We should look to other factors to explain why we don't regard the bowl and sphere as a working system—why it would be patently absurd, for example, to ask of this "system" *how it works*. Even if the "behavior" of this "system" did happen to promote some good—e.g. perhaps it is a large rock formation, and there is a hole in the bottom of the bowl through which water can flow, such that the system serves to distribute the flow of water in a way that is beneficial to plants below—it would be clear that it is not a genuine functional system.

To begin with, there is not plausibly any real "it" here at all, which is why one feels compelled to put the term "system" in quotes. There is no good reason for regarding the combination of bowl and sphere as constituting a single larger entity, since there is nothing uniting them other than purely accidental spatial proximity; they constitute a "system" only the physicist's loose sense, where we can look at any arbitrary collection of matter and analyze various forces, predict various outcomes, etc. So from the start there is *no real subject* to *do* anything in the sense in which a functional system *does* its *work*. Likewise, there is no genuine *type* of system of which this particular bowl and sphere constitute a token. Even if we were to regard them as constituting a genuine system, it could be considered only *as a particular* system, since there is no type of system to which it would be related in a non-accidental way, as it would have to be in order to be

43 When I say that nothing about the system that contributes to the homeostasis is functional, I mean of course *in this connection*; there may, of course, be independent functions in connection with the bowl and ball in the case where they are artifacts (the function of the walls of the bowl being to hold soup, for example).

44 Bedau (1992a) makes this move in explaining why such a system is not genuinely "goal-directed".

regarded as a genuine token of that type—at least if the type in question is a type of functional system.[45]

In the development of my account in the next chapter, I will explain in detail why the latter is a problem, but basically the difficulty with assigning functions to things regarded simply *as* particulars is that by leaving out the idea of standard roles played by types of parts in the working of a type of system, to which the behavior of a given token system of that type can be compared, one leaves no basis for non-arbitrarily evaluating the particulars in question; there is no way of distinguishing between a thing's proper functioning and its lucky malfunctioning: Any given result could as easily be attributed to the one as to the other. In this situation, where all we have to go by are particular causes and effects, there is nothing to set apart the sphere's returning to rest at the bottom of the bowl as an *end*, i.e. as the "system's" *work*; it is just something that happens, and there is nothing at all to distinguish it from the accidental effect of a malfunction (even if it is somehow beneficial), or the side-effect of some other function—the point being that since there is nothing to decide between any of these, it is therefore in fact none of these things. The concepts simply fail to apply here.

The fundamental problem involves the relations between the various parts, features or activities of the "system", on the one hand, and the various effects they have, on the other. Consider the parts and features that contribute to the homeostasis, i.e. the walls of the bowl and their semi-spherical shape, and the sphere and its spherical shape. (Actually, the term "part" is no less problematic here than "system". If there is nothing unifying the various objects into a single genuine system, then by the same token these objects do not really constitute *parts*.) If they were *functional*, making functional contributions to the working of the "system", then we could speak of their being *for the sake of* returning the sphere to its rest position at the bottom, this being the end. But it would be absurd to speak this way, and it is not hard to see why. It is just coincidence (or "accident") that the parts happen to be present and arranged in such a way that they tend to keep the sphere

---

[45] I discuss this issue further in connection with the problem of "accidental doubles", in section four of chapter six. It may be possible in some kinds of case for something to be a token of a given type simply by possessing certain current properties, regardless of how this has come about. But this is not so for functional entities, as shown by an example given in that later discussion.

returning to the same position, just as with various other things they may happen to do, such as trapping certain heavy gasses that might be present in the air, or vibrating at a certain frequency when the sphere is moved continuously. Where it is purely accidental that parts are present and arranged in such a way as to bring about a given effect (as in this case, *ex hypothesi*), their bringing about that effect can hardly be said to be *functional*, and it would be a mistake to speak of their being *for the sake of* the effect.

The contrast with organisms could not be more stark on this point. As I have tried to make clear, the principles of natural selection provide for just the sort of non-accidental connection that is missing in this case. That is, because the compresence and coherent organization of organic parts, features and activities is brought about by natural selection, it is *no accident* that organic parts, features and activities are combined and organized in such a way as to bring about a certain range of the effects that they in fact bring about; these are the effects that are ultimately relevant to certain sorts of genetic replication—the effects that figured into the natural selection history of the type of organism in question, and that are now to be regarded as genuine ends for that type of organism, according to the account I have offered. It is therefore easy enough to see why the various interrelated concepts I have identified as bound up with the notion of *working* do not apply to a mere bowl and sphere, and thus why we do not regard it as a working system. The fact that it displays a primitive kind of homeostasis, and that homeostasis is often a feature of functional systems, is dwarfed by the vast differences between such a "system" and either a machine or an organism.

It will be instructive now to modify the example, bringing it closer to being a working system, to see what sorts of factors seem to matter and why. So, for example, suppose that the bowl and sphere system is not just an accident on the coffee table or in the desert, but is a system designed for use in a physics class to illustrate the concept of homeostasis. Here a number of problems are solved right away. There is now a *non-accidental unity* to the system that was not present before; the bowl and sphere *belong* together as *parts* of this device, and together constitute a larger whole—a "gravity-homeostasis device", say. That is, they constitute a system that is genuinely a token of the type *gravity-homeostasis device*, and not just a set of two objects accidentally bearing certain close spatial relations to one another. Moreover, there is now something to set apart the effect in question from various other effects, as a genuine end: The device has been *designed* to return the sphere back to its resting position whenever it is disturbed; the curvature of the bowl contributes to this effect non-incidentally.

We should not, of course, jump to the conclusion that the fact of design is a general necessary condition for something to constitute a working system. As I've argued from the beginning, it is plausibly the case that the fact of design satisfies in artificial contexts a more general necessary condition which can equally be satisfied in other ways, which is why living organisms can be genuine functional systems, even on naturalistic assumptions. What matters is just that it be no accident that the system's parts and features are combined and arranged in such a way as to bring about the effects that are to be identified as ends (where the notion of *its being no accident* is subject to the qualifications mentioned earlier). This more general condition is met without the presence of literal design in the case of organisms, given that they are products of natural selection; the absence of literal design does not seem to present any obstacle to our considering them as functional systems. But let us return to the example at hand. We seem now to have a system that (i) possesses real unity, with genuine parts; (ii) is a token of a type of system to which it is non-accidentally related; and (iii) has parts and features that not only contribute to a certain outcome (returning the sphere to its resting position when it is disturbed), but are such that their being compresent and arranged in such a way that they do so is no mere accident.

Given all of this, we might legitimately speak of the effect in question as an *end* of the system, which is non-incidentally brought about. Likewise, we can speak of the walls of the bowl, for example, as being shaped as they are *for the sake of* returning the sphere back to its original position when it is disturbed, and we can judge parts or features of the system to be defective if they inhibit this (unless the problem lies in bad design, in which case this will be the criticism). On the other hand, what we have here is still a very minimal case of a functional system, perhaps a limiting case in terms of simplicity. There is so little in the way of *what it does* that, once we have an answer to *what it is used for* (e.g. illustrating homeostasis to physics students), there is little room for talk of how it accomplishes this, i.e. *how it works*—unlike in the case of something like a sewing machine, for example. It would be odd to ask how such a bowl and sphere system works because its "work" is simply to return the sphere to the bottom after it is displaced by some external force, and this is just a matter of its possessing such a structure that the force of gravity accomplishes this; that's all there is to "the operation of the system." This is therefore not in any case a very robust example of a working system.[46]

It would not take much, however, to bring this closer to being a central case of a working system (or sub-system), even if it is still relatively simple in its design and operation. Suppose, for example, that it is now a functional part of a larger machine—there is, say, a hole in the bottom of the bowl as in the case mentioned earlier, and the contraption serves as a valve governing the pressured flow of some liquid from a source below, such that the bowl fills to a certain height, after which the pressure above tends to force the sphere to the bottom, sealing the hole and preventing backwards flow.[47] Here the bowl and sphere sub-system has a function as part of a larger functional system,

---

[46] This is not to insist that talk of how such a system works is necessarily inappropriate. We can, after all, ask the question "how does it work?" even of simple things (i.e. things that aren't systems at all) that are used in certain ways, such as stain remover. The point is just that such talk in the case of the simple system considered above will be much less robust than in the case of something like a complex machine or organ.

[47] I am indebted to Robert Adams for this example. (It doesn't matter for the points being illustrated whether I've got the physics quite right, or what kind of larger machine this might be.) There are actually two interesting possibilities with regard to this sort of case. We might imagine either that (i) the valve system has been designed for this purpose, or that (ii) it was not designed at all (or was designed for a different purpose) but is at any rate being *used for* this purpose as part of the larger machine. The latter case would present the complication that it might well be an accident that the valve system possesses such an arrangement of parts and features that they bring about the immediate effects exploited by the designers or users of the machine. On the other hand, the valve system's *use* as part of the machine *would* make it non-accidental that it is present and has the effects it has within the context of the machine; that is, it is no accident that the machine contains such a part that has such effects within the machine (namely, regulating the flow of liquid as it does), inasmuch as this part has been deliberately integrated into the machine by virtue of its being such as to have such effects when so integrated. This would seem to be enough to allow for talk of its genuinely being a valve, its functioning well or poorly, its parts making certain functional contributions to the working of the machine, and so on. (There would, of course, be greater difficulties if the whole "machine" were just an accidental conglomeration. That sort of case would have to be treated in the way in which I treat the case of "accidental doubles" in chapter six.) For simplicity, however, I will assume that the valve system as well as the machine in general are products of design, and that their design matches their use.

rather than just being something that has a certain use in itself (e.g. to illustrate homeostasis). This embedding of the bowl and sphere contraption within a larger, more complex functional system makes it much more natural to speak in terms of *what it does* and not merely in terms of what it is used for, though it can of course be spoken of in these terms as well; what it does as part of the operation of the larger system—its *work*, i.e. regulating the flow of liquid, which it does through the use of gravity and the sphere—goes beyond merely returning the sphere to the bottom after disturbances, as in the previous case. This makes it much more natural to speak of it as a working (sub-)system, about which we might naturally ask *how it works*, i.e. how it manages to carry out the work of a valve. What we have here is not all that far removed from something like the Watt steam governor (a regulating device that uses spheres and centrifugal force), and it seems to have a similar claim to be regarded as a genuine functional (sub-)system, though it is still very simple.

Let us turn now to a different sort of case: a *natural* system where there are non-accidental relations between certain features and certain of their effects. Consider, for example, a silicate crystal with a structure that turns out to be partly the result of a kind of cumulative natural selection.[48] That is, there may originally have been a simpler type of crystal structure which underwent various mutations or "flaws" as such crystals replicated, resulting in a variety of different lineages; as some structures proved to replicate more prolifically in the given environment than others (perhaps growing and splitting more rapidly, due to certain properties resulting from their structures and chemical makeup), the former tended to increase in frequency over the latter, and so on with further mutations and selection, leading to increasingly efficient replicators.[49] Is there sufficient cause for speaking of a working system here?

What we have in the case of such a crystal is analogous *at best* not to an organism, but to an independent piece of DNA—or, if we are imagining a colony of such crystals, i.e. a macro clump, then the analogy would be to a clump of DNA: a number of copies of a single stretch of DNA all clumped together, one simple pattern repeated over

---

[48] This example will be discussed again in chapter seven, in connection with a criticism of Kitcher's (1993) view.
[49] See Dawkins (1987, pp. 151 f.)

and over again. There is thus no analogue of a living organism, or even of a genome—which is vastly more complex than a clump of multiple copies of a singe gene—in a mere crystal or crystal colony, and hence no particular reason to expect that we must have something like a working system here just because there is a natural selection background.

On the positive side, there is indeed a kind of non-accidental unity to a "crystal unit", by which I shall mean a crystal at the smallest scale at which the evolved structure is manifested; each unit is an instance of an evolved type of crystal, giving rise to more of the same as it replicates, and it is no accident that it exhibits such features as in fact enable it to replicate as it does. By extension, there might be a kind of natural unity at the level of clumps within a certain size range, insofar as the replicating success of the structure in question may have had partly to do with features exhibited at that higher level, so that it is no accident that the units tend to group together in clumps within that size range before splitting apart. But apart from this, there would not be such non-accidental unity at the level of clumps. How far, then, do we get if we try to think of a crystal unit or suitably unified clump—manifesting naturally selected structures and properties—as a working system? Is there any content to be given to the notion of its possessing various parts and features that play non-incidental roles in its working—non-incidental effects that interact to give rise to certain further effects that can legitimately be regarded as ends of the system?

In the case of an organism, we are dealing with a complex, hierarchically organized structure *built by* certain replicating molecules—a structure the organization of which is vastly more complex than that of the replicating molecules themselves, which enables it to carry out an almost unimaginably complex set of finely integrated activities aimed ultimately at the promotion of the replication of the molecules that built it. By contrast, what we are presently considering is a replicating molecule itself, not some phenotypic structure that does various things (constitutive of its *work*) to promote the molecule's replication. Thus, there seems to be nothing for *working* to consist in here, other than replication itself; there is no further activity carried out by a system that ultimately promotes the replication of the molecule, but just whatever happens in the replication process itself. Assuming that this amounts to little more than ions being attracted to each other and sticking together in a certain pattern, then breaking apart and attracting new ions, and so on, it is hard to see how we can speak of working here, or of parts or features having effects that interact cooperatively in the promotion of an end, such that they can be regarded as playing roles in the working of a system that might be said to be functioning well or malfunctioning, and so on. *Ex hypothesi*, the parts (certain ions) and features (their arrangement, and any resultant higher level properties) non-incidentally promote the replication of the molecule; but they do not do so in a way that makes it very tempting to apply to them the language of working systems.

In summary, even granting the existence of a natural selection background, a replicating crystal, unlike an organism, is simply not much of a *system* at all, with very little in the way of a cooperative endeavor by various interacting parts and features in bringing about replication, and it is this that motivates a reluctance to speak of teleology here. A given feature may be present because such a feature improves replicating efficiency, thus possessing a consequence etiology; but it is a stretch to think of it as playing a role in the working of a system, and this explains our natural reluctance to attribute a function to it—as we would be committed to doing according to standard etiological views, and even according to more restricted etiological views inasmuch as the etiology involves a natural selection background (and so likewise, it would seem, a "design background" in Kitcher's sense).

It is not part of my view that there should be any sharp dividing line between non-functional entities and genuine functional systems. Perhaps there is already some slight pull toward speaking of functions in the above case, and if not, at some point there will be as we modify the case to bring it closer to biological cases—predictably with some gray areas where intuitions become cloudy. To begin with, cumulative natural selection among various crystal lineages might give rise to crystals that "happen to catalyze the synthesis of new substances that assist in their passage down the `generations' . . . .[These substances] could be seen as tools of the replicating crystal lineages, the beginnings of primitive `phenotypes'."[50] These new substances might serve as

[50] Dawkins (1987, p. 156). In fact, it has been hypothesized (by Cairns-Smith) that long before the origins of life, organic molecules were among the "'tools' of . . . inorganic

protective coating for the crystals (this being very similar to the case of a virus, which is basically just a piece of DNA or RNA coated in protein), or they might help to attract minerals more effectively, or to dissolve other crystals whose minerals are then absorbed, and so on. At some point as the "phenotype" increases in complexity, either still in connection with such cases or in connection with more complex cases involving rudimentary life (i.e. where organic replicating molecules combine in mixed sets and build cells), there will be enough to the system in question to allow for natural and fairly robust applications of teleological concepts. This is admittedly somewhat vague, but such vagueness is in the nature of the subject—just as it is with concepts such as *life* or *welfare*, which are equally difficult to nail down in this area—and it is not cause for suspicion about clear cases, such as multicellular organisms.

---

crystalline replicators," leading eventually to the appearance and takeover of replicating DNA.

CHAPTER V

Development and Applications of the Account

Let me turn now to developing the account in chapter four by taking up a number of associated issues and exploring some interesting applications.

1. USE-RELATED BIOLOGICAL FUNCTIONS OF ITEMS OTHER THAN ADAPTATIONS

The connection between biologically functional entities and adaptations is not as direct as it might at first seem to be, though it is still very close. In some cases, such as that of a secreted hormone or neurotransmitter, which plays some standard role in e.g. growth or memory, it may be unnatural to speak of the functional entity in question as an adaptation. In such a case, however, both the process and the mechanism by which the entity is manufactured are adaptations, and it is thus natural to speak likewise of the entity's *presence in the organism* as constituting an adaptation. Of course, in other cases, such as that of a given type of vitamin that is ingested with food and plays some important role in metabolism, the entity might not even be a product of adaptations of the organism. Yet even here, both its presence in the organism in certain quantities and especially the ways in which it is used by the organism, in the sense in which it can genuinely be said to play a role in the organism's functioning, will be governed by adaptations, thus no less reflecting the shaping imposed by the principles of natural selection.[1]

The most interesting cases of function, however, involve entities which are themselves adaptations—whether organs, physical traits, organic processes or patterns of behavior. Any adaptation (again *type*, not token—I'll deal with attributions to tokens in section eight below) which falls within the phenotypic framework expressed by the co-adapted genome of a given type of organism will typically play some standard role in the working of such an organism—a role it was "designed" to play by the processes of natural selection that shaped the system. In cases involving processes or activities (e.g. the pumping of the blood), the activity will make a constitutive contribution to the working of the organism, being both constitutive of its functioning at one level, and promoting certain biological ends associated with the working of the organism at other levels (as the pumping of the blood promotes the distribution of nutrients throughout the body). In other cases, such as a coloration pattern that attracts pollinators or camouflages against predators, it may be a little odd to speak of the adaptation's having this effect as being *constitutive* of the organism's functioning (though I am not sure we shouldn't say this); but in any case, the crucial point is that the adaptation plays a role in the organism's functioning, contributing to the satisfaction of some end associated with the working of that type of organism.

2. BEYOND SELECTED EFFECTS: USE-RELATED BIOLOGICAL FUNCTIONS OF ADAPTATIONS

We should not, however, jump to the conclusion that the function of every adaptation is limited to the effect(s) for which it was selected. In

[1] Thus, an organism may be "designed" by natural selection to use largely ingested vitamin D for the absorption and metabolism of calcium and phosphorus in the production of bones, just as a boat engine might be designed to use the water it sucks in for cooling. One interesting difference between substances that are manufactured by the system in question and those that are not is that while the former might be said to be defective, i.e. if there is some malfunction in their production, the latter cannot. Blood or insulin can be defective, for example, along with synthesized vitamins, but not the vitamins or minerals that are simply ingested. (Similarly, if the water sucked in by a boat engine happens to be polluted in such a way that it fails to cool the engine properly, that does not make it defective water; it is merely unsuitable for the engine.) On the other hand, the *level* of a non-manufactured substance in an organism can certainly be *deficient*, as with dehydration or rickets, just as in the case of manufactured substances.

light of the above points about use-related biological functions, it is plausible that an adaptation can itself have use-related biological functions as well, quite independently of the effects for which it was selected. Suppose, for example, that sea turtles regularly use their tails for protecting their eggs, yet this is not the effect for which the tail or its traits were selected.[2] Can protecting eggs be regarded as part of the tail's function on the above account, assuming that this tends to contribute to reproductive success, and hence to gene replication? Or must it be regarded as a mere side-effect, at least as far as biological teleology is concerned, since the tail was not selected for its egg protecting abilities? The answer depends on the nature of the egg protecting behavior. If that behavioral trait is itself either an adaptation selected because of its ultimate effects on gene replication, or a characteristic way of pursuing a goal the having of which is an adaptation, then on my view it may indeed be said that the tail has the protection of eggs as one of its proper functions. Let me take up the two cases in turn.

First, suppose that the tendency to use the tail to protect the eggs is itself a behavioral adaptation—"hard-wired" as a result of natural selection acting on various alternative behavioral traits.[3] Certainly this behavioral trait would be a functional one, on the view I have presented: It would play a standard, non-incidental role in the working of the organism as a gene replicating system. But now since the tail plays a role in this functional behavioral trait, i.e. it is *functionally used for* protecting the eggs, then to this extent the tail plays a standard, non-incidental egg protecting role in the working of the organism, and can therefore be assigned such a function in addition to whatever function it has for which it was selected. This should not be surprising in light of what has already been said about such things as vitamins and minerals:

---

[2] I am indebted to Marc Lange for this example, and for helpful comments on the questions it raises.

[3] That is, suppose that it is what ethologists call an *innate motor program* (or series of programs). In making this supposition, we needn't imagine it to be a classical *fixed action pattern*, on the order of the egg-rolling behavior of greylag geese, which once "released" continues to the end even if the eggs are artificially removed, thus displaying extreme insensitivity to feedback. It may be more like the complex innate motor programs involved in the burying behavior of ground squirrels, the nest building of weaver birds, or the web-spinning of orb-web spiders, for example, which, although "pre-wired", nonetheless allow for significant adjustment in response to feedback, enabling the organism to deal with various contingencies. See Gould (1982, ch. 11).

They certainly play important roles in the functioning of the organism, and are hence assigned functions, even though they are not themselves adaptations selected for these effects. The point, then, is that adaptations with other functions can *also* acquire functions in the way that vitamins and minerals do. Just as the organism *uses* certain vitamins and minerals in certain functional ways, giving them functions in the life of the organism, so too the organism can *use* its body parts in certain functional ways, thus likewise giving those parts functions in relation to that use (in addition to whatever other functions they may already have).[4]

It is worth noting that this intuitively plausible position on such functions is not open to proponents of at least the most straightforward versions of the etiological view. Neander, for example, explicitly maintains that "biological proper functions are effects for which traits were selected by natural selection".[5] This would allow her to maintain that protecting the eggs is the function of the behavioral trait, but not that it is one of the functions of the tail, since the tail was not selected for this effect. The situation is less clear in Wright's case, though there are problems however his position is interpreted. Wright maintains that "the function of X is Z iff (i) Z is a consequence (result) of X's being there, and (ii) X is there *because* it does (results in) Z."[6] If we understand the second condition to imply (in biological cases) that X's doing Z is at least partly responsible for X's selection, i.e. for its coming to be a characteristic trait of members of the population, then Wright is no better off than Neander, and must likewise deny that the turtle's tail has an egg protecting function; for interpreted in this way, the tail "is there" *not* because of its egg protecting effects, but because of other effects for which it was selected. In fairness to Wright, however, the sense in which X must "be there" because it does Z is deliberately left rather vague, and is not obviously meant to require that X must have been selected for the effect Z. It may be enough on Wright's view if the protection of eggs has made at least some

[4] I will henceforth generally omit the qualification "non-incidental" when speaking of roles things play in the working of an organism, as it is somewhat cumbersome. Unless otherwise indicated, such references to roles should be understood as so qualified.

[5] Neander (1991a, p. 168). Cf. also Griffiths (1993, p. 412): "A feature will have a proper function only if it is an *adaptation* for that function. The trait must have been selected because it performs that function."

[6] Wright (1976, p. 81).

contribution to the *maintenance* of the tail in sea turtles, thus figuring at least in this way into a causal explanation of why the tail is present in current sea turtles. If so, he may be able to claim that the tail's helping to protect eggs is a function after all, at least if this has helped to maintain the tail against selection pressures tending to eliminate it.[7] But this position is still not satisfactory. There may in fact not have been any such selection pressures operating against the tail, in which case the etiologist would again be stuck with denying the plausible claim that it has the function of protecting eggs. And even if there were such pressures, the tail's other functions may have been more than sufficient to maintain it, so that the egg-protecting might still fail to figure into the consequence-etiological explanation of the presence of the tail, again leading to the denial of function.

Finally, even if there happen to have been such selection pressures, and the egg protecting use did make at least some genuine contribution to the maintenance of the tail, it does not seem that the question whether or not the tail has the function of protecting eggs should turn on the existence of such selection pressures (or on whether or not the egg protection actually made a contribution to the maintenance of the tail). It seems clear that the tail would have the egg protecting function even if there had been no such selection pressures against the tail or its relevant features, and so no past contribution on the part of the egg protection toward the maintenance of the tail or its relevant features (which were maintained for other reasons). It is, I think, a liability for etiological accounts that they cannot deal adequately with such cases, whereas the account I have offered can readily accommodate them. (More will be said about etiological accounts in chapters seven and eight.)

Let us return now to the second of the two kinds of case distinguished above, which is more complicated. In the case of animals capable of goal-directed behavior—perhaps even in the sea turtle's case—it is possible for there to be complex, non-accidentally adaptive behavior which is not simply "hard-wired", i.e. not simply a result of natural selection acting on genetic variation for a number of alternative behavioral traits, whereby one comes to be selected over the others. What I have in mind here is a case where what is actually selected is not

---

[7] Cf. Kitcher (1993, pp. 384 f.), who considers such versions of the etiological view explicitly. This comes up again below.

a particular behavior pattern, but the more general tendency to have a certain *goal* under certain circumstances, and to exhibit *some* appropriate behavior from the organism's behavioral repertoire, in the service of that goal (which behavior might vary greatly depending on the circumstances).[8] For example, it may be an adaptation in sea turtles to have the general goal of protecting their eggs from harm, and there are then a variety of ways in which that goal may be pursued, given the turtle's behavioral repertoire and the circumstances that obtain. Whenever a turtle thus employs some bit of behavior in its repertoire to protect its eggs, this behavior is non-accidentally adaptive, being a result of the more general adaptation in the given circumstances. And this is sufficient, I think, for regarding such behavior as biologically *functional*. Even though it is not an instance of any *specific* behavioral adaptation (as in the "hard-wired" cases), it is an example of the tendency to behave in ways conducive to the protection of the eggs, which *is* an adaptation; since the latter plays a standard role in the working of the organism, this is enough to make the particular behavior count as functional in nature.

Now whether or not we can speak of such behavior as conferring a general egg-protecting function on the *tail* will depend, I think, on whether such behavior is relatively characteristic of sea turtles. If it were something displayed only by a few individuals (the rest only using other parts of their bodies for this purpose, say), then while the

[8] It should be noted that in speaking of *goals* I am here introducing a distinct teleological concept that has not yet been discussed—a *psychological* teleological concept, rather than a functional teleological one. (This distinction does not, of course, imply that the two types of teleology are never closely related. Indeed, what I am considering in the above example is the possibility that some psychological teleological phenomena are also biologically functional, or that part of an organism's proper biological functioning may involve its having certain psychological teleological dispositions). In making use of this concept at this stage, I am not committing myself to any particular account of what having a goal consists in. It is irrelevant here, for example, whether a purely physicalist account is possible, such that having a goal can be explicated in non-teleological, physical terms, or whether (as I suspect) this cannot be done. If someone is concerned to avoid any teleological involvement that is not ultimately explicable in non-teleological, physicalist terms (in the way that *functional* teleology in biology is on my account, i.e. in terms of certain kinds of causal structures and non-accidental relations), then he may be wary of the use of such a concept here without any indication of the prospects for such an understanding of it. Even if so, however, nothing in the present discussion actually requires any very rich notion of having a goal, and the same points could be made by speaking simply of general dispositions, which might be thought less problematic.

improvised behavior in question would still be functional, it would not enable us to attribute the function of egg protection generally to the tail; for it could not naturally be said that the tail plays a standard egg-protecting role in the working of such an organism. But if it is in general quite natural, given their behavioral repertoire and the circumstances in which they generally find themselves, for sea turtles to use their tails to protect their eggs, then it will be plausible to say that the tail plays a standard role in the protection of eggs, and hence that this is one of its biological functions in sea turtles.

It is, of course, crucial to my position in both kinds of case discussed, that the *use* in question be *biologically functional* use. This is why it was important to assume that the behavioral trait (i.e. the tendency to use the body part in question) was either itself an adaptation or something motivated by a more general adaptation operating in the particular circumstances that obtain: Such a connection to adaptations is necessary for the behavior, and hence the *use* in question, to be biologically functional. Similarly, in the case of ingested vitamins and minerals it is crucial that the use in question be biologically functional; we attribute functions to them insofar as they are used by the body as part of its proper functioning, the various processes in question being adaptations, and so on. If the use to which a body part (or anything else) is put is *not* biologically functional use, connected up in the right way with adaptations, then it will not support any attribution of biological function—even though it may incidentally serve the same ends that are served by biological functions.

Suppose that the tendency to employ the tail (as opposed to other body parts) in protecting the eggs is something that is learned and socially transmitted from generation to generation. Could we still consider such use to be biologically functional in nature? On my view, we could. I do not see why the addition of learning and social transmission should get in the way of the fact that what results from the combination of natural selection and these other factors is a characteristic behavior performed in the service of a goal the having of which is an adaptation; this fact seems to be sufficient to establish the relevant non-accidental connection between the behavior (i.e. using the tail to protect the eggs) and the biological end in question (i.e. protecting the eggs, which is the end served by the turtle's having the goal of protecting its eggs, and so likewise by the behavior motivated by its having that goal).

3. BIOLOGICAL FUNCTIONS AND TOOL USE BY ANIMALS

The above points may have interesting implications regarding the teleology surrounding tool use. Consider an actual case involving chimpanzees. It is presumably an adaptation to have the goal of procuring food, which takes on various particular forms in particular circumstances, such as opening boxes containing (or at least believed to contain) food. With regard to such goals,

> chimpanzees are able to invent techniques and to transmit them to others. The use of sticks to pry open food boxes, [for example], was invented by one or a few individuals at the Gombe Stream Reserve, then evidently spread through the troop by imitation. . . . Each tool-using behavior recorded in Africa is limited to certain populations of chimpanzees but has a mostly continuous distribution within its range. This is just the pattern expected if the behavior had been spread culturally.[9]

On my view we should regard such use of sticks as biologically functional in nature, just as with the turtle's use of its tail to protect its eggs; again, what we have here is a behavior that is performed in the service of a goal the having of which is an adaptation, so that it is non-accidentally related to the biological ends served by that adaptation. On the other hand, there is a difference between the two cases. In the case of the tail, I said that if the use is regular enough, we might extend a function attribution to the tail itself in this regard, claiming that it is one of the functions of the turtle's tail to protect the eggs. By contrast, the sticks used by chimps to open boxes are not body parts, and it seems much less natural to attribute biological functions to them in relation to their use, even if this use were to become as standard as the turtle's use of its tail to protect its eggs.

Of course, spider webs and bird nests are not body parts either, yet it is natural to attribute biological functions to them. So that cannot be

---

[9] Wilson (1978, p. 32).

the crucial factor.[10] Perhaps an important difference is that the sticks used by chimps are not their "artifacts" in the way that webs are spider "artifacts". But this is not entirely clear, as it is possible that the sticks are sometimes broken off or trimmed for the purpose of such use as tools, making them primitive "artifacts" after all; and in any case, since what is at issue here is use-related function, it is not clear why the difference between artifacts and found objects should be crucial. A more relevant difference between spider webs and the chimps' sticks may be that whereas webs and nests figure into the natural selection history of spiders and birds, and so in this way are intimately bound up with the biological ends they serve, this is not true of sticks in relation to chimps; thus, it may seem that sticks are just too indirectly connected to the ends they are employed to serve (even granted that this employment is biologically functional in nature) to be assigned biological functions. But this can't be the whole story either. The same, after all, could be said about the turtle's tail, at least with regard to the end of egg protection: *Ex hypothesi*, the tail does not figure into the turtle's natural selection history in any way relevant to egg protection; it has simply come to be used regularly for that purpose—yet we can still plausibly ascribe a biological function to it in connection with that use. (I am here assuming the second of the two scenarios imagined above, and imagining that the use has become fairly standard.)

There are, however, two further differences between the case of the tail and the case of the sticks, which may explain why they are intuitively treated differently. First, despite the fact that the tail does not figure into the turtle's natural selection history in any way relevant to egg protection, it is nonetheless—as a standard part of turtle anatomy—a genuine *type* of entity associated with turtle life, which thus lends itself to function attributions. It seems unlikely that such a thing can be said about the sticks used by chimps for opening boxes, even if the practice were to become standard; that is, it is not clear that there is any genuine functional type here, comparable to the sea turtle's tail. (I discuss the importance of types in section eight below.) Second, the description of the function in the case of the tail—protecting eggs—is general enough that it can be considered one of the turtle's natural biological ends (which could in principle be satisfied by other

---

[10] I discuss such cases in section twelve below.

functional behaviors), even though the tail's use to protect eggs does not figure into the natural selection history; by contrast, the description of the function of the sticks—prying open boxes—is too specific to be so regarded, even though it is *in the service of* the general biological end of procuring food. This, I think, is largely what is behind the reluctance to speak of the sticks' function as a biological function, and the inclination to treat it as a different sort of use-based function.

It may not be hard to get around these problems by modifying the example slightly. For example, it seems possible for there to be some tool, such as a particular type of hollow reed, that comes to be so standardly used for one particular purpose, such as sucking insects out of logs, that it can likewise be regarded as a genuine type of entity that has come to be sufficiently associated with chimpanzee life (even though it is not a body part) as to be relevantly on a par with the turtle's tail. Perhaps in such a case we should not refuse to ascribe a chimp-related biological function to the entity after all; it may be enough that it plays a standard role in the biological functioning of chimps, much as certain vitamins and minerals do. I see no objection to this, though in both cases it would be a good idea to be explicit about the fact that what is at issue is use-related function, and that in the case of the sticks the use involves innovation and learning; as long as this is made clear, the function attributions need not come across as peculiar.

As we move away from such a case to more idiosyncratic cases, and more specific, circumstance-dependent descriptions of what the entities in question are used for, ascriptions of biological functions to such entities will be less plausible, and we will be inclined to speak only of use-related functions without regarding them as biological functions. But even here, where the *use* in question is still biologically functional in nature, we have a situation in which the use-related functions are intimately related to biological functionality. For example, even if we refuse to ascribe a biological function to the sticks in relation to chimp life, their use-related function—i.e. their standard use in prying open boxes—is something they have as a direct result of *behavior* which is *biologically* functional in nature (i.e. behavior that is motivated by a more general adaptation, namely, the possession of the goal of procuring food).

4. USE-RELATED BIOLOGICAL FUNCTIONS IN HUMAN LIFE

The connection brought out above, between a certain range of use-related teleology and biological teleology, would seem to be important to understanding the teleology associated with a significant range of animal behavior and certain aspects of the environment involved with it. It also raises a further interesting question. If the use-related functionality of (non-human) animal artifacts or tools is also a case of biological functionality—as I would say in the case of the spider web—or is at least intimately related to biological functionality—as in the case of the chimps' sticks—then is there at least sometimes a similar connection between the functional teleology found in connection with *human* artifacts and biological teleology? For example, is the artificial functionality of the fisherman's net (the function of which is to catch fish, paralleling the function of the spider web, which is to catch and trap insects) also a case of biological functionality, like that of the spider web, or at least intimately related to biological functionality, like that of the sticks?[11] In order to answer this question adequately, we obviously need to take up the difficult matter of biological teleology in human life. That is the topic of chapter nine. I will, however, say something briefly about this here, with the aim of providing at least a tentative answer to this question.

It is not always a simple matter to determine whether a given sort of use is biologically functional in nature. Consider, for example, the use of the human nose to support eyeglasses.[12] At first, this might seem clearly *not* to be a case of biologically functional use, even though it may incidentally contribute to gene replication (by making the wearer less likely to walk in front of trucks, etc.). For using the

---

[11] The comparison of spider webs to fishing nets is made by Woodfield (1976, pp. 27–8). His claim is just that the former, no less than the latter, should be considered to have artificial functions, at least insofar as spiders may plausibly be said to use their webs to achieve their goal of catching flies. My claim (discussed in section twelve below) is that the function of the spider web is a *biological* function regardless of whether or not we can legitimately speak of spiders as having goals (in the psychological sense). That is, even if we can, so that the use-related function of the spider web involves spiders' psychological goals, the having of those motivating goals is still plausibly an adaptation, so that this case of use-related function (or artificial function) is still just a special case of biological function.

[12] This example is mentioned by Gibbard (1990, p. 63) and by Neander (1991a, p. 179, fn. 8). I was helped in thinking about it by comments from Marc Lange and Bob Adams.

nose to prop up eyeglasses is obviously not a behavioral adaptation (i.e. there was never natural selection for the tendency to use the nose to prop up eyeglasses), and neither is it the simple result of a more general adaptation, on a par with a turtle's using its tail to protect its eggs, or a rat's using its nose to move an obstacle to food. Unlike the turtle or the rat, we typically have *reasons* for wanting the end served by the use (in this case, the end of seeing better), which are at least partly independent of the influences of natural selection, often extending well beyond anything of relevance to gene replication; someone may want to see better in part so that she can once again enjoy reading fine-printed mystery novels, for example, and we may in general value clear vision simply because of the joy we take in visual beauty. This tends to remove our use of the nose for supporting eyeglasses from the realm of biologically functional teleology, suggesting that it is more on a par with the sort of use-based teleology that is typically involved when we use artifacts in the service of our various independently held ends—as e.g. someone might use a bookmark for keeping her place in a novel.

At the same time, however, it is surely plausible that at least part of our desire to see better has an evolutionary basis. Surely a general interest in seeing what is around one is an adaptation in a variety of animals (think of cats), and the same is plausibly true of us. Thus, to the extent that our using our noses to support eyeglasses is partly motivated by such a general interest in visual information and desire for adequate vision, which have an evolutionary basis, this case of use is relevantly similar to the cases involving the turtle or the rat; the contribution to gene replication is to this extent non-incidental, and there is thus reason for regarding the use of the nose to support eyeglasses as biologically functional in nature. What we have here, I think, is simply an interestingly *mixed* case. The use of the nose to support eyeglasses is neither simply a matter of biologically functional use (like the turtle's use of its tail to protect its eggs), nor simply a matter of use that is relatively independent of the influences of biologically functional dispositions (like the use of a book mark to keep one's place in a novel); it is partly one and partly the other. We may expect there to be a range of such cases, some closer to central cases of biologically-oriented use-based teleology, some closer to central cases of independent use-based teleology. At some point as we move in the latter direction we become far enough removed from evolutionary influences that it seems clear that what is involved is simply artificial teleology or technique-based teleology—as in the use of the fingers to fret guitar strings (even if this might happen to promote gene replication by tending to attract potential mates).[13]

It is worth noting that the latter sort of use-teleology, which applies to bodily parts but goes well beyond biological teleology, can plausibly ground certain familiar kinds of teleological claims that ought to be distinguished from biological-teleological claims. We may speak, for example, of the *function* of the thumb of the left hand in the context of classical guitar technique; and in relation to the ends associated with guitar playing, an aging player might complain that his fingers no longer *work* as well as they used to.[14] The standards of good working here are set not by biology, but by the ends to which the fingers are put by the player—though obviously there are limits here, so that one cannot complain that one's arms are not working well because, however hard one flaps them, they just won't provide sufficient lift for flight; the end must be within the realm of plausibility, even if—as is likely the case with any accomplished classical guitarist—it involves capacities that go well beyond the norms set by biology. The aging guitarist, for example, may even now have extraordinarily dexterous fingers—fingers which certainly could not be said to be defective or malfunctioning in terms of biological function—and yet he may be fully justified in saying that they no longer work as well as they used to, again applying norms of functioning or working that go beyond biology (while remaining within the realm of plausibility, at least for gifted performers).

Returning now to cases involving use that is both partly biologically functional in nature, and partly independent, i.e. partly based on motivations that cannot be traced significantly to biologically functional dispositions: What should we say about the functionality of the used objects themselves—whether artifacts, found objects or body parts? It is clear, for example, that the use-based function of glasses is to correct vision; but can this be considered a matter of *biological* function in relation to human life, inasmuch as the use of glasses is plausibly in part a matter of biological functional use? And similarly, what about the use of the nose to hold them up? The implication of my

---

[13] These issues are further discussed in chapter nine.

[14] I owe the latter sort of example to Robert Adams.

view is that this is likewise a matter of mixed teleology. The functionality is partly biological in nature, partly independent—lying somewhere between the functionality of the chimps' sticks or reeds (at the biological end) and the functionality of the book mark (at the relatively independent end).

Even if we refuse to speak of the function here as a *biological function*—which would indeed be misleading without plenty of explanation—the point is that it will still be partly biologically-based, sharing something in common with the case of the sticks, which is not shared with that of the book mark. Something similar can be said, I think, with regard to the fisherman's nets mentioned earlier. They are certainly not simply on a par with spider webs, but there is more continuity with such cases than is generally thought. That is really my point here—something taken up further in chapter nine. Of course, one important difference between eyeglasses or nets, on the one hand, and non-human "artifacts" like spider webs, on the other, is that the former are not merely used for certain purposes, but are *designed* for them as well. Perhaps the function of glasses, for example, must therefore be understood as having two aspects—a use-based aspect (what glasses are regularly used for) and a design-based aspect (what glasses are *designed to do*), even though these may coincide perfectly. In that case, only the use-based aspect bears comparison to cases of non-human animal "artifacts" and is affected by the partly biologically functional nature of the use.[15]

In the case of the nose, there is of course no issue of a design-based aspect; if there is to be any function attributed to the nose in relation to eyeglasses, it will be purely use-based. As in the case of the turtle's use of its tail to protect eggs, however, this use must be fairly standard before we can naturally speak of a *function* in connection with the nose's holding up glasses. And, of course, even if we do attribute a use-based function to the nose in this regard, it will again be at most partly biologically-based—falling somewhere between that of the turtle's tail and that of the guitarist's fingers (in relation to fretting strings). I'm not sure whether the use is standard enough to warrant a function

[15] Whether a similar point to the one made about use-related functions could be extended in some cases to the design-based aspect of artificial functions would depend on the extent to which the *designing of the object to perform that function* could reasonably be said to be motivated partly by more general biologically functional dispositions. This generally seems less clear than in the matter of use.

ascription, and this may account for a natural resistance to it; but if it is, then even though such a claim might sound very odd at first, I think it is within the bounds of plausibility. It will seem bizarre only if it is wrongly taken to suggest that the nose has this function in just the same way, for example, that the nasal cavity has the function of (among other things) warming the air on the way to the lungs. As long as we are careful to distinguish the ways in which these two things come to have their functions, being clear that the one case involves intentional use, unlike the other, and so on, there is nothing bizarre in making both function ascriptions, or even in claiming that the use-based function ascription is at least *partly* a matter of *biological* function as well.

In summary, there seem to be four general possibilities for use-related functions:

(A) Biologically-based use-related function, either:

(i) where the use does not involve psychology (e.g. biological functions of vitamins and minerals); or:

(ii) where the use involves psychology (e.g. the egg-protecting biological function of the sea-turtle's tail)

(B) Independent use-related function (e.g. the function of the book mark, or the function of the left thumb in the playing of the classical guitar)

(C) Mixed (e.g. functions related to the use of eyeglasses to correct poor vision, or to the use of nets to catch fish)

5. CHANGES IN FUNCTION

Before leaving this general topic, it is worth noting that the above discussion of use-related biological functions sheds light on the problem of *changes* in biological function over evolutionary time. First of all, it should be said that in many cases there is no real difficulty to begin with. Often what has happened in cases of apparent change in function over evolutionary time is not that one organic form has changed its function, but rather that an earlier form with one function has given rise to new, modified forms, with different functions. This seems to be the case with feathers, for example. Early bodily feathers apparently had the function of thermoregulation, as do contemporary bodily feathers; but along the way, some such feathers evolved into distinctive flight feathers, with the function of contributing to flight.[16] There is no real puzzle here even for proponents of etiological views of function, so long as we distinguish between bodily feathers and flight feathers. The various special features of flight feathers (which clearly set them apart as such) have indeed been selected for flight, so there is no difficulty presented by the fact that *ancestral forms* of feather were selected only for thermoregulation; flight feathers may be regarded as having the function of flight in a straightforward way, having been shaped by natural selection in relation to flight. Where things get more difficult is in cases where a *single form*, retaining at least most of its adaptational features, *changes* its function.[17]

Proponents of etiological views will see such cases as illustrating the need to construe the etiological condition in such a way that an effect need not have been responsible for the initial presence of the form in question (i.e. for its initial fixation in the population through natural selection) in order to count as a present function; such cases may be accommodated, for example, by construing the etiological condition in such a way that the effect need only have figured in the *maintenance* of the form in question, thus allowing that the new effect is the present function of the form in question, even though it was originally selected for some other effect.[18] I have already indicated why such a move seems to me to be unsatisfactory: There may not have been any actual contribution to maintenance (though admittedly, in the case where the old function is completely lost, it is generally likely that the new effect will have had to have contributed to the maintenance of the trait or part), and even if there was, this is dependent on there having been negative selection pressures of the relevant sort, and it seems very odd to make the present function of the trait or part in question dependent on *that*. In any case, my suggestion, in line with the above discussion

---

[16] This example (though not its treatment) is borrowed from Kitcher (1993, p. 384).

[17] I discuss the issue of complete loss of function in section seven below.

[18] See Kitcher (1993, pp. 384 f.). By restricting the condition to require that the maintenance in question be relatively *recent*, the effect for which the form was originally selected can be denied present functional status if it has not recently contributed to maintaining the form, thus allowing us to say that the function has changed from one thing to another.

and the account of biological function I have presented, is different. On my view, it is sufficient that—often in connection with the development of various other adaptations relevant to the new effect—the trait in question has come standardly to be functionally *used* by the type of organism in question in relation to the new type of effect, just as the turtle's tail has come to be functionally used for protecting eggs.[19]

6. FUNCTIONAL TENDENCIES AND NON-FUNCTIONAL OR DYSFUNCTIONAL SIDE-EFFECTS

Once we begin thinking about cases where a certain fairly general tendency is said to be functional, as in the above discussion of the turtle's adaptive tendency to protect its eggs, a complication arises in cases where this tendency results in various more specific tendencies, not all of which seem to be functional. This is really just the same type of problem that is already familiar to biologists in discussions of *adaptiveness*, which may be illustrated by a simple example.[20] Moths have a general tendency to approach bright light sources in the dark, and this appears to be an adaptation. Let us suppose, for the sake of simplicity, that this is because for moths such light sources will very often be escape holes from enclosed spaces, such as caves, so that the tendency to fly toward them is adaptive and has been selected as an adaptation in moths. Now it turns out that this general tendency to approach such light sources brings with it the derivative tendency *to fly into candle flames* when they happen to be present; for moths are not generally able to discriminate between escape holes and flames. The question, then, is how we are to regard such a derivative tendency and such token behaviors.

---

[19] The above is sufficient, I think, to deal with similar cases presented by Prior (1985, pp. 315–16)—one of which comes from Boorse (1976, p. 76)—which were offered as objections to the etiological view. Again, I am not convinced that the modified etiologist's appeal to maintenance is sufficient to answer them, at least in an intuitively plausible way. My solution seems to avoid these problems, though it is to be expected that there will still be difficult cases where it will not be entirely clear whether or not the new use in question is non-accidental, biological functional use.

[20] See Dawkins (1982, pp. 36–7). The example is attributed to R.D. Alexander. For a variety of other classic examples where adaptations involving simple "releasing mechanisms" give rise to non-adaptive tendencies in the presence of relevantly similar stimuli, see Gould (1982, pp. 37, 43–5).

Clearly it would be wrong to say that flying into candle flames (or the tendency to do so) is *adaptive* for moths; obviously self-immolation is anything but adaptive. Thus, we would not say that it is an *adaptation* in moths to tend to fly into candle flames, even though this is a sub-tendency of an adaptational tendency; for we do not speak of adaptation in connection with behavior characterized in a specific, non-adaptive way.[21] On the other hand, it is no mere accident that moths fly into candle flames. It is the straightforward result of an adaptational tendency exercised in certain circumstances, and is intelligible in light of that background. Thus, even though flying into candle flames is not adaptive, the behavior can still be explained by appeal to the natural selection of the more general adaptive tendency from which it stems, along with some account of why there is no capacity to distinguish flames from such things as escape holes. (For example, as Dawkins notes, it might be that such a capacity has never evolved because the overhead costs would be too high; also, self-immolation might be relatively rare, so that there is not much selection pressure to develop the means to prevent it).

Similar points can be made with regard to *functionality*. On my view, the general tendency to fly toward bright light sources in the dark, which we are here supposing is an adaptation, would indeed be functional; it plays a role in the life (i.e. in the biological working) of moths. It would clearly be wrong, however, to say that the derivative tendency to fly into candle flames has a biological function. For this obviously does not play any role in the life of moths; on the contrary, it severely disrupts their proper functioning, and may thus be regarded

---

[21] This is not, of course, to suggest that the derivative tendency's being adaptive would be *sufficient* to allow us to speak of it as an adaptation. Suppose, for example, that flying into *light bulbs* were somehow of great benefit to moths (e.g. the sudden burst of heat proved to energize and strengthen them, without being so intense as to kill them, as with flames). Flying into light bulbs would thus be adaptive, but this would be a sheer coincidence, having nothing at all to do with the behavior. The tendency to fly into light bulbs is again (we are assuming) just a result of the more general tendency to fly toward bright light sources, which is an adaptation in relation to typical navigational benefits. I think it is clear, then, that we would no more speak of the tendency to fly into light bulbs as an adaptation than we would of the tendency to fly into flames, since it is a sheer accident that the former is adaptive. What is *non*-accidentally adaptive is just the more general tendency to fly toward bright light sources, and that is why we speak of adaptation when the behavior is characterized in this way.

as a dysfunctional derivative of a functional tendency.[22] On the other hand, when a moth does fly into a candle flame, this is not just an accident, nor is it merely the exercise of a specific non-functional or dysfunctional tendency (i.e. the tendency to fly into flames): It is equally an exercise of a more general functional tendency, i.e. the tendency to approach light sources. Or to put it another way, a token act of flying into a flame is not merely a token of the specific, non-functional or dysfunctional behavior-type *flying into flames*, but is also a token of the more general, functional, behavior-type *flying toward bright light sources*. Given this fact, therefore, the token behavior should still be regarded as *functional in nature* in this sense, even though it leads the moth to disaster and so does not in any way promote gene replication, i.e. even though it is not adaptive (if tokens can be spoken of in this way); it is functional *as* an instance of the general functional type of behavior, though *not* as an instance of the more specific, non-functional or dysfunctional type of behavior.

Indeed, unless it is part of the moth's functional repertoire to distinguish such things as flames from such things as cracks in cave walls—and I have been assuming it is not—the token behavior is not only functional in nature (in the above sense), but is even *non-defective*; it is no more defective than the closing of a Venus Fly Trap leaf on a pencil, the difference being just that in the moth's case the effort is not merely wasted, but productive of harm. Insofar as a moth's flying into a flame is a non-defective exercise of a general functional type of behavior, it should thus be regarded as functional and non-defective—and simply unfortunate for the moth; insofar as it is also an instance of a more specific dysfunctional type of behavior, it may be regarded as dysfunctional (though still not as defective). There is no contradiction here, as long as we are clear about the sense in which it is said to be functional and the sense it which it is said to be dysfunctional.

Finally, as with evolutionary explanation, there is no *straightforward* functional explanation for the tendency to fly into candle flames (since this derivative tendency does not have a function),

---

[22] The point could equally be put in terms of behavior-types (rather than tendency-types): Flying toward bright light sources in the dark has a function in moth life, but flying into candle flames does not, even though this is a sub-type of the more general functional type of behavior.

but there is still a less direct form of functional explanation for it. It can be made sense of by appeal to the function of the more general tendency from which it stems *along with* an appeal to the lack of any mechanism for eliminating the non-functional derivative tendency. With regard to a *token* act of flying into a candle flame, since the token behavior can be regarded as an instance of the general functional type (flying toward bright light sources), there *is* a straightforward functional explanation for it, just as in other cases where we explain a token by appeal to the function of its type; but in order not to be misleading we will certainly have to add that the general tendency to approach light sources is functional by virtue of factors which do not obtain in the case of flames.[23]

The example of moths flying into flames is a rather extreme case, but the points it illustrates apply to a wide range of familiar cases. To take just one, consider the following complaint of Searle's against what he calls "excessively crude Darwinism":

> If it were true that every innate predisposition of an organism were the result of some selectional pressure [thus giving some evolutionary advantage to the organism], then I would have to conclude that my dog has been selected for chasing tennis balls. He has a passion for chasing tennis balls, and it is obviously not something he has learned, but that is no reason for supposing it must have some biological payoff.[24]

Now it is certainly true that chasing tennis balls is not an adaptation in dogs. There was never plausibly any genetic variation in a population of dogs that gave rise to variation with regard to their behavior toward tennis balls, which was then acted upon by natural selection over many generations, resulting in the eventual selection of the tendency to *chase* tennis balls. Thus, Searle is also quite right in claiming that there is no reason to expect that chasing tennis balls *must* have "some biological payoff" (i.e. be adaptive) for dogs. So much is obvious, and it would

---

[23] I discuss functional explanation in chapter eight. The above remarks are meant only as relatively general observations, compatible with different accounts of the nature of functional explanation.

[24] Searle (1992, p. 106).

take a crude Darwinian indeed to think otherwise. But neither is the matter as simple as he seems to suggest by leaving it at that.

It is entirely possible that a dog's disposition to chase tennis balls is an aspect of a more general disposition—i.e. to chase fast moving objects, especially within a certain size range—which latter *is* an adaptation. Perhaps this is an adaptation that makes dogs on the whole more likely to catch prey, and for whatever reason they simply haven't developed the tendency to ignore non-prey, like tennis balls; again, the overhead costs of developing such powers of discrimination and control might simply be too high.[25] If so, the ball chasing tendency would indeed be intelligible in light of considerations having to do with natural selection and the benefits of the more general tendency of which ball chasing is an aspect; so there is still a close connection with natural selection. And in terms of biological function, even though we would not speak of the specific ball chasing tendency as having a function in dog life, we would still properly regard instances of ball chasing behavior as functional behavior, in the sense that such behavior is the exercise of a general functional tendency; indeed, it may well be fully *non*-defective as such. It makes no difference that the dog's token ball chasing does not in itself serve any biological end: That was never said to be a condition for a token entity to be functional in nature. Such behavior would thus still admit of functional teleological explanation.

Finally, it should be said that it *may* even be the case, at least in principle, that the tendency to chase moving objects is an adaptation in a more general way, the point being simply to maintain and sharpen the dog's chasing skills, so that when it does come across prey, it will be that much better able to catch it.[26] In that case, chasing tennis balls

---

[25] Compare the egg-rolling response of greylag geese, as studied by Lorenz and Tinbergen. When a goose notices an egg lying outside the nest, it executes a certain behavior pattern whereby the egg is rolled back into the nest. This behavior pattern is no doubt an adaptation. But it turns out that it can be "released" (i.e. triggered) not only by the sight of an egg outside the nest, but equally by objects ranging from baseballs to beer cans. The criteria employed by the "innate releasing mechanism" are simply of insufficient precision to discriminate among the various objects, though this is not normally a problem in the gull's natural environment, where it is relatively unlikely to encounter items other than eggs that lie near the nest and possess the crucial "releasing" characteristics. See Gould (1982, pp. 38–9).

[26] Cf. Gould (1982, p. 177), on the importance of play in lions: "Lion cubs must perfect their pounces before they are left to fend for themselves, and the distinction we like to make between their fetching practice on windblown leaves or their parents' tails, and actual attacks on small animals, is probably an artificial one."

wouldn't be wasted effort at all (like a Venus flytrap's closing on a pencil); it would be adaptive after all, and a perfect example of just what the dog had been adapted to do in the way of maintaining and sharpening his skills. Here, although it would still be misleading to speak of chasing *tennis balls* as an adaptation (as if tennis balls figured into the dog's natural selection history, or as if there were something special about tennis balls), the truth is not far from this. The dog's chasing tennis balls would be a non-accidentally adaptive result of a more general adaptational tendency.[27]

7. THE PROBLEM OF VESTIGIALS

A crucial issue that needs to be addressed by any account of biological function is the problem of vestigial traits—trait-types which used to possess functions but no longer do. There are three possibilities here. First, a trait may be vestigial because due to regressive evolution it has atrophied to such an extent that it fails to have even the proximate effects which would previously have been identified as its function in the type of organism in question. Second, it may be instead that it still has these proximate effects, but because of various changes these effects no longer promote any genuine *end* (except perhaps accidentally) either because (i) the causal structure has changed so that the ends once served are no longer served by these more proximate effects, or (ii) the latter still lead to at least some of the same more remote effects that would earlier have been identified as ends, but these remote effects no longer constitute genuine ends.[28] In any case, a vestigial trait (type) is one which no longer makes any *functional* contribution to any genuine *end.* Which effects presently constitute genuine functions and ends (the latter being more remote, the former being more proximate) within a functional system? On my view, these are the effects presently associated with the working of a given type of organism.

There is no implication, then, that the sorts of effects on the system that were *originally* responsible for an entity's incorporation into it are *still* genuine functions or ends. Again, it could very well be that the organism-type has evolved significantly, retaining the trait as a

---

[27] Contrast this with the above hypothetical case of moths flying into light bulbs, where the benefit was *accidental*, having nothing to do with the general adaptation.

[28] Cf. Griffiths (1993, p. 417), who likewise notes that there can be non-atrophied vestiges as well as atrophied ones, both of which must be taken account of.

vestigial trait. Even if it still has some of the effects for which it was selected, if those effects no longer tend to promote the reproductive success of the type of organism in question or of its kin (either directly or by promoting survival), those effects no longer play a role in the working of that type of organism as the gene replicating system it fundamentally is. In that case, on my view, these effects are no longer genuine functions or ends with respect to the type of organism in question. And in the case of a vestigial organ like the appendix, which no longer has even the proximate effect for which it was selected, or any other effect that promotes the relevant sort of gene replication and hence tends to maintain it in the "design" of the organism (it remains simply because there is not sufficient selection pressure to eliminate it), it is even more clear that it no longer plays any standard role in the working of the type of organism to which it belongs, and I am therefore not committed to assigning it a function, even though it is present as a result of natural selection.[29]

It is important to stress that we are speaking here of *types*, not tokens—though if a certain type of trait is vestigial, then it follows that its tokens are as well. But now when exactly can we say that a type of trait (or other entity) *no longer plays a role* in the working of a given type of organism? I will argue in the next section, in connection with a problem raised by Neander, that it is not sufficient that many or even most tokens of that trait-type fail actually to perform their functions; an entity's playing a role in the working of a given type of system is not a matter of statistics. One implication of this may be seen in the fact that we do not speak of loss of function where the environment simply changes suddenly, or when organisms are taken out of their natural environments; there may be immediate loss of function*ing* in many or even in most particular cases, but not immediate loss of function in the sense in which a vestigial trait has lost its function; the trait-type in question still *has* its function, i.e. it still *plays a role* in the working of that *type* of system, in the relevant

[29] It is important not to confuse these claims about vestigial trait-*types* (and derivatively tokens of those types) with claims about *ordinary* trait-*tokens* in individuals that are past the age of reproductive significance, or who are defectively sterile (e.g. mules). I am not claiming, for example, that a particular heart that no longer makes any contribution to gene replication—because, for instance, the individual to which it belongs is too old to reproduce—is thereby non-functional. I will take up this issue just below, where functions in connection with token traits are discussed.

sense, though the sudden change of environment has made it impossible for tokens to carry out this role. Of course, this can change over time, but the point is just that something more than just a sudden change of environment is required. The notion of a role played by something in a type of organism must evidently be understood as involving an implicit reference to a standard environment—roughly, I think, the one in which that type of organism was shaped by natural selection to work as it does, i.e. the environment in which it has evolved and maintained itself.[30]

8. TYPES AND TOKENS

I have been using the distinction between types and tokens, but it is important now to bring out why it is crucial that an account of biological function be formulated in terms of the roles played by *types* of things within *types* of systems. The reason for this becomes clear when we consider the case of defective or malfunctioning particulars. As Neander and Millikan have pointed out, it is entirely possible for a particular part to fail actually to make any contribution to the working of the particular system to which it belongs (e.g. if it is defective) and yet still to *have* the function of making some such contribution; and a satisfactory account of function must obviously allow for this.[31] Thus, it is crucial to avoid saying that the function of any token entity is the contribution it actually makes to the working of the system to which it belongs; it may be out of commission and fail actually to play any role in the (likely sub-optimal) functioning of the system, and yet still *have* the function that belongs to that type of entity. The solution is to

---

[30] It is, of course, no objection to my view, as it would be to an ahistorical view, that the notion of a type of system that works in a certain way involves an implicit reference to normal environment, which in turn—as Millikan (1989, p. 300) argues—involves an implicit reference to history. How exactly the details surrounding vestiges should be spelled out beyond the sorts of general observations made in the text—e.g. just how long it takes for traits to become vestigial—is something that is probably best left to biologists to determine, partly as a pragmatic matter. Griffiths (1993, p. 417) gives a plausible suggestion in this connection, which would be compatible with my view of function (though his view of function in general is not).

[31] Neander (1991a, pp. 181–3), Millikan (1989, p. 296); cf. also Woodfield (1976, pp. 112–13). This is another reason that Cummins' (1975) account of function fails, at least as an account of proper function, as both Millikan (p. 294) and Neander (p. 181) point out: There is no room in his view—as there would have to be in a satisfactory account of proper function—for a token entity to have a function which it nonetheless actually fails to be disposed to carry out.

attribute functions primarily to types of things in types of systems, in terms of the roles played by those types of things in the working of those types of systems, and to attribute functions only derivatively to particulars, simply insofar as they are tokens of those types. An idiosyncratically non-performing part still *has* the function it has because it is an instance of a presently functional type. A heart valve that doesn't open and shut properly is still a heart valve, and still has the function of regulating blood flow in certain ways because that is the present function of that type of part in that type of system; the particular valve is simply defective or malfunctioning (in the latter case it may be that there is a problem elsewhere that is making it impossible even for a non-defective valve to function properly).

At the beginning of chapter one we noticed that any entity, type or token, may in fact *function as* a ϕ-er without having ϕ-ing as its function (e.g. the rain functions as a street cleaner), though of course any *type* of entity that has ϕ-ing as its function must *function as* a ϕ-er. What we have now added is that not only can a *token* entity function as a ϕ-er without having ϕ-ing as its function, but it can have ϕ-ing as its function even though it doesn't in fact *function as* a ϕ-er.

My solution to the problem raised above is different from Neander's. Her move simply involves a direct appeal to etiology: The token may be said to have ϕ-ing as its function, even though it actually fails to ϕ, because ϕ-ing comes into the consequence-etiological explanation of how the token came to be present; that is, the token is present because past tokens of that type ϕ-ed, causing that type to be favored by natural selection, etc. By contrast, while my solution does involve an indirect appeal to history in establishing the function of the type, the crucial move with regard to the problem of defective or malfunctioning particulars is the assigning of priority to types over tokens in the account of function, rather than looking directly to the historical-causal background of the token and being concerned with the type only insofar as it figures into this (as in Neander's case).[32] Nor do

---

[32] In this focus on types I have been influenced by Michael Thompson, though the details of our views differ significantly. Neander (1991a, p. 174) herself maintains that at least in biology "proper functions belong primarily to types and only secondarily to tokens," the reason being that "natural selection does not operate on individuals or their biological parts and processes" (by which she means that these are not the units of selection). But this priority is not central to the move she makes to solve the problem of malfunctioning particulars, since it is just a direct appeal to etiology, which can be made

I fall into the sort of error that she thinks tends to plague non-etiological accounts, namely the mistake of characterizing *standard roles* in *statistical* terms. As she and many others have correctly pointed out, it is entirely possible for things of a certain type to have a certain function even though most of them fail to perform it—whether through lack of opportunity, as in the case of sperm, or through widespread dysfunction, perhaps caused by epidemic diseases.[33] But nothing I have said suggests otherwise. The notion of a *type* of entity playing a standard role within a certain *type* of organism does not in any way imply that most tokens of that type actually make such a contribution in their token systems. What *is* implied about tokens, I suppose, is that enough tokens have actually made such a contribution throughout the evolutionary history of the organism-type in question to result in the incorporation of the entity into the "design" of the organism-type through natural selection; obviously if almost no instances of a certain coloration pattern had ever attracted pollinating insects, for example, the trait would never have been favored by natural selection. But in any case there are certainly no general present implications about what most tokens are actually doing. When we speak of a type of entity playing a standard role within a type of organism, we are not making a disguised statistical remark about tokens, but simply a remark about a type of working system that has been put together by natural selection in a certain environment—a paradigm *way* in which matter and energy can be, and sometimes is, organized into a complex gene replicating system, in a certain natural environment.[34]

It is important that this primary focus on types in an account of function is not just an *ad hoc* move to avoid the sort of problem raised by Neander and Millikan, as Neander claims is the case for others who

---

either with respect to the type or with respect to the token (though again, reference will be made to the type even in the latter case). By contrast, the move to types *is* central for me, together with the appeal (at least in biological cases) to types of system in explicating the function of a type of entity. A token worn out organ still has ϕ-ing as its function because it is still a token of a certain *type*, which latter has ϕ-ing as its function within a certain type of organism.

33 Cf. Neander (1991a, p. 182), (1991b, p. 459, fn. 7), Millikan (1984, p. 34), Thompson (1995), Hsu ("On the Natural History of Artifacts"—unpublished manuscript).

34 On the irreducibility of such appeals to types, cf. Thompson (1995).

have made this move.[35] The feature of functional discourse to which Neander and Millikan point reflects something central to the logic of functional discourse: Particular entities fall within the domain of functional discourse not *as* particulars, but *as* tokens *of a certain type*, in relation to which they are evaluated as good or defective specimens. We do not approach a particular heart valve, for example, simply as a physical object in a certain particular physical setting, and inquire as to its function *as such*. It is not clear how we could even arrive at a notion of working for a particular system regarded simply *as* a particular bunch of ever-changing matter and energy (as opposed to being regarded as a token of a particular type of organism). Even if we could settle on identity conditions for it over time, considered only in its particularity, how would we distinguish in a non-arbitrary way between what belongs to its working and what are incidental side-effects or cases of malfunctioning? Similarly, in the case of the heart valve itself there would be nothing to distinguish its proper functioning from some malfunction that might happen accidentally to contribute to whatever it was we decided constituted working for the particular system to which it belongs. Instead, when we are interested in function we approach something like a heart valve *as* an instance of a functional type of thing, and the system of which it is a part *as* an instance of a functional type of organic system. And, as I have argued, we arrive at non-arbitrary notions of working for such a system and of roles played by such an organ, by considering the organism-type as a product of natural selection, which has "designed" that type of system to work in certain ways, with various parts and features that play standard roles in that working. The motivation for putting the account of function primarily in terms of types is therefore deeply rooted in the whole approach I have taken to cashing out the notion of *working* in biology.[36]

---

[35] Neander (1991a, p. 182, fn. 11): "Thus these theories are also forced to acknowledge that biological proper functions apply primarily to types and only secondarily to tokens, but the fact does not fall neatly out of the logic of the theory; it is introduced to ward off a plethora of counterexamples." I maintain that the fact *does* "fall neatly" out of the logic of my account, as I'll go on to explain.

[36] The above points about the difficulties with applying functional concepts to systems considered in isolation from any functional type to which they belong are related to points made by Millikan (1989, pp. 292f.), Neander (1991a, pp. 178f.), Thompson (1995) and Hsu (manuscript) about the difficulties with applying functional concepts to (hypothetical) systems that actually lack any lineage or genuine type—as in the case of an ostensibly biological system that comes together spontaneously through some sort of

In summary, then, attributions of function are made primarily to types, but this doesn't prevent us from attributing functions derivatively to tokens. We may say that token instances of functional types, *as such*, have those functions within their particular systems, so that particular hearts as well as *the heart* can be said to have the function of pumping the blood, and can accordingly be said to be functioning properly or to be defective or malfunctioning if they are not. A *token* entity can have the function of ϕ-ing, even though it fails to play any role in the actual functioning of the token organism, so long as that *type* of entity currently plays a role in the functioning of that *type* of organism, giving that *type* of entity (and hence derivatively also tokens of that type) a present function. Returning to the issue of vestigial traits, there is thus a big difference between saying (i) that a certain *type* of organ is vestigial, and so no longer plays any role in the functioning of a certain type of organism—from which it follows that token organs of that type in token organisms of the relevant type lack a present function—and saying (ii) that a certain *token* organ of a presently functional type no longer plays any role in the functioning of the token system to which it belongs—from which it does *not* follow that this token organ does not still have a proper function.[37]

---

quantum mechanical accident. I will discuss such cases in more detail in section four of chapter six, in connection with a possible objection to the present account of function.

[37] The general point about types and tokens can be made in connection with artificial function as well. A defective spark plug that fails to make any contribution to the working of the engine to which it belongs does not thereby fail to have a function; its function is still to spark—since it is a spark plug, and that is the function of the spark plug—and this is precisely why we call it *defective* when it fails to do so. (A. Hsu makes a similar point in an unpublished paper, "On the Natural History of Artifacts", regarding the functions of whole machines. A defective adding machine may well fail to be disposed to yield sums, but that does not mean that it is not its function to do so, i.e. that it is not really an adding machine after all.) Thus again, an account of artificial functions should be put primarily in terms of types, with functions assigned to tokens only derivatively. (Cf. Millikan, 1989, pp. 294–6.) There may be certain special cases where functions may be assigned directly to particulars, but I am inclined to think that even in many cases that appear this way—such as the case of the parts of a singular invention—the item is assigned the function not *as* a particular, but *as* a token *of a certain type*, even if it is the only one ever to exist. For it falls within the domain of functional discourse not simply *as* e.g. a bit of metal and silicone, but *as* e.g. a "circuit-modifier" (I'm just making this up), of which others could in principle be made, even perhaps of different materials; and it is judged in functional terms according to how well it fills the role played by *such* a thing, not simply as a particular. Someone might argue that we *can* here speak of the particular part as receiving its function *as* a particular simply insofar as the designer *intends* it to do something. But it is equally

9. INCOMPLETELY FUNCTIONING SYSTEMS

Now if it is true that from a biological-functional point of view organisms are ultimately gene-replicating systems—that the ultimate end toward which lower level functions and ends are generally geared is inter-generational gene replication (by way of reproduction or promoting the reproduction of kin)—then surely many organisms that are still very much alive nonetheless fail to function fully. Consider, for example, a token organism that is past reproductive age and belongs to a species without a significant repertoire of "kin altruistic" traits.[38] It is no longer functioning *as* a gene-replicating machine, even though its heart is still beating, it is still breathing, and so on. We must conclude, then, that it is no longer functioning fully, given the type of thing it is. And that, I think, is plausible. What would be *im*plausible would be to say that because of this its heart and lungs no longer have any function, or are no longer making any functional contribution to the organism's functioning. But on my view we are not led to say either of these things.

First, the fact that a token heart no longer tends to make any eventual contribution to genetic replication does not for a moment imply that this heart no longer has a function. As I have just pointed out, even if it made no contribution to the organism's functioning at all it would still *have the function* of pumping the blood, since it is still a heart and that is the heart's function. And in any case, it is clearly false that this beating heart makes no contribution to the organism's functioning. It makes no contribution to *gene-replication*, since the organism is no longer functioning as a gene-replicator, but the fact that the highest level of an organism's proper functioning is no longer being carried out or promoted does not imply that its various lower level activities cease to count as functioning. This heart's pumping of blood is still functional, making functional contributions to the working of the organism; the difference is just that the working of the organism is somewhat truncated—no longer promoting the *ultimate*

---

plausible, I think, to say that by intending this part to have a certain function in the machine, the designer made it to be an instance of a certain *type* of thing (at least if he was moderately successful), and its function belongs to it *as* such a thing—whether or not the type happens to have any name. In any case, even if I am wrong here and there are such exceptions to the point, this is irrelevant to function in biology, where intentions do not come into the picture.

[38] Cf. Hull (1982, pp. 305–6).

end for that type of organism, but only various subordinate ends, including warding off death. Such subordinate ends have in this particular organism lost their ultimate point, as it were, which involved contributing indirectly to gene replication. But again, all that means is that what we have is a partially functioning system, no longer functioning in such a way as to promote its proper ultimate end; it does not mean that the system is not functional at all, or that the lower level ends cease to be genuine ends.

An analogue in the realm of artifacts may be helpful here. Suppose someone designs a diamond synthesizing machine, which synthesizes diamonds from pure graphite by mixing it with liquid iron and compressing and heating the mixture over a period of two years. The ultimate end of the system, to which various lower level functions and ends are directed is the production, once every two years, of a synthetic diamond chunk of a given size. That is, at lower levels, the machine's functioning involves such things as pulverizing the graphite and adding heat to a certain chamber, aimed respectively at the subordinate ends of producing a uniform graphite powder and maintaining a certain high temperature; and the ultimate point of this lower level functioning, and of the realization of these lower level ends, it to produce synthetic diamonds—which is also how we would describe the work of the machine (or what working consists in for it) at the highest and most general level. Now suppose that due to various limitations in the materials of which the machine is constructed and in the ingenuity of its design, certain parts wear out before others, so that e.g. the compressor gives out long before the pulverizer, heater, and mixer do. If such a machine is set up and allowed to run indefinitely, there will thus come a point after which it continues to carry out many of the lower level activities it always did, though it no longer produces diamonds. Its *function* hasn't changed, but the nature of its actual functioning has: It is no longer functioning fully as a diamond synthesizing machine. That means that its lower level activities no longer actually contribute to the ultimate proper end; but if they still make contributions to lower level, proximate ends—e.g. the heater still keeps the chamber hot—then they are still making functional contributions to the (partial) working of the system. We are certainly not forced to say either that such parts and activities of such an "aged" machine no longer have functions or that they are no longer functional. They have the same functions they always had and their continued operations are still functional in nature. And the same is true of the heart and its blood pumping in the aged organism considered above.

10. THE PROBLEM OF STERILE HYBRIDS

A slightly more difficult case, which might initially be thought to provide a serious problem for the view I have offered, is that of mules.[39] If living organisms are fundamentally gene replicating systems from a functional point of view, what are we to say about mules? Surely in some sense they are *not* gene replicating systems, since they are sterile (and unlike worker bees, have not been "designed" to work toward the reproductive success of kin). Does that mean that they are not living organisms?

Such a claim would be extremely implausible, I think, and is not implied by my view.[40] As we have seen, it is entirely possible for particular organisms to fail actually to function as gene replicating systems, just as it is possible for particular diamond synthesizing machines to fail actually to function as diamond synthesizers. This may be a result either of standard degeneration or of defect (as in the case of a syndrome that involves sterility). On the face of it, then, it is natural to say that mules are simply *defective* gene replicating systems. They have a defect at the level of reproduction, which affects their functioning at the highest levels—parallel to a diamond synthesizing machine with a defective compressor—but this does not keep their lower level activities from being functional in nature.

There is, however, a complication here. The type *horse* is a type of gene replicating system (or more accurately, it encompasses two interdependent, closely related but importantly different types of gene-replicating system, namely *male horse* and *female horse*). Thus, if a particular horse has some syndrome that makes it sterile, it is a defective gene replicating system insofar as it is a defective horse. But there is no direct parallel to this for mules. Mules are not defective

---

39 Cf. Hull (1982, pp. 305–6).

40 Though I believe Michael Thompson (1995) may actually maintain that mules are not strictly speaking living things, since he takes species membership—or the possession of a "life form"—to be essential to life. His position will depend on how exactly he understands the relevant notion of species or "life-form", which is not made sufficiently clear to answer this question. I have more to say about Thompson's views in chapter six.

members of a species, such as *horse* or *donkey*; they are species-less hybrids, and so cannot be said to be defective Fs, where "F" stands for some species term. Thus, if we say that mules are defective gene replicating systems, it is so far unclear what type of gene replicating system is at issue; again, they are not defective horses or defective donkeys.

One possible response here would be to drop the claim that mules are defective gene replicating systems, and to say instead that mules are non-defective tokens of the functional type *mule*, which latter simply happens not to be a type of gene replicating system. This, however, would indeed run counter to the view of organisms for which I have argued. The view certainly allows for token organisms that fail to carry out the ultimate function belonging to their types, but it does not allow for types of biological-functional system that do not have gene replication as their ultimate biological function. I am therefore committed to denying that *the mule* is a genuine type of biological-functional system on a par with *the horse*.[41]

Instead of saying that mules are non-defective tokens of the functional type *mule*, which latter simply happens not to be a type of gene replicating system (or perhaps happens to be a *defective type* of gene replicating system, if such a notion even makes sense), it is more natural and plausible to keep the claim that mules are defective gene replicating systems, and simply to give a more general specification of the type of gene replicating system that is at issue, such that mules may indeed be regarded as tokens of that type. Mules might be said, for example, to be *defective equine mammals*.[42] There are, of course,

[41] This denial does not extend to hybrids generally. In some cases, hybrids are fertile and go on to enjoy significant reproductive success. For example, while the first-generation hybrids resulting from the crossing of a radish and a cabbage were almost completely sterile (since the incompatible chromosomes could not pair properly during meiosis), the second-generation hybrids—formed from rare, mutant gametes each containing the full diploid set of chromosomes—were fertile and bred true to the new kind, which resembled a radish on top and a cabbage in the roots. A similar phenomenon has occurred in nature. A certain very successful species of marsh grass, *Spartina townsendi* (which has 126 chromosomes), turns out to have evolved as a result of the hybridization of two related species (one of which has 56 chromosomes and the other of which has 70). (Villee and Dethier, 1971, p. 286) Clearly, then, sometimes hybrids *do* perpetuate themselves in reproduction, and even come to form what are standardly recognized as *species*. Here I think we should grant that we have a genuine new type of biological-functional system—unlike in the case of a mule.

[42] I owe this suggestion to Robert Adams.

limits to how much can be specified in the way of norms at such a general level; obviously there is much less to be said about the requirements for being a non-defective equine mammal than there is for being a non-defective horse. But it certainly seems that we can say at least that a non-defective equine mammal is capable of reproduction, and that is enough to brand the mule as defective.

Even though mules do not plausibly instantiate any type of biological-functional system *at the level of species*, they nonetheless share a great degree of the functioning of the two genuine species-level types (*the horse* and *the donkey*) from whose tokens they are descended. I therefore see no reason why we should not be able to speak of the pumping of the mule's heart, and so on, as obviously functional in nature, just as in the case of a horse that is defectively sterile; the difference is just that the sterile horse is a defective token of a species-level type, lacking a bit of functioning that belongs to it *as* a horse, whereas the mule is a not a token of any species-level type, but only a defective token of a higher-level type—an individual which simply happens to possess a limited repertoire of functioning in the first place, derived from two genuine species-level types. The functions and ends in a mule will thus still reflect the shaping of natural selection, even though *as instantiated in the mule* these functions will never promote gene replication.

A parallel in the domain of artifacts would be a case of a manufacturing confusion that results in the production of a cross between two slightly different models of diamond synthesizing machine, M1 and M2; the hybrid machine does almost everything but deliver the final product (because, for example, the combination produced a compressor that failed to provide adequate force). This machine is neither an M1 nor an M2, but it is still plausibly a diamond synthesizing machine—a *defective* one, since it does not actually produce diamonds. We would still refer to the various things the hybrid machine *does* manage to do—pulverizing the graphite, mixing it with iron, heating it, etc.—as *functioning*, even though none of it tends to produce diamonds in this particular system where the functioning is truncated; for it is certainly no accident that these things are done, inasmuch as the hybrid was formed through the same manufacturing processes that go into producing M1s and M2s (only mixed up a bit, resulting in some loss of function). This is, I think, basically the same situation we face with regard to the truncated biological functioning of mules.

11. THE DUAL FUNCTIONAL RELATION BETWEEN GENES AND ORGANISMS

Returning now to the general account of biological function, I said that probably the most familiar and interesting examples of function in biology involve adaptations, such as the heart and its pumping of blood, or the magnificent structures of the eye. But if such phenotypic features of organisms are paradigmatic functional entities because of the roles they play in the working of organisms, as this has been cashed out, then surely it is equally proper to speak of the genes that code for these phenotypic features as functional entities as well. Each of these genes plays a role in the construction or modification of the organism, making some contribution to its working. As in other cases, these roles can be considered either at relatively proximate levels—e.g. the effects the genes have on the structure and activities of certain cells—or at more remote levels, such as the resultant effects the genes have on the quality of certain tissues or organs; either way, the crucial point is that if the phenotypic effects play functional roles in the organism, then obviously so do the responsible genes. For example, there is a certain gene on chromosome 7 in human beings that is responsible for promoting the manufacture of elastin; when it is missing, the individual exhibits Williams syndrome. If elastin can properly be said to have a function in the human body, then so too can this gene: Its function (or at least one of its functions) is to promote the manufacture of elastin.[43]

---

[43] There is an interesting complication here when we consider the issue of types and tokens discussed above. In the case of an organ, such as the kidney, we assign a function primarily to the organ-type, and derivatively to tokens of that type. The function (or one of the functions) of the kidney is to excrete metabolic waste, and so the function of any particular kidney is to excrete metabolic waste—whether it actually does so or not. But with genes the situation is more complex. The function of the gene for elastin—meaning here the gene-*type*—is to promote the manufacture of elastin. But it seems odd to claim that every token gene of that type thereby has that function; for we have such tokens in every cell in our bodies, and they are *active* in only a small fraction of these cells. The difficulty here is not simply that most token elastin genes are never expressed, as most token sperm cells never fertilize an egg; that is not itself a difficulty at all, since function ascriptions carry no statistical implications about the actual behavior of tokens. (The fact that most sperm cells never come close to an egg

Now it might seem that this kind of talk contradicts a central claim made earlier about the relationship between organisms and genes. Recall Dawkins' remark that "organisms are tools of DNA rather than the other way around." In my terms, eliminating the metaphor: Organisms are functional systems whose ultimate end in each case—the ultimate end for the sake of which all lower level functioning occurs—is the replication of certain germ-line genes. Organisms work for the sake of genetic replication, rather than genes' existing for the sake of organisms' welfare. Yet now we are admitting that genes can be said to have functions within organisms, making certain contributions for the sake of certain phenotypic effects. How are these two pictures to be reconciled? In fact, there is no real contradiction here at all. Individual genes or gene complexes are said to have functions within the functional system of which they are parts, inasmuch as they make some contribution to the working of that system; but the working of that system must ultimately be understood as being for the sake of the replication of genes of the types represented in the organism's co-adapted genome (and not for the sake of the organism's flourishing, for example). Thus, each gene makes some functional contribution to a certain "cooperative project", but the point of that project is nothing other than the replication of copies of the "cooperating" genes. It is in this way that we can say both that genes have functions within organisms, doing certain things for the sake of certain higher level effects within the organism, while still insisting that organisms do the various functional things they do ultimately for the sake of gene replication.

The account of function I have offered thus allows us to incorporate Dawkins' significant insights about genes and organisms without giving up the plausible idea that the parts of organisms—right down to their genes—are to be understood as having functions in relation to the

---

does not prevent us from assigning the function we do to the sperm cell and thus derivatively to each token sperm cell.) With genes, however, it is not simply a matter of chance that some are expressed and others are not: It is part of the very functional structure of the organism that most token genes of a given type are never expressed. It would thus be somewhat odd to attribute the standard function of the gene for elastin to tokens of that type located in cells where—as a matter of the cell's proper functioning—those tokens are inactive. What we have in the case of genes, I think, is just the interesting situation where there is no automatic inference from a type-wise function ascription to token-wise function ascriptions; we must look at the wider context to see *which* tokens actually possess the functions of the type.

working of the whole organism. This is far preferable to the extreme position—which Hampe and Morgan attribute to Dawkins—according to which "DNA has no function for the organism".[44] I do not believe that Dawkins ever meant to make such a claim, though it is true that he says things that strongly suggest such a reading. In one place, for example, he writes:

> Birds' wings are obviously `for' flying, spider webs are for catching insects, chlorophyll molecules are for photosynthesis, DNA molecules are for. . . . What *are* DNA molecules for? This question takes us aback. In my case it touches off an almost audible alarm siren in the mind. If we were to accept the view of life that I wish to espouse, it is the forbidden question. DNA is not `for' anything. If we wish to speak teleologically, all adaptations are for the preservation of DNA; DNA itself just *is*."[45]

While this sounds like the claim that genes cannot be attributed functions within organisms, it can also plausibly be read as an unfortunately misleading way of expressing the general point that I have conceded, which is Dawkins' real concern: namely, that organisms have been "designed" by natural selection ultimately to preserve and replicate their DNA, rather than DNA's somehow having come about to serve the independent interests of organisms (i.e. to promote their welfare). In any case, Dawkins certainly does not *need* to take the extreme position that genes do not have functions within organisms. His interesting and important claims about the process and results of natural selection are fully compatible with the recognition that genes have functions within organisms, *as long as this functional framework is fleshed out ultimately in gene-centered terms*, as I have done. By continuing to speak of genes as having functions within organisms, we preserve a natural and coherent aspect of functional discourse, which (as mentioned above) seems inseparable from ordinary talk of *phenotypic traits* having functions within organisms; we also avoid problematic implications that a rejection of such talk would have—in particular, the implication (according to Hampe and Morgan) that it would "no longer

---

[44] Hampe and Morgan (1988, p. 119).

[45] Dawkins (1982a, p. 45). Dawkins' view of functional discourse will be discussed in more detail in chapter seven.

be possible to use the language of information theory in the description of the function of DNA in the sense in which it has been used", which "would change fundamentally our view of DNA".[46]

12. FUNCTIONS OF EXTENDED PHENOTYPIC EFFECTS

Finally, it is worth noting how Dawkins' idea of extended phenotypic effects, mentioned in chapter four, bears on the account of function I have offered.[47] I have said that adaptations which fall within the phenotypic framework expressed by the co-adapted genome of a given type of organism will typically play some standard role in the working of such an organism. But as illustrated by the case of the Nosema parasite discussed earlier, sometimes such adaptations—whether they are standardly called "adaptations" or not—are found not in the organism possessing these co-adapted genes, but in another organism. It will be recalled that in this case certain genes in the co-adapted genome of the Nosema exert phenotypic effects in flour beetle larvae, resulting in the trait of giantism in those larvae; they do so, of course, only by way of more immediate effects in the body of the Nosema itself, but that does not change the fact that their more remote and equally significant effect is the production of a certain trait in flour beetle larvae. (Actually it takes millions of Nosema to manufacture the required chemical in sufficient quantity to produce this trait in the larvae, but that doesn't affect the point.) Now inasmuch as this effect is a trait of an organism—albeit a different organism than the one carrying the responsible genes—and not merely some more general effect on the environment (such as is produced in any case of genes exerting phenotypic effects), it is fairly natural to attribute a function to it. The difference from an ordinary case, of course, is that this trait of flour beetle larvae has a function in relation *not* to the life of flour beetles, but to that of Nosema. It plays a role in the working of Nosema, making an eventual contribution to the replication of *their* genes by securing an environment for these parasites that is better suited to them than a normal flour beetle larva would be.

---

[46] Hampe and Morgan (1988, pp. 119, 126).

[47] Cf. also Griffiths (1993, p. 416).

We can plausibly treat certain other cases of extended phenotypic effects in the same way. For example, it seems appropriate to attribute functions not only to such effects when they constitute traits in organisms, but also when they constitute "artifacts" of animal behavior. I have already mentioned that secreted substances such as hormones and neurotransmitters are typically functional, as the adaptively significant products of processes that are adaptations, and there is no reason why the same should not be said about products of higher levels of adaptational organic activity (i.e. products of "behavior" in the common, restricted sense), such as spider webs and bird nests. Indeed, it is already common practice to do so. Anyone will agree, for example, that the function of the spider web is to trap insects for food; this is no more controversial than talk of the function of the heart. The account of function I have offered readily accommodates such cases. The spider web has the function it has because as the adaptively significant product of certain adaptational processes and activities—namely, the secretion of scleroprotein and the spinning of the web—it plays a standard role in the working of the spider as the gene replicating system it is, promoting the securing of food which enables the spider to maintain itself in existence and to carry on the various processes that lead to reproduction. Other "artifacts" of animal behavior, at least insofar as they are the adaptively significant products of adaptational behavior patterns, may be dealt with in the same way.

13. BIOLOGICAL TELEOLOGY AT HIGHER LEVELS: GROUPS AND ECOSYSTEMS

Finally, let us consider the possibility of biological functional systems at levels higher than that of an organism or its sub-systems. Perhaps the most obvious and clear-cut example of such a system is something like a colony of bees or ants, so I'll examine this first. To begin with, each member of a honeybee colony, for example, is an instance of a certain type of functional system that has evolved through natural selection. It is, of course, an oversimplification to speak just of *the honeybee* here, as if there were just one type of functional system associated with the species. In fact, there are at least three, corresponding to the three castes into which honeybees in a colony are divided: A queen is an instance of one type of functional system, a worker is an instance of another, and a drone is an instance of a third.

The three types of functional system are closely related and share much in common, all being examples of the more generic type: *the honeybee*. But there are important differences between them, and it is important to keep them distinct in thinking about the functional teleology associated with various members of a colony. On the other hand, we can understand each type of functional system only in the context of the others, since it is only in such a context that each type of functional system is able properly to carry out the full range of functioning that it was "designed" by natural selection to carry out.

This is most obvious in the case of a worker bee. As with any organism, the various biological functions in a worker bee are geared ultimately toward the inter-generational replication of certain germ-line copies of genes. But the copies in question do not reside in her body; in fact, there are no germ-line copies of genes in her body, as she is sterile. The worker bee is a functional system "designed" by natural selection—particularly by a heavy dose of kin-selection—to do various things ultimately to promote the inter-generational replication of germ-line gene copies *residing in the reproductives* of the colony (i.e. the queen and the worker's reproductive siblings).[48] Thus, divorced from the reproductives, the ultimate point of everything else workers do would be missing; they function as gene-replication-promoting functional systems only in the context of the rest of the colony. Similarly with drones and queens, neither of which can carry out the full range of their biological functioning without the other; and of course their reproductive efforts would come to naught without the many activities of the workers—their construction, maintenance and protection of the nest, and their caring for the queen, drones and brood. In a honeybee colony, then, members of different castes are instances of different co-evolved types of biological functional systems, all "designed" by natural selection to carry out their proper functioning together, as a whole colony. And all of this functioning is geared ultimately toward promoting the reproductive success of the reproductive members of the colony, and hence the replication of copies of their germ-line genes (rather than being geared, for example, toward the democratic welfare-promotion of members of the colony, or "the good of the colony").

So far, this is all just spelling out certain implications of the view of biological function for which I have argued. But we can take a step

---

[48] See chapter three.

further by recognizing that this makes possible talk of members of each caste *playing a certain standard, non-incidental role*—or *function*—in the *working of the colony*. Workers build, maintain and protect the nest, and care for the others; the queen lays eggs, some of which become more workers, others of which become reproductives (going on to continue the inter-generational spread of genes represented in the colony); drones eventually fertilize an outside female (again, continuing the spread of genes represented in the colony). What we are recognizing here is that the various functional activities of the various members of the colony are tied together in such a way that they each promote their ultimate end in concert with the others, by making certain contributions to a "cooperative endeavor", which is what we are referring to in speaking of the functioning of the colony.[49] The notion of *the functioning or working of the colony* is a derivative one: The colony may indeed generally be regarded as a higher-level gene-replicating system, but its functioning just consists in the integrated combination of the biological functional activities of its various members, the end being the same, i.e. the reproductive success of the reproductives in the colony.

This may be expressed by expanding on what was said earlier about the dual functional relation between organisms and genes, to include something like a bee colony as well. Certain genes may have certain biological functions within the body of a given bee (e.g. to direct the manufacture of the stinger), and certain bees may have certain biological functions within the colony (e.g. to protect the nest), but the biological functioning of the colony must ultimately be understood in terms of the replication of genes—since it is derived from the biological functioning of its member organisms, which is understood in genetically-oriented terms—bringing us around full circle. There is thus no incompatibility between the view of biological function presented in chapter four and the recognition of biological functions at higher levels, such as the functions of a worker bee within a colony.

---

49 This is not entirely true, however, as illustrated by the case of sex-ratio manipulation discussed in chapter three (though that case involved ants rather than bees). The "genetic interests" of the queen regarding sex-ratio among reproductive offspring may conflict with those of sterile workers (since the queen is equally related to male and female offspring, whereas the workers are much more closely related to their sisters than to their brothers, due to certain peculiarities in hymenopteran genetics), and it is thus possible for conflicting behaviors to evolve here.

There will likewise be higher-level functions among other social animals, though it is important not to make the mistake of extending the special features of something like a hymenopteran colony to organized groups of conspecifics generally. A bee colony is a single large family whose members instantiate functional structures that have been dramatically shaped by kin-selection, such that reproduction takes place as a cooperative endeavor at the level of the entire colony. It is for this reason that it is possible for the most part to treat a bee colony as a higher-level gene-replicating system, with a relatively unified end to which its functions are directed. Such a thing is not true of groups of conspecifics in general—even where there is a high degree of cooperation among them. Consider, for example, a pack of conspecific but non-kin cooperative hunters, whose cooperative tendency is an adaptation enabling each of them to do better than he would on his own (since together they can subdue large prey that would be beyond the reach of any individual on his own). Each individual is ultimately a gene-replicating system whose functions are directed to the replication of germ-line copies of his own genes; each can be expected to do everything in its power to maximize its own reproductive success, which includes cooperating with others in the hunt, but does not involve making any net sacrifice in personal reproductive output for the sake of another's reproductive output; indeed, it may well involve competition with hunting partners over mates, for example. There is not, then, any group-level "consensus" on any single end beyond the catching and sharing of prey, so that it would be wrong to think of the society here as a single gene-replicating system, with anything like the unity of function displayed by an organism or a bee colony.[50]

On the other hand, there is no objection to speaking of the organization into packs as being biologically functional behavior, or of certain behaviors as having certain functions in coordinating or in maintaining the integrity of the pack, or of certain individuals as playing certain functional roles in the pack, and so on. Such biological functions are only to be expected in the case of social animals, with social behavioral adaptations, and they may again be fully understood within the general account of biological function that I have defended. An individual may have a certain function within a pack, but the pack is itself a functional entity (within the framework of biological

[50] Cf. Dawkins (1989, pp. 255 f.).

teleology) only insofar as such a social structure plays a non-incidental role in the functioning of its members—ultimately making some contribution to each member's reproductive success, as the pack does by enabling each to acquire more food than it would on its own. What we have here is thus the interesting situation in which a given organism is itself a functional system, and it plays a certain functional role in a larger system (i.e. the pack or society), which in turn gets its functional nature by virtue of playing a role in the functioning of the organism and others like it. The primary level of biological functionality is still the organism, and we are simply recognizing that in social animals this tends derivatively to give rise to biological functionality at higher levels as well—all of it being ultimately genetically-oriented.[51]

A similar situation obtains for "societies" composed of symbionts, such as ants and aphids. Like the societies of cooperative hunters considered above, such a society would certainly not be a single gene-replicating system; the ants and aphids have different "genetic interests", and cooperate in the way that they do only because and to the extent that such cooperation tends to contribute positively to the spread of their own genes. Still, inasmuch as the behaviors on both sides are adaptations—i.e. the ants' protecting the aphids being an ant adaptation, while the aphids' providing food for the ants is an aphid adaptation—and are hence biologically functional, co-evolved to constitute a functional relationship serving both kinds of members, there is no objection to speaking of the ants and aphids as performing various distinctive *functions* within the society they form. The crucial point, however, is that the functionality here is ultimately genetically-oriented, as in the other cases examined above, belonging to ordinary biological teleology; it is *not* as if the welfare of the society is some kind of end-in-itself, functionally promoted as such by the ants and aphids.

There are other interesting cases one might consider here, but the above observations should suffice to indicate how we might proceed in dealing with them, without positing any special kind of teleology in

---

[51] It should be kept in mind that I am not yet addressing the case of human beings and human society, and the possibility of biologically-independent, artificial social functions (nor will I discuss human beings until later chapters). I am certainly not suggesting that all social functions must be understood in terms of biological teleology, but only that they are in the case of such things as bees and hyenas, and perhaps generally so throughout the non-human biological world.

biology in addition to the kind I have discussed. Again, my claim is that our application of functional teleological concepts at levels higher than organisms—at least in the case of such things as colonies or other social organizations of conspecifics—is just part of the same biological teleological discourse we have been considering all along.

What threatens to lead to the positing of a distinct kind of teleology within biology, at levels above that of the organism, are not thoughts about societies of conspecifics or genuine symbionts, but certain thoughts about even higher-level entities such as ecosystems. Dawkins, who rejects any such move, refers to this as the "BBC Theorem", and describes such thinking as follows:

> [In pop-ecology commentaries] there was something called the "balance of nature", an *exquisitely fashioned machine* in which plants, herbivores, carnivores, parasites and scavengers each played their appointed *role for the good of all.* The only thing that threatened this delicate ecological china shop was the insensitive bull of human progress, the bulldozer of . . . , etc. The world needs the patient, toiling dung beetles and other scavengers, but for whose selfless efforts as the sanitary workers of the world . . . , etc. Herbivores need their predators, but for whom their populations would soar out of control and threaten them with extinction, just as man's population will unless . . . , etc.[52]

There are in fact two issues here: (i) whether there exists a rich network of relationships within an ecosystem, such that a gross disturbance of some element within it will tend to have significant and widespread ramifications throughout, and (ii) whether this justifies the above sorts of teleologically laden claims, where "the good of the ecosystem"—or its harmonious "functioning", or the collective good of the populations that make it up, etc—constitutes an end-in-itself, functionally promoted by the various activities of the various populations.

It is obviously possible to take a negative position on the second issue without denying the claim regarding the first. Indeed, it is only to be expected that ordinary natural selection occurring within populations of various species sharing a common environment will give rise to such complex relationships. But nothing follows about any one

[52] Dawkins (1982, pp. 236–7), my emphasis.

population, or its members or their activities, being *for the sake of* other populations, or *for the good of* the ecosystem as a whole (where a "healthy" ecosystem might be thought of as one well disposed to support a wide variety of species). One population might, of course, *need* even relatively unrelated populations in the sense that the existence of the latter is necessary to the balance of the ecosystem, which in turn is necessary to the continued existence of the first population. If without dung beetles, for example, an important part of the ecosystem would collapse, leading to the extinction of oak trees among other things, then there is obviously a sense in which oak trees can be said to need dung beetles: Oak trees would depend on the continued existence of dung beetles for their own continued existence. But this does not automatically imply that it is part of the proper function of dung beetles to promote the good of the ecosystem or the welfare of oak trees, as such.

Dawkins sums up his objection to the "BBC Theorem" as follows:

> A network of relationships there may be, but it is made up of small, self-interested components [i.e. organisms]. Entities that pay the cost of furthering the well-being of the ecosystem as a whole will tend to reproduce themselves less successfully than rivals that exploit their public-spirited colleagues, and contribute nothing to the general welfare.[53]

The point here—though it needs to be qualified to allow for some legitimate forms of group selection (as discussed just below and in chapter six, section two)—is that organisms have not generally evolved to be "public-spirited" contributers to the good of an ecosystem, any more than they have generally evolved to be selfless contributers to the good of their own species or population; as illustrated in chapter three, they have evolved—in the context of their environment—to be efficient promoters of the inter-generational replication of certain copies of the genes that built them. The more "public-spirited" models would at least largely have tended to be replaced by the more "selfish" ones—i.e. those dedicated to maximizing their reproductive success in relation to that of others in the population, without devoting resources to promoting "public goods" within the population or ecosystem. In short, organisms

[53] Dawkins (1982, p. 237).

have been "designed" by microevolution through natural selection primarily to promote the spread of their genes through maximizing their reproductive success, *and* inasmuch as this long process of "design" has taken place in a common environment, where each population forms part of the environment of the others, a complex web of relations has developed as a derivative aspect of this process, such that various types of organisms that share an ecosystem are generally well-adapted to thrive in the context of the others.

There are, however, at least two further issues that must be considered in this connection, the first of which leads to one of the important qualifications referred to earlier. First, it would seem to be possible in certain cases for there to be a kind of higher-level selection of the following sort. If T1 and T2 are "rival" traits that may be possessed by members of a given species, with one or the other tending to predominate in any given group, and (i) the possession of T1 by members of a group tends not only to promote the short-term reproductive success of its possessors in the given environment, but also to be favorable to the continuation of the group in the given environment, while (ii) the possession of T2 by members of a group tends to lead to the extinction of the group in the given environment, even though it promotes the short-term reproductive success of its possessors, then the differential survival/extinction of groups on this basis can account for the general prevalence of T1 in members of the species.[54] This is especially relevant where T1, no less than T2, is "evolutionarily stable" in the sense that it can spread within a group and thereafter resist "invasion" by rival traits.[55] But this is not absolutely necessary. As Sober and Wilson have stressed, while it is true that higher-level selection in favor of T1 will tend to be countered by ordinary selection against it in cases where T1 is invadeable by T2 (i.e.

[54] Thanks to Robert Adams for stressing this possibility to me.

[55] On the possibility of higher-level selection between alternative evolutionarily stable traits—i.e. due to the extinction of populations where one has predominated, and the survival of those where another has predominated—see Dawkins (1989, p. 321). His example involves two rival behavioral strategies, labelled "cheat" (which involves exploiting the altruism of others) and "grudger" (which involves discriminating altruism). Each of these is evolutionarily stable (unlike undiscriminating altruism), but populations in which the former has come to predominate are more likely to go extinct than populations in which the latter has come to predominate. Thus, part of the explanation of the general predominance of the latter trait among members of the species generally will be in terms of this higher-level selection.

T1 is less individually optimal), this countering force is not necessarily sufficient to undermine the effect of higher-level selection; given the right sorts of group dynamics—e.g. a certain rate of periodic dispersal and reorganization of groups—the problem of "subversion from within" can be overcome and T1 can evolve.[56] For simplicity, however, I'll focus on the first type of case, where both traits constitute "evolutionarily stable strategies."

In one sort of case, all that is at issue is a difference in how well each trait promotes its possessor's reproductive success in a group dominated by individuals who also share the trait—with T2 perhaps doing so in a way that proves insufficient to keep the group going for very long, thus leading to the predominance of groups whose members possess T1, and hence to the predominance of T1 in such creatures.[57] Still, it could arguably be said in such a case that the type of organism in question has been "designed" by natural selection—considered at *both* levels—to be such that the end for such an organism is not merely the inter-generational replication of certain copies of its genes, but this *in a manner that allows the group to be sustainable in the long term in its environment.*

Another, more interesting, sort of case would be one in which the difference between the two traits has to do with something like impact on environmental resources—with T2 perhaps leading to unsustainable use of resources in the long run. (An example involving different clutch sizes among swifts in different groups will be considered in detail in chapter six, section two.) After the impact of higher-level selection in favor of T1, it seems that we will again have to allow that organisms of such a species have been "designed" by natural selection (considered at both levels) not merely to promote the inter-generational replication of certain copies of their genes, but to do so in a manner that allows the group to be sustainable in the long term in its environment—where this involves certain limitations in the impact on the environment. And in that case, we might legitimately speak of such "conservation" of the

---

[56] Sober and Wilson (1998).

[57] This is what is going on in Dawkins' example of grudgers and cheats (above). A population of cheats is more likely to go extinct simply because *even though* the trait in question is evolutionarily stable (i.e. it can spread, given the right initial conditions, and then resist invasion by rival traits, such as "sucker" and "grudger"), it is hard for cheats in a population dominated by cheats to survive and reproduce at a rate sufficient to keep the population going in the long run.

environment as *part of the biological end* of such organisms. This, then, is a further qualification to the account in chapter four. Having said that, however, note that this additional element would not be some kind of end-in-itself—as if predators in a given group, in manifesting T1 and not "overfishing", were working for the good of the prey population in its own right. That would be a complete misunderstanding of T1's teleological significance. Rather, the environmental conservation is merely instrumental to the end of sustaining the organisms' own group.

The second general issue referred to above is the following. Suppose we grant Dawkins' claim (as I understand it) that the network of relationships within an ecosystem is simply the result of the co-evolution of various populations of different species that all form part of the others' environments—where the primary evolutionary driving force is just the usual microevolution through natural selection within populations, leading generally to organisms that tend to be good at promoting the inter-generational replication of certain copies of their genes, in their environment. It does not immediately follow from this that there is no higher level functional teleology of any sort within ecosystems. For the same point can be made about a "society" of symbionts—such as a society of ants and the aphids they farm—and I argued above that there *are* legitimate higher-level teleological claims to be made here, though they must be understood to belong to the usual biological teleological framework. That is, the fact that the harmonious relations are just the result of the evolution of "selfish" adaptations on both sides does not prevent us from recognizing the social roles to be functional—speaking, for example, of certain ants as playing a special functional role in the ant/aphid society; such talk is legitimate, so long as we recognize that the well functioning of the ant/aphid society is only a subordinate biological end, geared ultimately to the genetically-oriented ends of both ants and aphids. The question remains, then, whether at least this sort of teleology may exist also more generally at the level of ecosystems.

This amounts to asking whether the organisms making up an ecosystem, or perhaps the various populations, are to be regarded quite generally as standing in genuine symbiotic relations with one another. To answer this, it is necessary first of all to make an important distinction. It is one thing to recognize that as a result of the fact that different species have evolved in a common environment—which they each in part constitute—each will tend to be well adapted to co-exist with the others in that environment, resulting in a certain "delicate balance" within the ecosystem. It is quite another, however, to suggest that each has actually evolved certain adaptations through benefitting various other elements of the ecosystem which in turn benefit them—in the way that ants have evolved the adaptation to protect aphids, or that aphids have evolved the adaptation to provide food for ants. Let us restrict the term "symbiosis" to refer only to the sort of phenomenon appealed to in the second of these two claims, since they are very different and it is the second sort of phenomenon that is at issue in central cases of symbiosis. There is no dispute, I take it, over the first sort of claim; the question is whether there is any merit to the second.

I do not know enough about ecosystems to know how much of the network of relationships is a result of a general manifestation of the second kind of phenomenon (i.e. a kind of general symbiosis), but it seems plausible on the face of it that a great deal of it could be explained simply in terms of the first—where the various benefits that organisms of a given species confer on organisms of other species are to be understood simply as side-effects of their ordinary adaptive activity. This would, of course, influence the further evolution of the other organisms, as any environmental factor does, and similarly the other way around, so that it would in a certain sense be *no accident* that they all tend to be well adapted to live in this common environment which they all help to constitute; but this is still very different from suggesting that the conferring of benefits on the other organisms is in each case an adaptation (resulting from the fact that conferring such a benefit also brought sufficient payback to help the relevant genes spread over other alleles), as in cases of genuine symbiosis.

In any case, to the extent that ecosystem harmony *can* be attributed to genuine symbiosis, I have already indicated how this can be understood in terms of ordinary biological teleology. To the extent that it is to be understood simply as a derivative aspect of the evolution of the various species in a common environment that they all influence, it does not seem appropriate to speak of any higher-level teleology, whether understood in terms of ordinary biological teleology or in some other way. Again, we can concede that it is in some sense *no accident* that populations of organisms within an ecosystem exist in a certain balance, with a high level of interdependency, each population doing certain things that tend to maintain the balance. But the non-accidentalness at issue here is just a derivative aspect of ordinary intra-population natural selection in a common environment, without the special relations characteristic of genuine symbiosis, and hence does not seem relevant to supporting any teleological claims of the sort we might make in connection with genuine symbionts.

Consider the following fanciful analogue. Suppose that the noises made by the heart of a given type of predatory organism, as a side-effect of its ordinary biological functioning, happened to scare away certain mutants of a certain prey species, thus benefitting them. At first this would be entirely accidental, but if the tendency to fear the noise wound up spreading through natural selection and becoming an adaptation, then this would no longer be so: It would be *no accident* that they tend to benefit from the noises made by their predators' hearts.[58] What should we say about this? Certainly we should say that the tendency to fear and to flee from the noise is biologically functional on the part of the prey. But as far as the predators are concerned, the production of heart noises is merely a side-effect of the heart's proper functioning; the benefits to the other species are purely incidental from this perspective—certainly not some functional contribution to their good within some broader teleological context! The same things can be said, I think, in connection with "contributions to the ecosystem", at least to the extent that this is a result not of genuine symbiosis, but simply of species evolving in each other's environments. And again, even insofar as genuine symbiosis may be involved, we certainly do not arrive at a picture of the ecosystem as a functional system whose good or health is some kind of final end, which organisms or populations have been "designed" ultimately to promote.

14. THEOLOGY AND TELEOLOGY

Biological teleology is a form of functional teleology that is plausibly found in connection with all known living things, and it gets its gene-oriented character from the fact that all known living things are instances of organism-types that have been assembled and shaped over

[58] This example is based on a similar one suggested to me by Sean Foran.

evolutionary time through natural selection. It is worth pointing out, however, that I have not claimed that all *conceivable* living things will manifest this sort of biological teleology. I do not want to commit myself to the view that being an instance of an organism-type that has evolved through natural selection is a necessary condition for being alive. There are, it seems, conceivable objects that we might plausibly regard as alive even though they would not satisfy this condition. Imagine, for example, objects perfectly resembling actual cats in makeup and behavior, but created in a lab or by a supernatural creator, or at least tracing their ancestry back to such artifacts. (Actually, a supernatural creator would itself be another example, since it would plausibly be alive and yet would not be a member of a species that has evolved through natural selection.)

If there were such biological objects—either artifacts created directly by an agent, or offspring of such artifacts—they would be instances of biological types designed by the agent(s) in question, rather than shaped by natural selection. They would manifest a certain sort of functional teleology, but this would not necessarily share with ordinary biological teleology any special orientation toward the inter-generational replication of certain molecules—in fact, presumably it would not, unless the creator were rather perverse.[59] Thus, my use of the expression "biological teleology" as a label for the sort of gene-oriented natural functional teleology found in connection with actual known living things may be somewhat misleading. It is not simply by virtue of their being *alive* (i.e. "biological") that they manifest this form of teleology, but by virtue of this together with the fact that they have been shaped by natural selection—something that is perhaps only a contingent feature of living things, manifested by all actual known living things, but not necessarily by all conceivable ones.

If there can in principle be other kinds of living things, then there is no single kind of functional teleology that applies to all conceivable living things as such. To be precise, then, we might treat the expression "biological teleology" as a broad category, encompassing both "natural teleology"—i.e. the sort of functional teleology that is manifested by all natural living things on the assumption that they are indeed products of natural selection—and "artificial biological teleology", which would apply to such things as living artifacts of a

---

[59] I will consider just such a hypothetical example, for another purpose, in chapter six.

divine creator. I will not do so, however, and will continue to use "biological teleology" and "natural teleology" interchangeably, to refer to the first of the above two categories. The reason is that functional teleology found in connection with possible living artifacts or their offspring would still really just fall under the category of *artificial* teleology, deriving its character from the creators' designs, just as with other artifacts. It does not, then, seem necessary to speak of "artificial biological teleology" in such cases, since the fact that the hypothetical machines in question are *alive* would seem irrelevant to the nature of the functional teleology at issue. It would just be an example of artificial functional teleology that happens to involve a living entity, as is already the case with artificially bred traits, for example. By contrast, while I have conceded that it is not simply by virtue of being alive that actual, known living organisms manifest what I have called natural or biological teleology, I am assuming that there is in fact an intimate connection between life on earth and natural selection; that is, the former has in fact arisen and evolved through the latter, so that all actual, known living things, including human beings, are instances of organism-types that have evolved by natural selection. I will therefore continue to use the expression "biological teleology" or "natural teleology" to refer to the sort of teleology found in connection with actual organisms and their potential progeny under this assumption.

It is worth mentioning one further possibility. Suppose that living things are indeed products of natural selection, but it turns out that the mechanisms and circumstances of natural selection—perhaps even the background laws that make it possible—were put in place by a divine creator with the intention of thereby creating the various forms of life over the course of evolutionary history, culminating in the creation of human life. This is presumably what many theists who accept the theory of natural selection believe. If this were the case, then the sort of natural biological functional teleology I have been discussing would have to be understood to be embedded within a larger *psychological* teleological framework, creating a complex functional teleological situation for living things that is neither simply natural nor simply artificial. Is such a possibility consistent with the general claims I have made about biological teleology in earlier chapters?

I believe that it is, inasmuch as my account of biological function—along with the neo-Darwinian evolutionary theory to which it appeals—is meant to be a *secular* account, remaining neutral on such larger theological issues, rather than being an *atheistic* account, committed to a denial of such broad theological possibilities. What *is* denied, of course, are certain theological claims that would be inconsistent with the evolution of living things through neo-Darwinian selection processes—such as the claim that each species was designed directly by a divine creator. But the theological possibility raised above is not in this way inconsistent with our best biological theories about the origins of living things, or with accounts of biological function based on them, which can and should be pursued in isolation from such theological issues. Nothing I have said about biological teleology has been meant to imply any commitment to a rejection of such a theological possibility. Talk of the "ultimate end" of any living thing, for example, has been meant to be understood as a claim within the biological teleological framework, which leaves theoretical room for there being further ends within a different framework, no longer the concern of biological teleology.

The theological picture described above would involve at least two teleological commitments with regard to natural selection and living things that go beyond any commitments of a secular account of biological teleology. First, on such a theistic view, evolution by natural selection would itself be a teleological process, as a tool designed and used to bring about certain long-term divine goals—its function being to yield increasingly complex forms of life, culminating in humanity. By contrast, on secular accounts there is no commitment to treating evolution as a teleological process. In fact, it is often explicitly denied that it is, but we have to be careful here. What is properly denied from the secular perspective is just that anything about natural selection itself (or about any scientifically accessible factor related to it) makes it teleological; it is rather the source of the biological teleology manifested in living things, and there is nothing within the secular picture that would contribute to its being teleological itself. It would be a further step to deny categorically that natural selection is a teleological process, which would involve denying the above theological possibility; and again, this would not seem to be the business of a secular theory.

The second teleological commitment involved in the theological picture, but not in the secular one, is the idea that living things that were ancestral to human beings would have had ends beyond what I have identified as their ultimate biological ends. On such a view, the ultimate biological end of any such organism would be subordinate, within a broader framework, to further ends of a different sort, having to do with the teleological process of evolution by natural selection, leading to the fulfillment of the divine goal. In the human case, things would get even more complicated. We have, it might be supposed, reached the upper end of the evolutionary chain used by the creator to generate various life forms culminating in humanity, so there is no further end *along that chain* to which human biological ends would stand in an instrumental relation. The various biological ends human beings have, as well as the ways in which they are structured, are on such a view the result of a tool (i.e. natural selection) the creator was using ultimately in the service of certain higher ends involving human beings. And that tool has now served its purpose, having yielded human beings, who have a biologically-independent teleology stemming from the creator's higher purposes in addition to our biological teleology. Thus, our biological ends would seem, on such a view, to be nothing but residual products of our natural selection history, relevant to our further ends only insofar as they genuinely promote them as part of the creator's plan. And in that case, certain biologically functional traits relevant to promoting gene replication might fail to be relevant at all to our higher ends, while biologically functional traits that have tended to promote gene replication by promoting survival may now be relevant to our higher ends *directly*.

These are no doubt interesting issues to explore, but I shall not further pursue them here.

CHAPTER VI

Welfare and Natural Teleology

1. INDEPENDENT PROBLEMS FOR AHISTORICAL WELFARE-BASED ACCOUNTS, PART I: GENERAL CONSIDERATIONS

The argument for the view of biological function I have presented began with a principle stating the relevance of causal history to the teleological structures of functional systems. The principle was:

> P: With regard to a given type of functional system S, whatever it is that is ultimately causally responsible for the compresence and organization of its parts and features—in such a way as to make it the case that it is *no accident* that the system possesses *such* parts and features and *such* an organization among them that they interactively bring about a certain special subset of the *effects* they do, thus allowing us to distinguish non-incidental effects or contributions from incidental ones—is thereby also responsible for determining the specification of working for S; that is, it equally determines the nature of the functions and ends served by these parts, features and activities so organized, and the way in which these functions and ends are related at various levels, i.e. how exactly they fit into the hierarchical teleological structure of S.

This was meant to apply both to machines and to organisms. In the case of machines, it is the designer that plays the crucial causal role, while in the case of organisms, it is the natural selection history behind the evolution of the organism-type that plays this role. This principle thus led me to conclude that it is natural selection—"the Blind Watchmaker"—that is responsible for shaping the biological functional structures of organisms.[1] And as argued in chapters two through four, this means that organisms must generally be understood ultimately as complex gene replicating systems—functional systems that have for their general and ultimate biological end *the replication of certain germ-line copies of the genes represented in their co-adapted genomes into the next generation*, with lower level functions and ends all geared toward this end; moreover, as summarized in section six of chapter four, this certainly need not generally coincide with the promotion of organismic welfare, and indeed often diverges from it.[2]

The natural move for an opponent who wishes to hold on to some sort of welfare-based view of natural teleology will therefore be either (i) to reject P from the start, going on to argue for an alternative welfare-based account of natural teleology, or (ii) to insist that there are two different kinds of natural teleology associated with living things, one of which is captured by P and the account that follows from it—and is perhaps characteristic of discourse among biologists in particular—while the other, which is equally legitimate, familiar and objective is quite independent of P and admits of an ahistorical, welfare-based account. Call the first the "exclusive approach" and the second the "complementary approach".

On the exclusive approach, no less than on the complementary approach, it will be granted that organisms have been put together as the coherent and integrated systems they are by causal factors having ultimately to do with natural selection. (I am not here concerned with an opponent who denies the theory of natural selection.) But it will be claimed that functional concepts have application to organisms not in relation to any of this as such, but rather in relation to species-typical promotion of *welfare*, or the satisfaction of species-typical welfare-

---

[1] This expression from Dawkins (1987, pp. 4–5) is an apt one: On the one hand, the metaphor of a watchmaker brings out the fact that the natural selection history behind the evolution of a given type of organism plays the same causal role with respect to that type of organism that is played by a designer with respect to a certain type of machine; on the other hand, the limits of the metaphor are indicated by the specification that the watchmaker is blind, meaning that the forces of natural selection have no foresight—no capacity for planning.

[2] Again, this formulation of an organism's ultimate end within the framework of biological teleology is only an approximation, and is subject to the qualifications discussed in chapters four and five.

related *needs*. One obvious difficulty with such a line, however, lies in motivating the rejection of the parallel between organisms and machines. After all, it will presumably be granted that P holds in the case of a machine, where the causal factor that is ultimately responsible for giving the system its particular collection and arrangement of parts and features, and establishing certain non-accidental relations among them and certain of their effects will surely also be what normally determines its functional teleological structure. So the position will have to be that this is just a peculiar fact about artifacts and the role of literal design, with no bearing on the case of organisms, despite the deep parallels I've brought out with regard to exactly the sorts of relations that are at the heart of functional teleology. But why should that be?

One potential motivation behind such an insistence on a fundamental asymmetry between machines and organisms is the suspicion that we cannot identify genuine *ends* in natural systems except by appeal to the notion of *good* or *welfare*: How else could we distinguish between genuine ends and mere effects? In the case of machines, it will be said, we can identify the effects that are ends by relating them to the *intentions* of the designer and of users; but since there are no such intentions involved in the case of organisms, we must look to the notion of *good* to make the distinction. A type of effect is a genuine biological end *only if* its realization promotes (instrumentally or constitutively) the good of the organism, or perhaps of its group. The circulation of blood, for example, is a genuine end served by the heart's pumping, because it promotes the organism's good; by contrast, the production of noises does not do any good, and hence is not an end, but just a side-effect. And if it is present welfare-promotion that is relevant to teleology, it is not clear why we should be concerned with history at all.

This is misguided, however. It is true that there are no intentions behind the structuring of the parts, features and activities of organisms, but there are principles behind it nonetheless—principles that explain how organisms have non-accidentally been put together as coherent, hierarchically organized systems, and what it is, as a result of this, toward which their activities are non-incidentally directed. By examining these principles of natural selection, and the organic systems to which they give rise, it is not hard to distinguish in an intuitive and non-arbitrary way between genuine functions and ends, on the one hand, and mere effects, on the other, as I have done in earlier chapters. The former play some non-incidental role—even if very remote—in the coordinated inter-generational replication of certain copies of the genes represented in the organism's co-adapted genome, while the latter do not. The distinction between the circulation of the blood and the production of noises, for example, is no less clear with regard to gene replication than it is with regard to welfare. It is perfectly clear that the former, but not the latter, makes a standard and non-incidental contribution to coordinated, naturally selected activities culminating in the replication of the organism's genes into the next generation; in fact, in many cases this question will be significantly easier to settle than the question whether the trait makes a standard contribution to welfare, since the relevant notion of genetic replication is far easier to nail down than is the notion of welfare, about which our intuitions are often quite unclear.

To this it might be objected that while we can indeed distinguish between two classes of effect—those which play some role in the coordinated replication of genes and those which do not—it is not clear why this should be thought to capture the distinction between genuine ends and mere effects. Why should effects of the former class be regarded as *ends* (or functions), as opposed merely to being effects that are of relevance to natural selection? The answer is that it is not merely because they are of relevance to natural selection that the former effects are singled out as special; this *by itself* would indeed be a poor reason for calling them *ends*, as I will emphasize in distancing myself from etiological views in the next chapter. Rather, they are special because as a result of the natural selection history behind a given type of organism, they fit into a non-accidentally produced, coordinated hierarchy of effects, without which it is doubtful that the system could intelligibly be spoken of as a coherent whole that *works* in a certain way. This way of distinguishing ends from mere effects is thus far from an arbitrary choice reflecting a scientistic overemphasis on genes and evolution. If both machines and organisms ostensibly constitute working systems, and the notion of working can be understood in the case of machines without reference to the machine's welfare (an obvious point, since there is no such thing), then there is so far no reason to think that the same cannot be true of organisms. We have seen no good reason to doubt that the distinction between genuine ends and mere effects in organisms can be made in a plausible and intuitive way by appeal to coordinated gene replication, and thus no good reason to think that an appeal to welfare is necessary.

Perhaps many will still insist—as I believe Foot would—that it is just intuitively obvious that biological teleology must be understood ultimately in terms of welfare promotion, claiming that there is simply a conceptual connection between function in biology and the promotion of welfare or the satisfaction of welfare-related needs. In response to this, I'd like first to consider a thought experiment that may help to remove the temptation to suppose that welfare promotion *must* come into the functional teleology of living things, weakening the opponent's confidence in such a conceptual connection; then I will say a few things about why it might seem that there is such a conceptual connection, even if there is not.

First, suppose there is an extremely clever inventor who has discovered in the laboratory how to design and create organic machines that look and act much like natural organisms, maintaining and reproducing themselves through very similar methods, including many of the same bio-chemical reactions, and so on. Let us imagine that the similarity is close enough that these machines are conceded to be alive, and to be such that we can genuinely speak of their welfare (unlike any machines that have actually been created so far). This inventor, however, is rather perverse: He is single-mindedly obsessed with the propagation of the replicating molecules in these machines which correspond to genes in natural organisms—call them "shmenes"—and cannot be persuaded by his more conscientious colleagues to place first priority on their welfare as he designs them. Instead, he designs his machines simply to work ultimately toward the most efficient replication of their shmenes, however this affects their welfare; some of the things they are designed to do also promote their welfare, while others detract from it—just as with natural organisms "designed" by natural selection. We might even imagine that he designs machines that are virtually indistinguishable from living things of certain kinds, appearing exactly as if they were products of natural selection.

Now it seems clear to me that what he has created are primarily and ultimately shmene replicating machines. The functions pertaining to these things are generally and ultimately related to the end of shmene replication, not to the promotion of the machines' welfare. The fact that they have *welfare* at all is incidental. They were not designed around its promotion, but simply to be effective vehicles for the preservation and replication of the shmenes in question, regardless of how this might bear on their welfare. But now if this is right, then clearly there *can* be living things whose functionality does not have generally or ultimately to do with the promotion of their welfare, but has to do merely with (in this case) shmene replication. There is not, therefore, any conceptual connection between function in living things and welfare promotion.

Someone might try to respond to this by weakening the claim to say that there is at any rate such a conceptual connection in *natural* cases, not involving design. But this is a groundless move. Why should such a restriction be thought to *help* the case for such a connection? The perverse designer was deliberately described in such a way as to bring out the fact that he is simply duplicating the effects of natural selection in the design of his organic machines. The welfare of organisms is no more taken into account in natural selection than it is by the gene-obsessed designer, and it is therefore unclear why the restriction to natural cases should be thought to help. Why should welfare come into the picture here any more than in the artificial case? Perhaps it will be claimed that the designer's intentions are sufficient to make shmene replication the ultimate end of his organic machines, but that in the absence of such an intention in the natural case, the causal similarities between the two cases are of no significance, and everything about function is instead determined simply by facts about welfare and need-satisfaction. While I have not yet said enough to rule out such a response, however, it obviously has no force without some argument for it.

Why is there such a strong tendency to think that the promotion of welfare must be central to biological function—to the point where this may even seem to reflect a conceptual connection? The reason has already been touched upon in previous chapters. First, we invariably learn to apply functional concepts to organisms long before we learn anything about genes and evolution, focusing on relatively accessible traits whose contributions to gene replication do indeed involve contributions to survival and reproduction that can plausibly be regarded as contributions to organismic welfare. It is natural in this situation to assume that organisms are functional systems geared toward the promotion of their own good, or that of their offspring, or sometimes that of their group or species—and that this connection with welfare is the very essence of function, at least in living things; what else, after all, would functions pertain to if not the *good* of the organisms in question? Second, these rather vague thoughts about welfare and function tend to continue even after we learn about natural selection, partly due to sloppy thinking about natural selection (e.g. talk of traits being selected because they are "good for the species")—which, far from bringing out the problems with this view of biological function, serves only to reinforce it—and partly due to continued focus on a fairly narrow range of cases which do indeed involve the promotion of welfare or the satisfaction of welfare-related needs. Again, if these errors in thinking about natural selection are not cleared up, and the range of cases considered when thinking about biological function is not expanded, it will naturally seem that function in biology has generally and ultimately to do with welfare promotion. It will not even be apparent what alternative there could possibly be, and it will thus probably just seem to be a conceptual matter that function in biology pertains to welfare—a matter no longer open to serious question. If considerations about genes and natural selection are then raised (as I have done), it will *appear* to someone in this position that this is just missing the whole point, confusing the issue of teleology with something else.

As mentioned earlier, there is also a weaker position for my opponent to take: the complementary approach. This does not require rejecting P or the parallel between machines and organisms, and is compatible with granting my account of natural teleology—though only as an account of one of two alleged kinds of natural teleology. The question, however, is why we should believe in the second kind of natural teleology—the kind that's allegedly just as legitimate and objective as the one I've identified, but is independent of facts about causal history, and can be grasped simply by considering the ways in which welfare is characteristically promoted and needs characteristically met, however this might relate to the structure of non-accidental relations among traits and their effects that I've explored in connection with organisms. Does such teleology also exist, right alongside the shmene-oriented teleology, in the example of the living artifacts discussed above? How do we come to know of such teleological facts over and above the merely statistical facts about what things tend in fact to promote the organisms' welfare?

There is no epistemological problem on the view I have defended, apart from the usual problems with knowing historical facts; given the historical facts, we have an argument for why certain teleological facts thereby obtain, and what sort of structure they have. But on a view such as Foot's, it is entirely mysterious how we're supposed to come to know the teleological facts as such. This is especially true on the complementary view, where these facts are supposed to exist in *addition* to the whole slew of teleological facts that have already been accounted for on my view—which includes just the sorts of examples Foot herself appeals to in supporting her intuitions: the sight of an owl, the behavior of bees, wolves and lionesses, and so on. Why think that she's doing anything more than just fixing on a subset of the teleological facts I've accounted for—namely, the facts about functions that involve welfare-promotion as a proximate end—and taking a narrow view of them, simply ignoring the further ends beyond welfare-promotion? Obviously this will be the suspicion unless some good reason is given for thinking that she's discovered a distinct set of differently structured teleological facts superimposed over the others; and her discussion contains nothing in the way of such an argument.[3]

Again, one needn't look far to find a plausible theory of error to explain why it may *seem* as though there are two forms of natural teleology, rather than just the one I have accounted for. Consider, for example, that for most practical purposes we are certainly *interested* only in welfare-related aspects of biological teleology, ignoring functions and ends that don't pertain to welfare promotion—in particular, ignoring the final end of gene replication. Welfare-related aspects of biological teleology make up such a central and large part of it that unless we are careful ultimately to return to the larger context, and to view these aspects within that context, they might well falsely

[3] See Foot (1994, 1995). Her discussion is explicitly in terms of the *evaluation* of particulars according to natural species-typical standards grounded ultimately in facts about how species-typical needs are properly met. But as noted in chapter one, this is plainly an appeal to natural teleology, the evaluations—such as judgements of natural defect or non-defectiveness—being natural *function*-related evaluations, with the meeting of welfare-related needs coming in as the alleged ultimate *end* to which functions are related. Thus, the burden is on Foot to argue for the view of natural teleology to which she is implicitly appealing. And again, that is what is lacking.

appear to constitute a special sort of teleological framework in its own right.

Perhaps someone will wish just to *stipulate* a certain use of functional concepts in the context of living things, according to which we treat species-typical welfare-promotion as a general and final end. I have no objection if someone wishes to speak this way for certain purposes—for example, in medical contexts in connection with pets. But it is important to see that a crucial threshold will have been crossed—something that would certainly be unacceptable to Foot given her larger purposes: We will no longer be dealing generally with *objective* teleology in biology, but with a partly artificial framework shaped by our own interests. There will therefore be a loss in the objectivity of our teleological judgments, at least in every case where they differ from the judgments made within the biological functional framework I have defended. If a pet is claimed to be dysfunctional, for example, because it spends so much of its time fighting (suppose that this is an adaptation that tends to increase the reproductive success of such a creature, but does not promote any relevant welfare—as in the case of elephant seals considered earlier), this judgment must be admitted simply to be relative to the interest we take in the pet's welfare; in relation to the objective teleological framework pertaining to such an organism, it is not dysfunctional at all, and any drug that took away its fighting disposition would actually be inhibiting its proper functioning, even if it made the animal better off.[4]

2. INDEPENDENT PROBLEMS FOR AHISTORICAL WELFARE-BASED ACCOUNTS, PART II: SOME BIOLOGICAL EXAMPLES

There are even worse problems for ahistorical welfare-based views that become especially clear through the consideration of some salient examples. (I will consider the compromise of an *historically informed* welfare-based position, which fares no better, in section three.) What these cases illustrate is that such accounts cannot distinguish between genuine ends functionally served by a trait and effects that, while beneficial, are only *incidental* results of the traits in question; and this obviously applies to both the exclusive and the complementary

[4] Cf. the point about neutered male cats, in chapter three, section four.

approaches described above. On some versions, where the appeal to welfare is very general, the problem will be especially clear. But even more sophisticated versions (such as Foot's) that try to specify the level at which welfare comes in—i.e. whether functions are to be understood as promoting the good of the species, group, or individual—will run into difficulties. By ignoring the historical background, they will still wind up committed to treating what are plainly *incidental* benefits (i.e. benefits achieved simply by accident) as ends functionally served by traits. Since the distinction between function and accident, or end and mere effect (even if beneficial), is central to functional teleology, these problems are serious ones.

Consider, then, the following example. Suppose we notice that typical clutch size varies among different groups of swifts in proportion to the scarcity of food resources in their environment: Clutch size tends to be reduced from 3 to 2 in environments where food resources are more scarce. After some investigation, we discover that the food scarcity in the group with the reduced average clutch size is not so severe that it just automatically prevents females from manufacturing the higher number of eggs; they have the nutritional resources necessary for making 3 eggs, but nonetheless, their clutch size is reduced from 3 to 2. In light of this, it is natural to wonder what *function* this might serve in the life of swifts; indeed, let us suppose that we are told by field biologists that it is in fact an adaptation, though we are told nothing more about what its function might be. After further investigation, we may discover that such "birth control" significantly reduces the risk of group extinction, since it prevents the group from over-exploiting and endangering the relatively scarce food supply; without such reduction, the group might well do itself in. A natural thought, then—at least for proponents of ahistorical welfare-based accounts of biological function—will be that surely this is the function of the reduction of clutch size (laying 2 eggs rather than 3): to conserve group resources in order that the group avoid the risk of extinction. Swifts have smaller clutches in such environments "for the good of the group", it might be said. As Foot might put it, *this is how they manage to conserve their resources and guard against extinction* in times of food scarcity, each couple doing its part in the conservation effort. Why should we even be concerned with the historical origins of clutch size reduction, when what is at issue as far as functional teleology is concerned is simply the function it serves in such organisms at a given time? And isn't the latter plain enough?

The answer is that it is not, and this becomes clear once we consider the complex issues that arise in thinking about the evolutionary story behind this particular trait. In fact, there are at least two very different evolutionary possibilities here. The first is that the trait—i.e. an individual's having a smaller clutch size than it could have, i.e. "birth control"—has evolved through group selection of the sort defended by Sober and Wilson (1998). That is, even if a clutch size of 3 is always biologically optimal from the individual organism's point of view, groups of swifts that exhibited this "selfish" trait even in conditions of scarcity wound up straining their resources and ultimately doing less well than groups of swifts exhibiting the altruistic "birth control" trait, so that the latter phenotype ultimately came to be more widely represented in the population. For a long time this was dismissed as a possibility because of the familiar problem of "subversion from within". Any group of individuals who lay fewer than the "optimal" number of eggs is inherently unstable: The gene pool will tend to be invaded by genes that tend to make their possessors lay closer to the optimal number of eggs, leading to a population of optimal breeders. We should not, then, expect to find groups of individuals who exhibit such restraint, "voluntarily" laying fewer than their optimal number of eggs.[5] But as Sober and Wilson have pointed out, this problem is not necessarily fatal, provided roughly that those exhibiting the altruistic trait (i.e. the swifts restraining their individual reproduction, hence helping to limit the strain on group resources) are concentrated into temporarily isolated groups and that those groups periodically disperse before the "invasion" by the more selfish trait goes too far, with the altruistic individuals again concentrating into groups, and so on.[6] In other words, group selection and individual selection will pull in opposite directions, the first favoring the altruistic trait while the second favors the selfish trait; but the latter will not automatically win out, as is often assumed, and given the right group dynamics, the altruistic trait might well evolve in this way.

Now *if* this happens to be the actual evolutionary story behind the swifts' behavior in our example above, then—as I've already indicated

---

[5] See Dawkins (1976, pp. 121–6).

[6] See Sober and Wilson (1998, ch. 1).

in section thirteen of chapter five—this sort of resource conservation will indeed count as part of the swifts' natural end; on my view, it will figure into the right sort of framework of non-accidental relations to count as a genuine end. So *if* this is how things happened, then the above proponent of the ahistorical welfare-based approach will have gotten lucky: She will have declared a certain benefit to be an end where it is indeed such. But the point is precisely that this will have been luck. For there is a second evolutionary possibility that can equally give rise to the same behavior in the swifts, and if that is the history that obtains then we shall see that treating the group benefit as an end served by the trait will be entirely misguided. And of course the point is that by ignoring causal history she will have no idea which is the case, making her teleological claims altogether unreliable and, when true, only accidentally so.

The second evolutionary story is a more orthodox one, and shows how such behavior can evolve without appeal to group selection.[7] For swifts in any given environmental situation, there is one optimal clutch size in terms of *maximizing personal reproduction*—a number such that given the parents' resources and the costs of laying eggs and caring for chicks, laying that number of eggs maximizes the probable number of *viable* offspring for that set of parents (i.e. offspring which are themselves likely to survive to the point of reproduction). Now it was assumed in the previous account that the optimal clutch size (in the above sense—i.e. from the individual's point of view), even in conditions of scarcity, is 3, and the argument was that a reduced clutch size could nonetheless evolve through group selection. But in fact, it may be instead that the optimal clutch size in conditions of scarcity is really 2 after all.

It is in fact plausible that the optimal number is lower here than in environments where resources are more plentiful because the competition for what few resources there are will result in fewer resources for each set of parents; this means that they have fewer resources to put into making viable offspring, so that the number they can *successfully* produce and rear will be lower. So the story may have gone like this: Parents who laid more than the optimal number of eggs for that environment—for example, those who laid 3 eggs—wound up

[7] The evolutionary account that follows is based roughly on Lack (1966, ch. 1 and appendix) and on Dawkins' discussion of Lack's (1954, 1966) view.

spreading what resources they had too thin, resulting in a less than optimal number of viable offspring; thus that tendency was disfavored by ordinary natural selection. By contrast, parents who carried genes tending to make their clutch size closer to optimal in this environment, i.e. reduced in comparison to more favorable environments, tended to have more (viable) offspring, who in turn tended to carry the same genes, and so on. That is why this tendency has been favored by natural selection to the point of now having become a characteristic trait of members of populations in such an environment.

Now if all of this turns out to help groups in less favorable environments to reduce their risk of starvation, that may be fortunate for them, but it is merely *incidental*. The process is not driven by factors having to do with group benefit. It is not even driven by factors having to do with individual benefit in the sense of personal survival or the long-term survival of descendents. It is driven simply by natural selection acting on each generation of swifts, with the result that whatever heritable traits in an individual tend to promote greater reproductive success relative to other members of the population naturally spread in the population. If individuals arose who were genetically endowed in such a way as to be both able to command greater than average resources and disposed to have larger than average clutch sizes, they would out-reproduce their peers, and their genes would spread. This is true despite the fact that these genes would be dangerous to the whole group, and therefore ultimately even to these individuals' descendents, who are after all members of the group that would be threatened with extinction; in fact, if the threat of extinction posed by the spread of such genes were imminent enough as to fall within the original individuals' lifetimes (which is unlikely, but theoretically possible), their "greedy" genes would spread even to these individuals' *own* ultimate detriment. The genes, as sifted by natural selection, do not "care", as it were, about anything as abstract as the good of the group or the long-term good of individuals—or *even* the long-term survival of their own copies; they simply spread or don't spread depending on their effects on the reproductive success of the individuals carrying them, one generation at a time, and—according to the present story—this explains why we find the traits we do in current specimens in various populations.

The reason, then, that all of this is problematic for ahistorical, welfare-based accounts of function is that by ignoring causal history, one would have no idea which of the above two evolutionary stories underlies the phenomenon in question; and yet if it is the second, then the treatment of the group benefit as an end would plainly be groundless. How could we possibly be justified in treating something that is entirely incidental—in the way that group benefit or long-term individual benefit is on the second evolutionary story, having no non-accidental connection to the trait in question—as an *end for the sake of which* the trait is manifested? If, from an evolutionary point of view, these effects are nothing but *side-effects* of a trait that is present for altogether different reasons, how could we rightly continue to treat them from a teleological point of view as genuine ends, *functionally*—and hence *non*-incidentally—served by the trait? Functions must, after all, be distinguished from mere accidents—even where the accidents happen to be beneficial; but it appears that it is precisely the latter that we are led to embrace as if they were genuine functions, if we look simply to the welfare that is in fact promoted and ignore the history behind the trait.

Indeed, if we are allowed arbitrarily to drag in such beneficial effects as ends functionally served by adaptations, then what would stop someone from going on to speak of adaptations' being for the good of the *species*—or even for the good of the *genus*, *family* or *order*?[8] What is to stop someone from speaking of them as being for the good of the local prey population (which is spared extinction from overfeeding by too many swifts), or for the good of the local ecosystem—something Foot herself certainly rejects? It is easy to see how one would be led to such claims if history is ignored. Consider the thought that the clutch size adaptation is for the good of the species. It might be suggested that we can speak of a species being benefitted insofar as groups of its members are made less likely to go extinct, thus making the species that much less likely to go extinct; so why should one not go ahead and say, in that functionally loaded way, that "*this is how the species (The Swift) helps to keep itself from going extinct*"? My point, of course, is that if only the second evolutionary story obtains, this is equally absurd, since such "species benefit" is just as incidental to the etiology of clutch size regulation as group benefit and long-term

[8] Dawkins (1976) poses a similar question.

individual benefit were; but if this fact is ignored, what is to stop someone from citing this as the end, and similarly with the other cases mentioned?

Finally, such arbitrary appeals to welfare—made even where its promotion turns out just to be a lucky side effect of traits that are present for altogether different reasons—raise a problem for another central feature of teleology. Functional teleological judgments are supposed to constitute genuine *explanations* of a certain sort for the traits in question. For example, the fact that the heart pumps the blood *in order to* circulate it (rather than in order to cool it) is supposed to answer the question "*why* does the heart pump the blood?", understood in a certain sense. But how could such seemingly arbitrary teleological judgments as those mentioned in the last paragraph possibly constitute genuine *explanations* for the traits in question? Even if it is maintained that functional teleological explanations are different in kind from consequence-etiological explanations—which I think is true, as explained in chapter eight—how could an appeal to group or species benefit, for example, possibly be thought to provide *any* kind of *true explanation* for clutch size reduction in a case where the second evolutionary story lies behind the trait—which, again, is a factor to which the ahistorical approach is oblivious? In such a case, how could it possibly be correct to answer the question of why swifts have reduced clutch sizes in conditions of scarcity to say: "Swifts in certain environments have a reduced clutch size *in order to* conserve group resources and thus to help prevent group/species/prey population extinction"? I do not see how it could, and this further illustrates the inadequacy of such appeals to welfare with no consideration of causal history. By contrast, I do think we can correctly answer the question in such a case with: "Swifts in certain environments have a reduced clutch size *in order to* maximize their personal reproductive output"; for if the second evolutionary story is what underlies the trait, then this is indeed the end toward which this trait is non-incidentally directed, by virtue of its having been incorporated into the swift's overall functional structure by natural selection.[9]

---

[9] Of course, if the first evolutionary story obtained, then we could appeal to the group-level benefit in a teleological explanation of the trait, since this would then be a genuine end promoted by the trait. The point is just that such an attempt at teleological explanation will be bogus if the second evolutionary story obtains, and yet an ahistorical

Let me emphasize that nothing in my argument turns on whether the group selectionist or individual selectionist story is actually true in this example. That makes no difference to the criticism of the ahistorical welfare-based view of natural teleology. What matters is just that the teleological facts depend crucially on the historical causal facts, whatever they are, so that any view that just ignores the latter will wind up at sea with regard to the former. Where the judgments are true—as the initial judgment about the swifts would be if the group selectionist story turned out to be true—they will be so only through luck, with no basis for understanding *why* they're true; and very often the judgments will just be false, attributing teleological relations between traits and beneficial effects where there is no basis whatever for positing such relations—as would be the case for the initial judgment about the swifts if the individual selectionist story turned out to be true.

At least *some* of the problems raised above would be avoided if it could plausibly be argued that the appeal to welfare should be restricted such that functions relate to the satisfaction of the welfare-related needs of members of a species as such, and to larger units such as groups only insofar as it is often through group level phenomena that such needs are normally met (as with social insects, for example). That is, for a given species F, functions would be identified simply in connection with *how Fs get the food they need, how they avoid predators, how they attract mates, how they reproduce*, and so on; the functions associated with a given species would be understood in relation to the promotion of the good of members of the species, and not in terms of contributions to the good of the group or species *per se*, or to the good of any other entity. Thus, there would be no commitment to speaking of the promotion of the good of the group, species, genus, or ecosystem as *ends*; nor will the good of members of *other* species count as ends in an account of the functions pertaining to a given species.[10]

---

view has no way of detecting which is the case, so that there is nothing to stop it from making such misguided teleological claims.

[10] This seems to be roughly Foot's (1994, 1995) view. When we speak of *what Fs need* in this sense, it is understood that we are speaking of what they need within the context of their given "way of life". The fact that wings would help mice escape from cats doesn't mean that mice need wings; and while something like the ability to hop several feet in a single bound might in many cases prove in some sense "necessary for the avoidance of harm", it cannot be said to be something mice need in the relevant sense, as such an ability is not within the mouse "repertoire"—i.e. it is not part of "the life of the species." Foot's account of *function*, then, appears to be this: For any biological species F, various

Even if such a restriction could be defended within an ahistorical welfare-based framework, however, the approach still runs into the same basic problem illustrated by the above example. Perhaps there will no longer be any problematic appeal to the notion of group benefit *per se*, which runs into difficulties when there is no supporting group selectionist story behind the trait. But since the view in question ignores etiology and just considers the ways in which needs are met in a species, it still commits its proponents to such claims as that the practice of cooperative "birth-control" among swifts in conditions where resources are scarce is the swifts' functional way of *meeting their need to conserve resources*, ensuring that their food sources will not be endangered and that their nutritive needs will therefore continue to be met; this is how such needs are met in the context of the swift "way of life"—part of *how they manage to survive in conditions of scarcity*. Indeed, we would presumably be committed as well to the corresponding evaluative judgment that swifts who do not "cooperate" and instead lay a greater number of eggs, thus not doing their part in the conservation effort, are thereby *defective*. (How could such judgments be avoided, given the very limited and vague resources of the ahistorical view?)

But again, these claims will just be false in cases where the second evolutionary story is what lies behind the trait, with no way for the ahistorical view to avoid the error. In such cases, the reduced clutch size is not really a case of "cooperative birth-control" at all, but is just a matter of *competitive reproducing*. If this happens to promote individual survival inasmuch as it happens to contribute to preventing mass starvation (which would include the starvation of the individual), then that is merely an incidental benefit which can hardly be cited as the end for the sake of which the trait exists. And if a given swift can manage to command the resources to raise 3 viable chicks rather than 2, so that 3 is its "optimal" number, then there is nothing *defective* about it for doing so, even if this happens incidentally to threaten the whole group, as discussed earlier. On my view, of course, a swift might be judged defective if it lays *more than its optimal number*, but in that case the judgment of defect is relative to a failure of function pertaining to

---

traits of Fs have functions just in case they play a part in the satisfaction of the needs of Fs, in the above sense; the function of any given trait is what it does that contributes to the satisfaction of such needs.

personal reproductive success, not relative to some supposed function pertaining to cooperative resource conservation. We cannot, then, escape the earlier difficulties simply by relating function to the satisfaction of the needs of individuals within the context of "the life of the species" or "their way of life", ignoring etiology.

It might be objected that the proponent of the ahistorical, welfare-based view is not after all committed to the problematic functional claims above, since coordinated population control for the sake of general resource conservation is not really "in the swifts' *repertoire*", and so may be ruled out when considering the function of the trait. But of course the question that would have to be answered is how we are to determine what is, and what is not, in an organism's repertoire or "way of life" without looking to history—something about which Foot gives little guidance. The swifts *do* in fact keep their population down in conditions of scarcity, which thereby helps them (or at any rate, may help them) to avoid extinction; on the face of it, then, this would indeed seem to be part of their "way of life" (something that will be even more clear in the next example). It is only when we look to history that we see that what they are really doing from a functional point of view is, for example, just *maximizing personal reproductive output*, rather than making a self-sacrificing contribution to overall population control (that is, if the second evolutionary story is the one that obtains). The proponent of the ahistorical view cannot just help herself to this insight and then proceed as if history were irrelevant. And I have already argued at length that once we do consider history, and its implications for the nature of biological teleology, we must reject the welfare-based view.

I hope it is clear that the considerations I've raised against the ahistorical approach are very general ones—and indeed, hardly surprising upon reflection. But it is worth considering some further examples, since they are easy to come by and will dispel the possible feeling that the points have relied on something special about the above example, which would be a needless distraction. So, for instance, consider a dominance hierarchy among elephant seals. It is natural to look at such a thing and to wonder what its function is. It is certainly part of their "way of life", if that expression is to have any meaning at all, and if one ignores etiology and focuses simply on contributions to welfare, it will seem plausible to conclude that the function of the dominance hierarchy is to promote group harmony, minimizing daily conflict within the group—not necessarily for the good of the group as such, but for the good of the individuals who make up the group. "*This is how they keep intra-group fighting to a minimum*", it will be said, "*thereby avoiding a great deal of injury and living more flourishing lives, perhaps even living longer and having more offspring.*"[11]

If the orthodox neo-Darwinian account of the phenomenon is correct, however, not only is this not a plausible account of the function of dominance hierarchies, but it is not even plausible to speak of their *having* a function. From an evolutionary point of view, according to this evolutionary story—which I shall assume is correct for present purposes (though as before it makes no difference whether it ultimately is so or not)—the dominance hierarchy itself is just a group-level side-effect of certain patterns of behavior on the parts of individuals. These behavior patterns—i.e. fighting for control of a harem, but subordinating oneself to the winner if beaten, and continuing to do so at least until a better opportunity arises, such as the dominant male's becoming sick and growing weaker—are adaptations that have been selected simply because of their genetic consequences, as discussed earlier; genes making individuals behave this way have spread in the gene pool because such behavior in each case is an especially effective way of getting those genes replicated into the next generation. The fact that the existence of the hierarchy, once it is established, tends to promote group harmony, reducing the incidence of regular fighting among members of the group, making them on average that much better off, is incidental.[12] But now once again, if it is merely incidental from an etiological point of view, how can we possibly regard it as a genuine *end for the sake of which* dominance hierarchies are

[11] Again, it is Foot that I have in mind here, this sort of remark being characteristic of her approach.

[12] See Dawkins (1976, pp. 88–9). Note that at least some of the benefit enjoyed by an individual as a result of his own behavior—namely, his avoidance of injury due to refraining from further challenges to the dominant male until a better opportunity arises—would not be incidental, since this indeed figures into the etiology of the adaptation; avoiding pointless injury increases the chances of continued survival and reproduction, and so on. But it would still be *instrumental*: It is relevant not in its own right, *as* benefit, but only insofar as it promotes the relevant sort of genetic replication. And in any case, other benefits resulting from the hierarchy—such as the benefit to a dominant male of not having to deal with constant attacks from others he has previously defeated, or benefits to females who are less likely to be caught in the middle of a fight, or the increased comfort that comes from less fighting—are straightforwardly incidental.

manifested? How can we possibly answer the question "why do elephant seals form relatively stable dominance hierarchies?" by saying that they do so "*in order to* promote harmony among members of the group, enabling them to live longer and more comfortably", and so on? In light of the evolutionary background, such teleological judgments and explanations just seem straightforwardly false. The benefit appealed to is not an end functionally served by the dominance hierarchy, but just a side-effect of behaviors that are manifested for very different reasons, and it explains nothing.[13]

Again, nothing turns on whether or not the evolutionary story I've assumed for the purposes of this example is in fact correct or not. It is enough that it could be true, and if it were, an ahistorical welfare-based view would wind up counting an incidental benefit as an end for the sake of which a certain trait exists. And again, the more general point is that this is exactly the sort of problem we should expect to arise for such a view. There will naturally be cases in which a trait is selected for a certain genetically-significant effect, though it happens to provide certain unrelated benefits to organisms as a side-effect; and these benefits may remain completely incidental. In such cases, then, an approach such as Foot's cannot avoid certain specious teleological ascriptions. By contrast, on the view I have defended, there is always a principled way of making the necessary distinctions.

Finally, there is a slightly different kind of example, involving parasitic flukes and snails, that is quite interesting and worth considering in this connection. The flukes manage to manipulate the shell-secreting cells of snails into diverting resources to build thicker shells. A plausible explanation is that fluke genes for such manipulation are rewarded over alleles in other flukes, since the thicker shell makes the snail more likely to survive and hence to serve as a host to the fluke. In Dawkins' terms, the thicker shell of the snail is an extended phenotypic effect of certain fluke genes.[14] Now this parasitic behavior

---

[13] I am not denying that in other kinds of cases group harmony might come in as a genuine end (though still only a proximate one). It certainly might—for example, in cases involving cooperative hunters, as discussed in section thirteen of chapter five. My claim about the elephant seals is just that if this is the correct natural selection story about the traits in question, then in this case group harmony is just an incidental effect, even though it occurs as part of the "way of life" of elephant seals; and this fails to be appreciated by a view such as Foot's.

[14] Dawkins (1989, pp. 241–2). See also chapter four, section one, and chapter five, section twelve.

on the part of the fluke against the snail works to the detriment of the snail's normal functioning as a gene replicating system. For in shifting the snail's natural and presumably relatively "optimal" balance of resources away from reproduction toward personal survival, it is reducing its natural efficiency as a gene replicator. This is characteristic of cases of parasitism. But what is interesting about this case is that while this is true, it is also true that the individual snail itself *benefits* from all this. It may wind up a less efficient gene replicating system, but because of its thicker shell it is more likely to survive and will tend to live a longer life—and hence is plausibly made better off. We might even imagine hypothetical cases where the parasite's effects on the host not only make it live longer, but make its life more *pleasant* in various ways. In any case, such benefits to the host are either just lucky side-effects (in the imaginary case of enhanced pleasure) or are at best of purely instrumental significance to the *parasites*, promoting *their* increased reproductive success; the benefit is not in any way for the sake of the snail as an end in itself.

The point of this example is to illustrate once again how we will tend to be misled if we ignore history. If we insisted on thinking about functional teleology in abstraction from the above considerations about natural selection history, we might very well come to the erroneous conclusion that this is a case of *reciprocal altruism* or *symbiosis*; that is, it might seem that as part of a functional relationship, the snail provides a home for the flukes, and in return the flukes make it better able to survive, both of these being *functional* benefits, as with the aphids and ants considered in chapter five. In fact, however, as we have seen, this is not what is going on here at all. The relationship is one of parasite and host, and the only plausible *function* of the increased shell thickness is to help to provide a better home for the flukes, the benefit to the snail not being an end in its own right, but only a means of providing a better home for the flukes. To continue to speak as if the benefit to the snail were an end in its own right, functionally served by the relation with the flukes, would make a mockery of the function/accident distinction. Yet this is precisely where we are led by ahistorical, welfare-based views of function.

These difficulties, incidentally, apply equally to Thompson's (1995) related discussion of what he calls "natural historical judgments" pertaining to "life forms" (i.e. judgments with a special "logical form", providing a normative specification of the form of life of a species), at least if it is held that these can be arrived at without taking careful account of natural selection history. It is impossible to know, in the absence of an understanding of the natural selection history behind a given type of organism, whether or not various things that such organisms do, or various features that they have, really figure into the distinctive sort of judgment he is trying to isolate. Consider, for example, the judgment: "Snails (of a certain species) harbor flukes (of a certain species) in certain parts of their bodies." This might at first *look* like a paradigm "natural historical judgment"—especially if we are told that the flukes contribute to thickening the snails' shells, which protects them from harm; it seems to be on a par with one of Thompson's own examples: "The Texas blue-bonnet harbors nitrogen fixing microbes in certain nodes on its roots." As soon as we look into the functional situation, however, through looking at the natural selection background, we see that it is very different; it is not plausibly a "natural historical judgment" at all (or if it is intended by someone as such, then it is false), but is instead on a par with merely statistical judgments like "domestic cats harbor fleas", and certainly does not belong in a description of the "natural history" or "way of life" (to use Foot's phrase) of the snail. By contrast, the judgment: "Flukes parasitize snails" *would* be a true natural historical judgment about the *flukes*—part of a functional description of their "way of life".

To take another example, we recognize that judgments like "moths fly into candle flames" are not natural historical judgments possessing a special logical form, but just statistical judgments about the creatures in question. Why? Because we recognize that flying into candle flames does not, as such, come into the description of proper functioning for moths, but is a non-functional by-product of a broader, functional tendency. Thompson's natural historical judgments, I believe, are properly thought of as derived from *functional* judgments and judgments closely bound up with them (e.g. about natural environments), which equally possess, and indeed are the *source* of, the special logical features he identifies as characteristic of his natural historical judgments. If this is right, then we cannot look to natural historical judgments to shed light on functional teleology and function-related evaluative judgment, as Foot (1994, 1995) suggests, but must instead develop a careful account of functional teleology before we will be in a position to evaluate and to apply natural historical judgments with any confidence. The general point is that Thompson's appeal to a special "logical form" in trying to isolate natural historical judgments still leaves unanswered the question of the truth conditions for such judgments; for that, we plausibly need an account of the truth conditions for functional judgment, and I have argued that this requires an appeal to natural selection history.

To return to the central point: The above examples are sufficient to establish that appeals to the promotion of welfare or the satisfaction of welfare-related needs, in the absence of historical considerations pertaining to natural selection, are unable to distinguish between genuine functions and ends, on the one hand, and accidents or incidental benefits on the other. Such appeals may sound plausible as long as we are ignorant of the causal history involved; but once it is brought out, they lose all plausibility as grounds for function attributions.

3. INDEPENDENT PROBLEMS FOR AHISTORICAL WELFARE-BASED ACCOUNTS, PART III: THE DISTRIBUTION AND RANKING OF ENDS

To this above general point we may add one more related observation. Not only are ahistorical, welfare-based views unable to distinguish between genuine ends that are functionally promoted and merely incidental benefits, but they are equally unable to make sense of the ways in which various functionally promoted benefits are *weighted* as ends—i.e. how various kinds of need-satisfaction on the parts of various individuals stack up as ends within the life of the type of organism in question, and why.

We have seen that some organisms do not simply promote their own survival and reproduction, but also do things that systematically promote the survival and reproduction of various family members (i.e. kin altruism), other members of a group (i.e. cooperation among conspecifics that are only distantly related), and members of other species (i.e. symbiosis). And these various benefits are promoted in ways that reflect definite priorities: They are not all just on a par, nor are they always weighted in the same ways. In some species, for example, an individual will easily give up its life to help a sibling, whereas in others this would happen only in comparatively rare circumstances, indicating that the two ends (i.e. personal survival and sibling survival) have different priorities in the teleological structures of the two species. And in any given species where kin selection has been a significant evolutionary factor, aid will generally be more forthcoming to closer relations and less so to more distant relations, indicating that the priority belonging to the end of the survival of relatives depends on the degree of their relatedness.

Now let's suppose that there is agreement in a wide range of cases that the behaviors in question are indeed a matter of proper functioning, so that such benefits are indeed promoted as genuine ends. The question is how any ahistorical, welfare-based view can account in any principled way for the complex weighting of ends in each case. Here is just a sample of the sorts of questions that naturally arise in this connection: Why does personal survival count less than the survival of siblings in one species (at least with regard to certain members and certain siblings, e.g. worker bees and reproductive siblings), but more in another?[15] In a given species, why are the ends of gaining personal nutrition and of aiding siblings weighted in such a way that a runt struggles against its siblings for limited resources up to just a certain point, but then gives up after that point even when it has some fight left in it? Why does the end of a son's survival at least sometimes seem to have priority over the end of a daughter's survival in elephant seals (recall the case of uneven resource distribution by parents)? Why does the promotion of relatives' welfare often have clear priority as an end over the promotion of the welfare of others in the local group? How does the promotion of welfare in various types of case compare as an end—i.e. is it supposed to be a *final* end in all cases, or is the promotion of the partner's welfare supposed to be merely an *instrumental* end in case of symbiosis, say, geared ultimately toward the organism's own welfare? Is there a difference in this respect between cases of kin altruism and cases of cooperation among conspecifics that are only distantly related? (Recall how different the etiological accounts are of

[15] To this we may add: Why in the former case does the situation suddenly change in certain circumstances, i.e. where a male reproductive sibling (a drone) has failed to mate? For in that situation, the very same worker bees that once helped to nurture him, putting his survival above their own as an end, now systematically work to kill him and expel him from the hive—indicating that in such circumstances his survival is no longer an end functionally promoted by them. Indeed, the new end is the bringing about of his death and removal, and it is pursued with the same impressive determination.

the two kinds of trait. Is it plausible to suppose that this has no bearing at all on such questions about ends?) Why is reproduction always given priority over survival as an end, so that, for example, the enhanced survival enjoyed by the snail in the above case, at the expense of reproduction, does not plausibly count as a functional benefit of a symbiotic relationship? And so on.

The historically informed view I have offered is able to provide clear, principled and detailed answers to all of these questions. By contrast, a view such as Foot's, which simply looks to *the ways in which members of a given species meet their various needs*, will not be able to say much of anything about this—about *whose* needs matter (in terms of their figuring into *ends*), with *what priority*, and *why*; as far as I can see, there will simply be an appeal to differences in "ways of life", where these are taken as given, with no principled account of why various ends should be weighted as they are and how more precisely they are to be understood. It will not do for a defender of such a view simply to concede that such details are indeed determined by natural selection, but claim that they are irrelevant to a general understanding of biological function. For what is at issue here is not just the priorities that are actually observed to be given to various aspects of need-satisfaction in the course of actual organic behavior—which could, after all, often be a matter of malfunction—but the crucial theoretical matter of the structuring of ends *as* ends within the teleological framework of a given type of organism. This is a normative matter, not simply a matter of what actually happens for the most part, and is central to any real understanding of the teleological facts. It is thus very unsatisfying for an account of function to have nothing to say about this. All of this serves, I think, to offer further support for P, and for the sort of account I have developed on the basis of it.

4. THE ARGUMENT FROM THE HYPOTHETICAL POSSIBILITY OF ACCIDENTAL DOUBLES

Perhaps the most intuitively powerful objection raised by at least some proponents of ahistorical views of biological function against historically-oriented views involves the hypothetical case of accidental doubles. This objection can be met in an intuitive way with the resources of the view I have defended, and it is instructive to bring this out—particularly by way of contrast with etiological views. Indeed, such examples can be turned around to raise problems for the very positions they have been thought to support.[16]

Suppose that through sheer accident a collection of organic molecules comes together in such a way as to produce something that is physically and behaviorally indistinguishable from a cat: It is made of the same material, organized in the same way, exhibiting the same kinds of changes at various levels that we ordinarily associate with life. There are a number of questions that might be raised about this incredible object. It might be wondered, for example, whether in the absence of any historical connection to cats it can properly be thought of as a cat; perhaps it must be regarded simply as a species-less impostor, even if it is capable of mating with cats and producing offspring. Likewise, we might well question whether the thing thumping in its chest is really a *heart*, and whether its various activities really constitute *breathing* or *eating*—or for that matter whether whatever it managed to do with a cat could ever really constitute *mating*.[17] The question that interests me, however, is whether the entity constitutes a functional system, and more specifically whether we can attribute biological functions to its various parts and features, parallel to the functions attributed to the corresponding parts and features of cats. (If there is thought to be some difficulty arising from the fact that there is only one such object, we may imagine instead that many of them have appeared simultaneously.) The objection posed by defenders of at least certain ahistorical views of function is that it seems obvious that this cat-like object *is* a functional system, and that its various parts and features have just the same functions as the corresponding parts and

[16] For discussions of this sort of case, see Boorse (1976, p. 74), Millikan (1989, pp. 292f.), Neander (1991a, pp. 178f.), Thompson (1995) and Hsu (manuscript).

[17] See Thompson (1995), who would answer these questions in the negative. In fact, he holds the radical (and interesting) position that such an entity isn't even *alive*, though it mimics living activities perfectly. The reason is that it does not possess what he calls a "life-form", as real cats do, though it is not clear what the conditions are for possessing a life-form. (Apparently the lack of historical connection to living things is sufficient for *not* possessing one, on his view.) It is not obvious from Thompson's discussion whether lack of a life-form is also supposed to prevent the entity from being *conscious*, for example, or from feeling genuine *pain*. On the face of it, these do not seem plausibly to depend for their identity on a non-accidental connection to types in the way that organs (e.g. hearts) or particular vital activities (e.g. breathing) plausibly do. But if we admit that the entity is conscious and can feel pain, etc., then I think this makes it rather difficult to maintain that it is not alive—particularly given the thoroughgoing physical similarities with living things.

features of cats; this fact, however, cannot be accommodated by views of function that make appeal to history, since the cat-like object has no such history.

According to etiological views of function, for example, facts about present function just *consist in* certain kinds of facts about causal history (or in these together with the fact that the trait still has the etiologically relevant effect).[18] It follows, then, that where such facts about causal history are absent, there will be no facts about function, i.e. no functions. On most current etiological accounts of biological function, the sort of causal history that is required for biological function would clearly be absent in the case of our cat-like object. For example, if the required causal history is one that involves natural selection, as on Neander's account, then it will obviously be absent in this case, as there is no natural selection history behind the presence of its various parts and features. Nor would it help to broaden the view to require only *some* kind of selection history, for this is equally absent. Likewise in the case of Millikan's version, which requires explicitly that the object in question have ancestors. One is thus forced on such views to answer the objection simply by biting the bullet:

> Such a double has no proper functions because its history is not right. It is not a reproduction of anything, nor has it been produced by anything having proper functions. . . . The thing that appears to be its heart does not, in fact, have circulating blood as a proper function, nor do its apparent eyes have helping it to find its way about as a proper function, and when it scratches where it itches, the scratching has no proper function.[19]

Now what is most striking about this answer, I think, is not the conclusion itself, but the nature of its justification, which relies *simply and directly* on the absence of a certain kind of causal history, making the conclusion implausibly clear-cut. According to etiological views, such causal history is the essence of functionality, and everything else that may seem important—for example, facts about the coherent organization of parts, features and activities into a complex,

---

[18] Etiological views will be discussed further in the next chapter and in chapter eight.

[19] Millikan (1989, p. 292). Cf. Neander (1991a, p. 180): "Without history, the usual biological/functional norms do not apply."

hierarchically organized system—is relevant only as a *probable indication* of the presence of such a causal history. Millikan, for example, compares these other properties—the properties that make such an impression on us when we examine the cat-like object—to color changes in litmus paper, with facts about function being compared to facts about acidity:

> Having the right sorts of current properties and dispositions is in point of *fact*, in *our* world, an infallible index of having proper functions. If you like, it is criterial—as criterial, say, as the red of the litmus paper is of acidity. But it is not turning litmus paper red that *constitutes* acidity, nor is it having the right sort of current properties and dispositions that *constitutes* a thing's having a purpose [i.e. function]. To the degree that each of these . . . descriptions in terms of current properties or dispositions is successful, each describes only a *mark* of [functionality], not the underlying structure.[20]

If someone came across the cat-like object, having no idea that it was not a normal cat, he would admittedly be strongly inclined to attribute functions to what he would regard as its heart, and so on. But according to Millikan, he should be regarded just like someone who mistakes "fool's gold" for gold, or Twin Earth "water" for water: He is taken in by certain properties that are normally very reliable indications of what really matters (i.e. the right kind of causal history), but in this case are misleading. If we continue to have intuitions about function even after learning about the bizarre circumstances surrounding the cat-like object, this will presumably be chalked up to a confused obsession with mere indicators, and a failure to focus on what really matters to functionality.

Such insistence on the direct and sole importance of causal history—to which everything else stands only as a likely indicator—seems very implausible, for reasons that I have already tried to bring

---

[20] Ibid., p. 293. What she means by "infallible" above is that such incredible accidents as I am imagining in the case of the cat-like object simply do not occur and will never occur, so that the presence of such organization may in fact be regarded as an infallible indication of the presence of some such causal history as natural selection or intelligent design. Strictly speaking, however, since the case is possible, even in our world, her view is that the presence of such organization is nothing but an extremely probable indicator or mark of the presence of some such causal history.

out in defending a system-oriented though still historically-informed account, and which will become even more clear when I defend a non-reductionist construal of functional explanation in chapter eight. There are indeed serious problems confronting any attempt to attribute functions to systems in ignorance of history (or in the absence of history, as in this case), as I have illustrated, but we can recognize the importance of history without simply *reducing* facts about present function directly to facts about history. That is, we can respect the crucial contribution made by certain kinds of causal history to the functionality of present systems while still maintaining that facts about present function are directly to be understood as facts about non-incidental roles currently played by certain entities in the working of such systems; this is a natural and intuitive position, which has the same virtues possessed by etiological views while avoiding the problems (as discussed in chapter seven). And I think this enables us to give a more satisfying response to the above objection regarding accidental doubles, and a more plausible account of the intuitions that motivate it.

How, then, should we deal with the above objection according to the view of function I have defended? My account ties functions directly to standard, non-incidental roles played in the working of systems of certain types. It is not, then, *immediately* obvious that there can be no functions in connection with the cat-like object, simply because there is no causal history of the usual sort. The crucial question is whether in this case we can speak in a sufficiently non-arbitrary way of things playing standard, non-incidental roles in the working of a type of system. In the end I think the answer is that we cannot, at least not in a very robust way. But even so, the difficulties are relatively subtle ones—at least more so than the simple lack of causal history—and it is easy to see how there could be strong intuitions in favor of speaking of functions, which are not simply a matter of being fooled about the presence of causal history or of being obsessed with its usual indicators. First, let me say why there are problems with attributing functions in connection with the cat-like objects, thus turning the objection around against ahistorical views.

The crux of the problem is well brought out in the following passage from Neander:

> Having stared at [the cat-like objects] in stupefied amazement for some time, we eventually begin to wonder about their wing-like protuberances on each flank. We ask ourselves whether these limbs have the proper function of flight. Do they? When we discover that [the cat-like objects] cannot actually fly because their "wings" are not strong enough, we are tempted to suppose that this settles the matter, until we remember that organismic structures are often incapable of performing their proper function because they are deformed, diseased, atrophied from lack of use, or because the creature is displaced from its natural habitat (they could perhaps fly in a lower gravitational field). On the other hand, often enough there are complex structures that have no functions, for instance, the vestigial wings of emus and the human appendix. The puzzle is where among these various categories we are to place the [cat-like objects'] "wings".

Generally, since everything about the compresence and arrangement of the parts, features and activities of one of these cat-like objects is merely accidental, it is hard to see how any of the above questions could be settled in a non-arbitrary way.[21] For the same reason, it is hard to see how we could possibly speak of a genuine *type* of system in relation to these objects. The fact that a particular system perfectly resembles and mimics genuine tokens of a given type of functional system is obviously not sufficient for it to constitute a token of that type of system, thus possessing the proper function of such a system. As Hsu points out, a malfunctioning adding machine, for example, may *behave* just like a quadding machine, churning out answers equivalent to quums rather than sums. Indeed, "it may [even] *look*, inside and out, just like a quadding machine: That melted wire [may look just like the] fuse [in quadding machines], that slipping gear [may look just like] a clutch."[22] Such a thing is still an adding machine, having the proper

---

[21] Actually, the claim about accidentalness needs to be qualified. I take this up below, at the end of this section.

[22] Hsu (manuscript, p. 1). The example comes from Kripke (1982, pp. 9, 34–5). A quadding machine is a machine we may imagine to be designed to calculate quums, where the quum of X and Y is the sum of X and Y when each is less than 57, and equal to 5 in all other cases. We are imagining that quadding machines (of a certain model) look just like adding machines (of a certain model), except that where adding machines have a certain gear, quadding machines have a clutch, and so on. When the gear in the adding

function of adding—which it is unable actually to carry out because it is defective; it does not suddenly become a non-defective quadding machine, with the proper function of quadding (though there is a loose sense in which it might be said that it has become *de facto* a quadding machine). And the same point holds in non-artificial cases: The fact that our cat-like object closely resembles real cats is not sufficient to make it a non-defective token of the type *cat*; we might just as well regard it as an interestingly deformed and malfunctioning token of the type *raccoon*.

On my view, then, there is no functional type to which the cat-like object is non-accidentally related as a genuine token. And if we propose simply to invent a new type, somehow derived from the particular (i.e. just stipulating that it is the type of which this particular is a token), all the same problems arise with respect to the identification of this supposed type. Since everything about the particular is accidental, so that there is no way of determining in a non-arbitrary manner what is a matter of proper arrangement and functioning and what is a matter of (possibly beneficial) defect and malfunctioning, there is nothing to determine *which* type is represented by this particular—i.e. whether the type in question is *cat* (of which this is a non-defective token) or *raccoon* (of which this is a deformed token). A given particular cat-like object is therefore not plausibly a token of any functional type, despite the fact that it perfectly resembles real cats, and this disqualifies us from speaking of standard, non-incidental roles played in the working of a type of system in this case. All we have are particulars *as such*, and that is not sufficient for the assignment of biological functions.

On the other hand, there is nothing to stop us from speaking of functions in the way that Cummins and Prior do, as mentioned in chapter one (and as discussed further in chapter eight).[23] We may certainly take a keen interest in how this complex, hierarchically organized system manages to keep itself going, i.e. to survive (if we are willing to speak of *life* here), and in relation to this we may speak of the various things that parts and features at lower levels do which contribute to various capacities and effects at higher levels, all eventually enabling the system to maintain itself, just like a cat does.

---

machine is defective and slips, causing it to malfunction, it resembles the clutch in the quadding machine.

[23] Cummins (1975), Prior (1985). Cf. Neander's (1991a, p. 181) discussion.

But there will necessarily be an element of *arbitrariness* here that is not present in genuine cases of biological function. I have argued that biological function is a species of proper function having to do with ends that are determined not arbitrarily, according to the interests we happen to take in biological systems, but in an objective way according to the principles governing the ultimate causal factors behind the assembly and conditioning of such systems. Gene replication is the ultimate end of any given biological system not because we happen to take an interest in it (which in fact most of us do not!), but because of the special place it has in the processes of natural selection responsible for putting together the systems in question. By contrast, when it comes to one of these cat-like objects, there is nothing to single out non-arbitrary ends, so that there is nothing left but the sort of interest-relative analysis of what various parts and features of the system "function to do" with regard to certain interesting higher level effects—this being a much less restricted, and less interesting, notion than that of proper function (as noted in chapter one).

The strong intuition that we must be able to speak of (proper) functions in this case, just as in the case of cats, is largely a result of recognizing that we can indeed give an identical Cummins-style analysis of how the cat-like object is able to maintain itself and to reproduce: The thing in its chest certainly *functions to* pump the stuff in its veins, which is relevant to its ability to do all the other things it does, and so on. But that was never in question. The question is whether we can speak of *proper functions*, which goes along with speaking of excellences and defects, proper functioning and malfunctioning, and so on; and it is clear from the kinds of considerations Neander raises, I think, that there is no non-arbitrary way to do that here.[24] We can account for the strong intuitions that there are proper functions in the case of the cat-like object as stemming from the recognition of all that is indeed shared in the two cases. Given such similarities with cats, there is an understandable temptation to treat the cat-like object as a genuine instance of a type of working system, and hence there is a natural inclination to want to speak of functions, as if it really were just another cat. It is only when we consider the sorts of difficulties that

[24] Cf. Millikan (1989, p. 296): "There is no way that any of [the cat-like object's] states or parts could be *defective* or might *fail*. That creature of accident, wonderful as [it] may be, falls under no norms."

have been raised that it becomes evident why these intuitions are misguided, i.e. why the cat-like object does not constitute a genuine working system after all. This, I think, is a more plausible treatment of the case than the one suggested by Millikan, and seems to me to be an adequate answer to the objection.

Having said this, a certain qualification is necessary. It is true that at the moment a cat-like object comes together, everything about the compresence and arrangement of its parts, features and activities is entirely accidental. After a relatively short period of time, however, this can no longer be said. Part of the reason why the object has the structure it has, whereby it is able to carry out such activities, is the fact that at earlier times it had such a structure and carried out such self-sustaining activities; if it had not, then it would have since collapsed into a mere heap and dissipated. There is thus a sense in which, as time progresses, we can say that it is no longer entirely accidental that it exists as a single thing, with such parts and features, organized in the way that they are, such that they carry out such activities. This then provides a sense in which we may speak of a certain subset of effects as non-incidental after all—namely, the ones relevant to the object's maintenance or "survival". And this puts at least some pressure on us, given the view I have defended, to allow the application of at least some concepts associated with the idea of proper function; in fact, there is clearly such pressure even for proponents of general etiological views, such as Wright's, which make no essential reference to a selection history. Perhaps we can indeed say that the function of the heart-like object, for example, is to pump the blood—meaning not just that this is what it functions to do, but that this is the non-incidental role it plays in the functioning of the system, where the latter is understood simply in terms of its maintaining itself in existence as a single, coherently organized entity.

This, I think, should just be conceded, though with the recognition that just as the non-accidentalness involved here is far less robust than the non-accidentalness resulting from cumulative natural selection over evolutionary time, so too should the teleology here be regarded as far less robust than that associated with actual living things. There may, for example, be a sense in which we can say here that the "heart" is *for the sake of* pumping the "blood", and does not merely *happen* to do so (as it did at the very beginning); but this is only a shadow of the sort of thought we have in the case where the heart, as a blood pumping organ, was put together a stage at a time through cumulative natural selection over millions of years, shaped at each stage by factors having to do with its blood pumping proficiency. One way in which this sort of discrepancy is reflected is in the fact that while we may wish to speak of functions in the case of the cat-like object, it is still doubtful whether we can legitimately speak of the heart-like object as a heart, for example, or make objective judgments about *defect* if it happens to pump in the way that would be defective for a cat heart, for example. How can a natural object be defective if it is not a token of any functional type to which certain non-arbitrary functional standards apply?

One might perhaps just make reference to the object's performance so far, and say that it is defective if it ceases to contribute to the maintenance of the system as efficiently as it used to. This would, however, be somewhat arbitrary, since there are no grounds for saying—as we could indeed say in the case of an actual cat heart—that such an object *should* continue doing what it has done so far; how, then, can the heart-like object legitimately be faulted for failing to do so? Of course, someone might just *stipulate* that this is how he shall use the relevant expressions: "Proper functioning" for such a system will be continuing to do those things that maintain itself in existence, with equal efficiency as in the past, and a part is "defective" if it fails to make the sort of contribution to this that it has made in the past. I have no strong objection to someone's speaking in this way, so long as it is recognized that this is only a pale reflection of the robust, non-arbitrary and richly grounded teleological discourse we can engage in with respect to actual living things. Ultimately, it counts in favor of the view I have defended that it allows us to make good sense of the intuitions that at least some kind of functional discourse is in place with regard to such cases, while showing how and why that discourse is importantly limited.[25]

---

[25] Similar points may be made with regard to the issue of reproduction: At some point, the cat-like objects may reproduce, and with regard to the offspring it will no longer be entirely accidental that they possess certain parts and features, organized as they are, such that they are able to bring about reproduction. We might then speak of such effects of the reproductive parts of second-generation cat-like objects as playing non-incidental roles in their functioning, which now includes reproducing as well as maintaining themselves in existence. Again, however, what we have here is still far less robust than what we have in

5. HISTORICALLY-INFORMED WELFARE-BASED ACCOUNTS

Suppose a proponent of a welfare-based account concedes the relevance of causal history to present teleology, giving up the ahistorical approach to the latter, but still insists that biological functions—at least in purely natural cases (i.e. excluding the hypothetical case of the shmene-obsessed inventor)—are generally and ultimately related not to gene replication but to the promotion of welfare. Basically, the idea here would be that the appeal to natural selection history is necessary to screen off accidental benefits of the sort illustrated in section two, and to isolate those which are non-incidentally—and hence functionally—produced; but the claim is that we can appeal to causal history in just this limited way without being led, by the arguments I've given, to a non-welfare-based, genetically-oriented account. On this proposal, those effects which would be recognized as ends would be a *subset* of the effects that I recognize as ends—namely, the ones whose realization promotes welfare as an instrumental part of promoting the relevant sort of gene replication. For example, the circulation of blood would count as an end, since it is good for the organism and its being good for the organism is instrumental to its promoting gene replication, so that it is not merely incidental. By contrast, the manipulation of the sex ratio among reproductive siblings (as discussed in chapter three) would not count as an end, since its realization does not promote welfare as part of its promoting gene replication; the adaptational tendency to manipulate sex ratio is just a consequence of natural selection, with no functional significance. Generally, the effect of this hybrid view will be to screen off both incidental benefits (answering the objections from section two) and effects that are relevant to gene replication but do not accomplish this by doing things that promote welfare.

While this view would avoid the problems I have explicitly raised so far, the first question to ask is what support there is for the kinds of restrictions it imposes, and indeed why we should find this more

---

actual cases, and there will still be much that is indeterminate—where there is nothing to determine the details about proper functioning (i.e. the structure of non-accidental relations among functions and ends, the relative weighting of various ends, etc.), in the way that there is for organisms informed by a cumulative selection history. One interesting question to consider is what we should say about a hybrid offspring of a cat and a cat-like object.

restricted view any more attractive than the much more straightforward one I have presented. (Recall that the former is really just a more restricted version of my view, recognizing as functions and ends only a subset of what I recognize as functions and ends, the restricting factor being relevance to welfare.) There are at least two likely answers here. First, it may again simply be taken for granted that there is a conceptual connection between natural, biological function and welfare, the concession now being that it is only by looking to causal history that we can discover *which* welfare-promotion is truly functional, and how various types of welfare promotion are weighted as ends. Second, it may be suspected that without such a restriction, there will be no way to distinguish between non-functional systems like tornadoes or crystals, and genuinely functional systems like organisms; by restricting functions and ends to effects relevant to welfare, we rule out erroneous teleological attributions to non-functional systems, whereas on the less restricted view, it is not clear how such mistakes would be avoided.[26]

I have already addressed the claim that it is simply a conceptual matter that functions in biology must relate generally and ultimately to the promotion of welfare. The hypothetical example of the shmene-obsessed inventor plausibly shows that there is no such necessary connection, and we have seen that it is not hard to come up with a plausible explanation for why it might have *seemed* as if there were be such a connection even if in fact there is not. Thus, this motivation is quite weak. The second reason mentioned above is more interesting, but again we have already seen that there *are* perfectly good ways to distinguish between functional and non-functional systems—at least in clear cases, where we ought to be able to make such a distinction—without appealing to welfare-promotion or the lack of it. So there is little to be said for the second motivation for making welfare promotion a necessary condition for biological function. Still, let us set aside the

---

[26] This argument is made by Bedau (1992b), though he goes on to argue that even the restriction to cases involving welfare-promotion is insufficient. In an effort to avoid certain counterexamples, he arrives at the suggestion that the promotion of welfare must enter *as such* into the etiology of the trait in question in order for the trait to have a genuine function (thus offering a hybrid etiological/welfare-based account). Since he thinks that natural selection etiologies do not involve welfare promotion *as such*, he winds up concluding—quite wrongly and unnecessarily, in my view—that there is no such thing as genuine biological teleology after all.

question of motivation, and turn to the view itself. It is not hard, I think, to show that it is untenable.

We may begin by recalling the example of the shmene-obsessed inventor. Suppose he invents a type of organic machine that perfectly duplicates an actual type of organism—the domestic cat, say. (Actually, he designs two sub-types, corresponding to male and female cats, and turns them loose to live just like cats.) It will be agreed, I think, that his "shmats" are complex shmene replicating machines; that is what they were designed to do, and it is shmene replication toward which all of their functions are ultimately directed. If "male" shmats exhibit a tendency to kill the already existing offspring of new mates, making way for the more efficient creation and raising of their own offspring, we will recognize this as perfectly functional behavior, geared toward spreading the male shmat's shmenes; it is a perfectly intelligible part of their functional repertoire, given their nature as shmene replicating machines.

Now suppose we consider the same trait in male cats. Recall that it occupies exactly the same position in a perfectly parallel hierarchy of traits and activities that is likewise non-accidentally geared ultimately toward gene replication. What could be more natural, therefore, than to regard this trait in cats as functional in just the same way? It did not matter in the case of shmats that it does not promote shmat welfare or satisfy their needs, so there is evidently nothing absurd in the very idea of *a trait's being functional within a living thing* even though it does *not* promote welfare. Why, then, should it matter in the case of cats that the trait does not promote cat welfare? Yet on the welfare-based view, we are forced to say that it does: The trait does not promote welfare in either case, but it is nonetheless functional in shmats, while it is *not* functional in cats; in cats, it will be said, the trait is just a non-functional—perhaps even dysfunctional—relic of their natural selection history.

On the face of it, this position seems both unnatural and groundless. As mentioned earlier, there is nothing about the case of cats that should make us think that welfare comes into the picture in any more central a way than it does in the case of shmats. The natural selection history behind the presence and organization of cat traits involves principles that are no more concerned with cat welfare than are the principles behind the inventor's designing of shmats. And the trait's promotion of gene replication in cats is every bit as *non-incidental* as the trait's promotion of shmene replication in shmats, so there is no cause for worry in this regard. There is thus so far nothing very intuitive about the above position, and therefore no reason—in the absence of some further argument—to take it very seriously as a rival to the view that the functional facts are exactly parallel in the two cases. The move from an ahistorical welfare-based view to an historically informed one may help to avoid the problems raised in section two, but it still leaves us with what seems to be a groundless restriction of functions and ends to effects that involve welfare promotion.[27]

The same basic problem can also be seen without recourse to hypothetical cases of living artifacts. The welfare-based view in question forces us to distinguish among the various traits of a given type of organism on the basis of whether or not they promote welfare (i.e. as an instrumental part of their promoting gene replication). Thus, some adaptations will be regarded as functional, while others will not, depending on their involvement, or lack thereof, in welfare promotion. But now consider two adaptations, one of which (A1) is thus judged to be functional, the other of which (A2) is not; and suppose that neither is the product of "renegade" genes, and for simplicity that neither involves kin altruism. A1 contributes to the organism's survival, say, and A2 contributes in some non-welfare-involving way to increasing its likely reproductive output. Now what seems implausible about drawing such an important line between these two traits is that A1 and A2 *are present in this type of organism for exactly the same reason*, in the following sense: Each is present because it makes a significant contribution ultimately to the *reproductive success* of the organism. So here we have a coherent system of traits all geared toward this common thing, and all present *because* of their typical contributions toward it, and yet on the welfare-based view some are regarded as having a function in the system, while others are regarded as not having any

---

[27] It is important to be clear about the limits of the above argument: I am not here arguing, for example, that the possibility of shmene replicating machines establishes that there could not be living things that had a welfare-based functional teleology. As I argued in section fourteen of chapter five, there could in principle be such living things—if, for example, they were created by a benevolent God (or for that matter, a *non*-shmene-obsessed inventor) concerned with their welfare as an end in itself. What the above argument helps to show is just that we have every reason to regard the functional teleology of actual living organisms—*given* that they are indeed products of natural selection—as being *gene-oriented*, rather than welfare-oriented.

function in it at all simply because they don't happen *also* to promote welfare; indeed, the latter may even be regarded as being *dysfunctional*, on this view, if they happen to detract from welfare.

The question is why welfare should be thought to be all-important here, especially considering how much traits like A1 and A2 have in common—properties that, in the case of a machine, for example, would certainly be understood to bear heavily on the issue of teleology. Indeed, traits like A1 promote welfare only in the course of doing the very same thing that traits like A2 do, i.e. promoting gene replication. They wouldn't be included in the system at all unless they did! On the welfare-based view, then, one is forced to maintain all of the following: (1) welfare promotion played only a purely instrumental role in the etiology of traits like A1; (2) what really mattered to such traits' coming to be present and coherently organized as they are in such organisms (namely, contributing to cooperatively engineered gene replication) is equally shared by traits like A2; (3) traits like A2 presently fit equally well into a coherent hierarchy of traits non-accidentally geared toward certain common results; (4) yet despite all of this, traits like A2 are to be denied functional status simply because they don't promote welfare. In the absence of some good argument for why welfare should be given such a special place in biological teleology, despite the first three considerations above, this position is plainly implausible.

Similarly, just as it seems unwarranted to deny traits like A2 functional status simply because they don't promote welfare, it seems equally unwarranted to treat the promotion of welfare by traits like A1 as an *ultimate* end. I do not myself deny that it is an end, of course: Surely helping the organism to avoid being killed by a predator, for example, is an end served by a protective trait such as camouflage. My claim is just that this end, like others, is a *subordinate* end, directed ultimately toward the end of gene replication—a state of affairs that reflects the ultimate causal principles behind the construction and conditioning of the organism. But on the welfare-based view, everything stops with welfare promotion. The ultimate end toward which biological functions are directed is the promotion of the welfare of members of the species, the satisfaction of their species-typical welfare-related needs (though as already mentioned, exactly *whose* needs matter, functionally speaking, and with *what priority* and *why,* is left vague, to be determined somehow in each case by the given "way of life"). The question is how this treatment of welfare promotion as an *ultimate* end is justified.

There is actually a difficulty here that I have not previously mentioned. The welfare-based view I am imagining in this section regards welfare promotion as non-accidental, and thus functional, when it is done as an instrumental part of the standard contribution to gene replication—hence ruling out the sorts of incidental welfare promotion considered in section two. But this will *include* the welfare promoted in organisms of different species as part of symbiotic relationships! One of the things an ant does in the course of promoting the replication of its own genes, for example, is to protect the aphids it "milks", thus promoting the aphids' welfare. Now on my view, it is obvious what to say about this. Promoting the aphids' survival is a *subordinate* end for the ants, geared toward the more remote end of maintaining their food supply (which the aphids provide), which in turn is geared toward the more remote end of promoting their own survival, which is geared finally toward the replication of their genes. Viewed in this way, there is really no great difference in kind between intraspecific welfare promotion and the promotion of the welfare of partners in symbiotic relationships (or for that matter, between personal welfare promotion and the promotion of the welfare of kin). Neither is an ultimate end; both are just subordinate ends geared toward the relevant sort of gene replication. But this is not how proponents of the welfare-based view (at least the neo-Aristotelian ones I have in mind, such as Foot) will tend to view that matter. They will want to say that the intraspecific welfare promotion *is* an ultimate end, while the promotion of the welfare of symbiotic partners is of only instrumental relevance to the promotion of the organism's own welfare, or that of other members of its own species. But the question is how they can defend treating the one kind of welfare promotion as an ultimate end, and the other as merely subordinate to this. Without some argument, the choice just looks arbitrary.

In any case, quite apart from the above question, the general treatment of intraspecific welfare promotion as an ultimate end puts proponents of the welfare-based view in a position no less awkward than their position with regard to the non-functionality of traits like A2. They concede that where a trait like A1 promotes welfare it also thereby (more remotely) promotes coordinated gene replication, and that it is only by virtue of its having this latter effect that the trait has come to be incorporated into the organism's phenotype to begin with; without the further effect of promoting gene replication, the promotion of welfare would never have done a thing to get the trait included as part of the system. Yet despite all this, it is nonetheless insisted that the promotion of gene replication is functionally insignificant, except insofar as it is used to help mark off non-incidental welfare promotion from incidental welfare promotion; functions are to be understood in relation to a hierarchy of ends that culminates in the meeting of welfare related needs, simply ignoring the further effects that had everything to do with the selection and organization of all these traits, and that are still cooperatively brought about by these traits so combined and organized. But again, this just seems entirely arbitrary. We have every reason to regard the promotion of welfare as nothing more than an end that is served only by a subset of functional traits, and even then only as a subordinate end, directed toward the ultimate end of inter-generational gene replication. There does not seem to be any good reason to regard welfare-promotion as a general and ultimate end in biological teleology, apart from appeals to intuitions and alleged conceptual connections, which can plausibly be explained away as vestiges of deeply ingrained, pre-Darwinian thoughts about living things.

6. WELFARE-BASED VIEWS AND TELEOLOGICAL NATURALISM IN ETHICS

Suppose that a proponent of the welfare-based approach were somehow to succeed in meeting the challenges I have posed to such accounts, making a convincing case for a special welfare-oriented teleology pertaining to living things as such, according to which functions in biology are generally and ultimately related to the satisfaction of species-typical welfare-related needs. Even if this could be accomplished, would it lend any real support to the sort of metaethical project advanced by Foot, as discussed in chapter one—i.e. the attempt to understand ethical normativity as a special case of natural teleological normativity, pertaining to the reason-involving dimension of human life? In fact, I think that there would still be serious problems with such a move.

First, much will depend on how the problems raised earlier are dealt with by this welfare-based view. If the problem of distinguishing functionally realized benefits from incidental benefits in biology is solved by looking to natural selection history, as with the view considered in section five, then this view of natural teleology will certainly not be suitable for any such metaethical project. For if the relevant kind of functional teleology in human life is taken to be a special case of natural teleology, then on such an historically-informed, welfare-based account of the latter, the only things that will count as genuinely functional—even in the realm of the will—are welfare-promoting adaptations and related traits stemming from them; and as I have already noted, it would obviously be misguided to try to understand ethics—as a justificatory scheme for human action—in terms of adaptations. If ethical discourse is to be understood in connection with some sort of functional discourse pertaining to human life, then that functional discourse is certainly not centered around human adaptations—which at best promote human welfare only as an instrumental part of promoting competitive gene replication, and do not (except perhaps accidentally) promote any further human good of the sort that would figure into any sane account of welfare-based ethics.

Secondly, even if this difficulty is somehow solved—perhaps because another solution is found to the problems raised in this chapter—it is clear that the rest of the biological world would present no real guidance in answering the questions, with regard to the natural teleology appealed to in human life, of *whose needs matter, functionally speaking, and with what priority, and why*. We have seen that not only are the *ways* in which organisms of a given species standardly meet their species-typical needs determined by the natural selection history behind the species in question, but equally so determined is the complex *weighting* of the needs within the teleological structure, with the satisfaction of some having higher priority as natural ends than the satisfaction of others. When someone speaks of *what F's do*, where this is supposed to be understood as belonging to functional discourse related ultimately to F-flourishing, this hides the complexity of the particular structure of ends that is involved in such flourishing. The meeting of needs in any given species is far from a simple, democratic matter, as I have tried to bring out with a host of examples. When we move to the human case, then, there are two choices. Either we again maintain that natural selection history determines this structuring of ends, or we do not. If we do, then once again we have obviously rendered the functional teleology in question unsuitable for the role it is supposed to play in the metaethical project: No sane ethical view will hold the structure of proper human ends (i.e. the weighting of the various needs that are functionally served by proper human behavior) hostage to the contingencies of our natural selection history. If we do not, however, then we are left without any clear specification of these crucial facts, and the real work remains to be done—again, something Foot has not even begun to provide (and similarly with Thompson). Moreover, if such facts about the structuring of ends are held to be determined by something other than what determines them in all other species (namely, natural selection history), then it is no longer clear that we would in fact be talking about a special case of natural teleology after all, as opposed to a distinct form of teleology associated with human life—perhaps one rooted in some way in the faculty of reason. I have no objection in principle to the latter possibility.

I want to stress the significance of the above point. It is easy to speak in very general terms about *what Fs do*, and *how they meet their species-typical needs*, as Foot does, and to imagine that we can just apply the same thoughts to human beings and come up with something reasonably determinate about *how human beings get on*, arriving rather swiftly at functionally laden judgments about the respecting of contracts and keeping of promises, for example. But this is all much too quick. Having looked more carefully at the biological cases, we have seen that there is much that needs to be said about *whose* needs are relevant to *what* functions in *which* particular organisms, and how the satisfaction of each need is *ranked* within the teleological structure pertaining to such organisms—all of which becomes clear once the natural selection history is understood. And this will vary greatly from species to species, as illustrated by some of the examples we have considered. When it comes time to move to human beings, therefore, it cannot just be assumed that we have some general intuitive understanding of these matters that falls naturally into place here, allowing us to speak in the normative teleological mode of how "we" get on, or of how "human good" is properly promoted. So even if a welfare-based view of natural teleology along the lines of Foot's picture could be defended, it would be so vague about the above details as to be of virtually no guidance in trying to figure out what it might say about ethics.

CHAPTER VII

Contrasts with Dawkins and with Standard Etiological Accounts

1. DAWKINS' TREATMENT OF TELEOLOGICAL DISCOURSE

I have relied heavily on Dawkins' work in theoretical biology in arguing for a genetically-oriented account of biological teleology. It is therefore useful to point out the crucial differences between his conclusions and mine. To begin with, it should be noted that Dawkins is primarily concerned not with developing an account of biological function, but with arguing for a certain general and unified understanding of natural selection—a gene-centered view, as opposed to a species-, group- or even individual-centered view—and hence for a certain understanding of organisms as products of natural selection. This is part of the reason why it is risky, as already mentioned, to attribute to him any such claim as that genes cannot be said to have functions in organisms, on the basis of his general remarks about genes and organisms from an evolutionary point of view. Still, he does think that his general argument has relevance for understanding the proper use of teleological language in biology, and in fact he often states his conclusions in explicitly teleological terms. This is true both with regard to his most general thesis about the nature of adaptations, and with regard to more particular conclusions about particular adaptations.

The general thesis is neatly summarized in the following passage from *The Extended Phenotype*:

> We look at an adaptation and want to say, 'it is for the good of. . . .' Our quest is for the right way to complete that sentence. It is widely admitted that serious error follows from the uncritical assumption that adaptations are for the good of the species. I hope I shall be able to show, in this book, that yet other theoretical dangers, albeit lesser ones, attend the assumption that adaptations are for the good of the individual organism. I am suggesting here that, since we must speak of adaptations as being for the good of something, the correct something is the active, germ-line [genetic] replicator.[1]

Now the first thing to notice about this formulation of Dawkins' thesis is that the language is clearly metaphorical. As Dawkins makes clear throughout his work, he is not suggesting that there is literally such a thing as the *welfare* of genes, as there is with organisms, any more than he is suggesting that genes are literally *selfish.* The talk of adaptations *benefitting* genes is just metaphorical shorthand for their promoting the genes' replication. Thus, when he claims that adaptations cannot generally be expected to promote the good of the species or the good of the organism, but only "the good of" germ-line genes (which often, but not always, involves promoting the good of the organism), what he means is simply that what adaptations can generally be expected to promote is the replication of germ-line genes—namely, germ-line genes of the types that code for the adaptations.

The next question, then, is how he understands the teleological "for" here: Why does he say that adaptations are really generally and ultimately "for the good of" genes (or equivalently, "for the sake of" genes—eliminating the metaphorical use of "good")? He doesn't just mean that they in fact promote the replication of genes. For he applies teleological language only where the effect in question is one that figures in the consequence-etiology of the adaptation.[2] Indeed, he seems to regard this as the main point of such language, i.e. to indicate which effects are the ones that really figure into the consequence-etiology of the trait, as opposed to other effects which are just incidental. This may be seen clearly in his discussion of particular

[1] Dawkins (1982, p. 91).

[2] As noted earlier, a trait has a consequence-etiology if the causal explanation of how it came to be present in current organisms involves an appeal to certain causally relevant effects that the trait (-type) has. This is further discussed in section two below and in chapter eight.

cases, such as the example of clutch-size regulation among swifts in conditions of scarcity, which we examined in chapter six, section two. After explaining what he takes to be the most plausible natural selection history behind this trait (the more orthodox, non-group-selectionist account), he formulates his conclusion as follows:

> Therefore, individuals regulate their clutch size for reasons which are anything but altruistic. They are *not* practicing birth-control *in order to* avoid over-exploiting the group's resources. They are practicing birth-control *in order to* maximize the number of surviving children they actually have (an aim which is the very opposite of that which we normally associate with birth-control).[3]

The reduction of clutch size tends to have the effect both of avoiding over-exploiting group resources and of maximizing individual reproductive output (since what limited resources the individual parents can secure are better invested in a smaller number of offspring, rather than being spread too thin among the usual number of offspring). But the teleological language is properly applied, on Dawkins' view, only to the latter, the reason being that this is the effect that—on the evolutionary story he is taking to be correct—was actually relevant to the selection of the trait in question, and hence to its coming to be present in the organisms in question—the other being merely incidental.

It appears, then, that this is how Dawkins views the proper use of teleological language in biology: It is properly used to indicate the etiologically relevant consequences of traits (and other entities, presumably); acceptable uses of teleological language point to true evolutionary stories, while unacceptable ones do not. But there is still a real question as to whether he views this use of teleological language as ever involving a *literal* application of the teleological concepts and expressions, or whether he regards it as involving only *metaphorical* applications restricted in the manner just indicated. To the extent that he might regard them as literal, his view is just a form of etiological

---

[3] Dawkins (1976, p. 125), my emphasis.

account.[4] But there are strong indications that he generally views them as metaphorical.

Much of Dawkins' use of teleological language is obviously intended to be metaphorical—as in the quote above, for example, when he speaks of maximizing the number of offspring as the swifts' "aim". He obviously does not mean that swifts actually have this intention, but only that they have been shaped by natural selection to behave *as if* they had such an intention (and that the effect in question—maximizing offspring—is in fact the effect that figures into the consequence etiology of this behavioral trait).[5] His use of the "aim" metaphor here suggests that the "in order to" in the above passage is likewise to be understood as a metaphorical application. Since it is restricted in the manner indicated, it is not *merely* a metaphorical way of talking, but a useful device for singling out evolutionarily significant effects of adaptations; but if it is to be understood within the framework of the metaphor suggested by the use of the term "aim", it is metaphorical nonetheless, not literal. The same point holds for his talk of the "aims" of genes: This is obviously metaphorical, and it suggests that the related teleological language used in connection with such cases is equally to be understood within the framework of the metaphor, rather than as involving literal applications of functional teleological language. He also sometimes exploits the machine metaphor: Because of the influence of natural selection, organisms behave at least in large part *as if* their behavior had been designed by a programmer for the purpose of maximizing their number of offspring; since the latter is the effect that indeed figures into the consequence-etiology of the traits in question, Dawkins will tend to use teleological language in describing such behavior. Again, while his use of metaphors certainly does not *commit* him to regarding *all* of his uses of teleological language as metaphorical, there is at least good reason to believe—particularly in the absence of any explicit argument to the contrary—that he does intend it all to be understood as usefully restricted metaphor.

---

[4] It would thus perhaps be close to that of Brandon (1981) or Neander (1991a). I discuss differences between my account and etiological accounts in the next section.

[5] The parenthetical remark is necessary, of course, since natural selection—even according to the orthodox story—has equally shaped them to behave *as if* they had the intention of conserving group resources. Again, Dawkins uses teleological language not merely where it makes for a useful metaphor, but only where the effect in question is the one that figures into the natural selection history of the trait.

Now whichever view he holds, it is significantly different from the account I have given of functional teleology in biology. First of all, the account I have offered is an account of what I take to be perfectly literal applications of terms and expressions such as "function", "end" and "in order to". I do not maintain that functional discourse in biology is a useful way of speaking metaphorically about the natural selection background of the traits of organisms, but rather that it is a literal way of speaking of their genuinely functional teleological nature—no less so than in the case of machines. Secondly, my account is not an etiological account, connecting the functions of individual adaptations directly to their individual natural selection histories, but an account identifying functions with standard roles played by types of entities (especially adaptations, but also other sorts of things, as described in chapter five) in the *working* of organism-types; the natural selection background is, of course, crucial, but comes in only indirectly. One important difference, then, is that on Dawkins' view, the function of any adaptation is ultimately the propagation of the genes that code for *it*. By contrast, on my view, while this is indeed the ultimate effect responsible for the trait's selection over its rivals, we should not understand *function* in this atomistic way: As far as function is concerned, any functional adaptation within a given organism-type has the same ultimate function, namely contributing to the replication of genes of the types represented in the organism's co-adapted genome. Also, on my view there is a perfectly straightforward sense in which genes themselves can have functions within organisms, doing what they do *for the sake of* various organismic ends (though ultimately these ends are subordinated to the final end of replicating copies of genes of the types represented in the genome). We can thus recognize not only that organisms work ultimately in the service of gene replication, as Dawkins stresses, but also that genes make functional contributions to the working of organisms, so that in a sense the teleological arrow points in both directions. I will now say more in general about the ways in which my account differs from etiological accounts.[6]

[6] Having contrasted my account of functional discourse in biology with what appears to be Dawkins' view, it should again be stressed that his primary objective is to arrive at a clear understanding of natural selection, and of the real parts played in this process by species, groups, individual organisms and genes; his treatment of teleological language as such is really a secondary concern. I intend my account of biological function to be fully compatible with the substance of his theses about natural selection, and therefore would

2. ETIOLOGICAL ACCOUNTS AND THE ISSUE OF REDUCTION

I have already noted one way in which my account differs from standard etiological accounts in the discussion of the example of the sea turtle's tail in chapter five: On my view, an entity X can have a function Z quite apart from whether or not Z enters into a causal explanation of how X came to be there; the turtle's tail, for example, can have the function of protecting the eggs, even if this is not the effect that enters into the natural selection explanation of the presence of the tail (whether of its initial fixation in the population, or even of its maintenance). This, I think, is a significant advantage, since such functional claims are indeed plausible, and so ought to be accommodated by a satisfactory account. There are other views that would yield the same result as mine here, against the etiological account, but they tend to achieve this only at the significant cost of capturing far too much, failing to draw a satisfactory distinction between genuine functions and merely fortuitous effects. For example, a crude welfare-based view that allows anything that regularly has good consequences for the organism (or its group, etc.) to count as functional will wind up counting as functional many things that are clearly just accidents or lucky side-effects.[7] By contrast, my account avoids such excessive liberality with function attributions: There must still be a close tie to adaptations, as I've stressed; it is, for example, as part of a behavioral adaptation that the turtle's tail is used for protecting the eggs.

What I want to do in this section is to draw attention to two further and more fundamental differences between my account and etiological accounts. There are, of course, various versions of the etiological approach to functional teleology in the literature. On some versions, an

---

expect that he might be relatively sympathetic to this account, even though it differs from his treatment of teleological discourse as far as it goes.

[7] A classic example involves heart sounds: As a side-effect of its functioning, the heart makes certain noises; the making of different noises thus tends to tip off physicians to potential problems: There might even be characteristic sounds indicating particular problems. The production of such sounds may thus be *good for* the organisms in question, but it can hardly on that account be deemed biologically functional, as if it were on a par with the pumping of blood; it is merely an incidental effect (i.e. incidental to the working of the organism, as I have argued this should be construed) that happens to have certain beneficial consequences (at least in an environment containing physicians). Wright (1976, pp. 78–9, 92, 99) offers several other examples along these lines.

entity has a function just in case it has a consequence-etiology—i.e. just in case it is present in part *because* (historical causal "because") of certain *effects* that type of entity tends to have, these effects thus constituting its function. On other versions, function attributions are restricted to cases where the consequence-etiology involves some sort of *selection*, whether intentional or natural; in some cases, a unified account is offered for both cases, while in others—particularly in cases where the author's concern is primarily just biological function—any such attempt at unification is either set aside or explicitly disavowed. In any event, etiological accounts are generally characterized by an atomistic approach to functions (though I will consider an exception below), and by an attempt to reduce facts about present function *directly* to certain facts about the causal history of the entity in question (or to a conjunction of such facts and the fact that the entity still has the effect that figures in that causal history), neither of which are features of the account I have offered. Let me take these up in turn.

In saying that etiological accounts approach functions in an atomistic way, I do not mean that they altogether ignore the larger contexts in which functions are situated. Indeed, in focusing on the causal history of the entities in question, they must often at least implicitly make reference to the wider context in which instances of the entities have been found throughout that history. For example, Neander's thesis that functions are selected effects requires us to consider these effects not in isolation, but in the context of the environment in which natural selection took place—including the rest of the traits of the type of organism in question, which form an important part of the background environment for a trait undergoing natural selection. Nonetheless, such an approach is still atomistic insofar as it ties the function of each individual trait (whether type or token) directly to its own natural selection history or consequence-etiology—rather than tying functions directly to contributions made to the coherent working of a certain type of system, and appealing only indirectly to natural selection history to determine the precise nature of that working and of those contributions.

The importance of this difference is nicely illustrated by the case of meiotic drive considered earlier, for which etiological accounts yield an implausible result. Segregation distorters and the phenotypic mechanisms by which they effect the distortion of meiosis have consequence-etiologies just as much as any other gene and trait that have been favored by natural selection. If we assign functions in biology simply on the basis of something's having been naturally selected, then we will have to conclude that it is the *function* of the segregation distorter and its mechanism to distort meiosis in the segregation distorter's favor. Indeed, we will have to say that it has this function *in the very same sense* in which the gene on chromosome 7 mentioned earlier has the function of promoting the manufacture of elastin, or in which the heart has the function of pumping the blood. There is, after all, a natural selection history behind the genes and traits in both kinds of case, and that is all that etiological accounts consider. Now this conclusion is, I think, very implausible. The problem is not simply that the effects of the segregation distorter are *bad for* the organism experiencing it; as I have argued, there are lots of functional traits that are either neutral or harmful to organisms. The problem with the segregation distorter is that it makes no contribution to the coherent working of the organism to which it belongs, as do the genes in the co-adapted genome that are responsible for constructing and conditioning the system into a coherent working unit. Indeed, it severely disrupts the working of the organism, pulling its functioning away from the high-level common ends being served by the genes within the co-adapted genome, toward its own enhanced replication. The segregation distorter and its mechanism thus belong in a different category from genes coding for organs like the heart and their phenotypic expressions; the former are really closer in many respects to viruses and their effects than to the "cooperating" genes in an organism's co-adapted genome and their functionally integrated phenotypic expressions.[8]

---

[8] It should be noted that while Neander (1991a) is indeed vulnerable to the above objection, given the obvious intent of her account, there is one place where the way in which her thesis is stated happens to enable her to avoid this problem—though this is really only due to an unintended oversight on her part, namely, her false assumption that all natural selection involves an item's contribution to inclusive fitness. Her thesis is fundamentally the atomistic etiological thesis that "the proper function of a trait is to do whatever it was selected for...i.e. whatever items of that type did that caused them to be favored by natural selection" (pp. 173–4); there is no mention of standard contributions to the working of a complex system, and so on. Yet when she gives the summary statement of her account, she puts the etiological condition in terms of past contributions to *inclusive fitness* (p. 174). Since the natural selection history of a segregation distorter does not involve any contributions to inclusive fitness, she would thus avoid this counterexample by formulating her thesis in this way. But again, this way of formulating her thesis is not in fact in accord with her obvious intent in the rest of the paper: It is the

My account of function thus has the advantage that it does not implausibly assign the segregation distorter a function on a par with the gene for elastin or genes involved in the proper formation of the heart. On the other hand, there is nothing in principle to stop us from regarding the segregation distorter and its mechanism as a type of functional system in its own right—a mini gene replicating system equally put together by natural selection, for which the rest of the organism is just part of the natural environment. In that case, we might speak of the function of the distortion mechanism within the context of this smaller system: In doing whatever it does to distort meiosis, it makes a contribution to the working of this system, the ultimate end of which is simply the replication of the segregation distorter gene. This would adequately account for our intuition that the meiosis-distorting mechanism must in *some* sense be regarded as functional, like other biological mechanisms. In principle it might, after all, be very complex in nature—just the sort of thing about which we could intelligibly ask "how does it work?", as with other complex biological mechanisms. By distinguishing between the relevant systems, we can account both for the sense in which the meiosis-distorting mechanism is a functional entity, and for the sense in which it certainly does not have a function *in the (working of the) organism*, as e.g. the gene for elastin does, instead falling outside of this. Still, I think this sort of case should still be regarded as a special case, involving an extension of the ordinary notion of biological function, rather than being a paradigm of biological function. Our ordinary notion of biological function involves the notion of roles played in the working of a certain type of *organism* as a coherent functional unit, and this should be reflected in an account of biological function even if we allow for natural extensions to cover other cases. Thus, I think the account given in chapter four should remain as it is, rather than generalizing it, as e.g. Griffiths does with his account, to accommodate such cases as the segregation distorter as central cases.[9]

Again, the standard etiological view does not have the resources to deal with these issues: By linking functions directly to individual

---

possession of a natural selection history that matters, on her view, not whether or not this involves contributions to inclusive fitness. Thus, it is only by a lucky mistake that her view, on this formulation, avoids the problem I have raised here for etiological views.

[9] Griffiths (1993, pp. 415–16).

consequence-etiologies, there is no room for drawing the above sort of distinction, at least if it is to be understood as firmly rooted in the theory, and not just as an *ad hoc* move; this is true whether the nature of the consequence-etiology is left open (as in Wright's case) or specified as involving natural selection (as in Brandon's case), and whether or not it is also stipulated that the entity must still have the effect for which it was selected (as in Wright's case). One could, of course, attempt to modify the basic etiological account and draw the distinction by pointing to differences in the natural selection histories—e.g. distinguishing between traits whose natural selection histories involve the selection of genes within co-adapted genotypes and those whose natural selection histories involve "renegade" selection. But this just illustrates how misguided the whole atomistic approach is to begin with: The distinction would simply be *ad hoc*, to avoid implausible results, with no real justification for so distinguishing among different kinds of natural selection history, since the heart of the etiological approach is just supposed to be consequence-etiology or natural selection-based consequence-etiology. By contrast, my justification for treating cases such as the segregation distorter differently from ordinary cases lies in the direct connection of the notion of biological function to the notion of a non-incidental role presently played in the working of a given type of organism—exploiting a parallel with the case of machines and their functional parts and features. The appeal to *causal history* comes in only insofar as it bears on this present fact about a certain type of working system, i.e. establishing the relevant non-accidental relations. Thus, the differences in causal history between the case of the distortion of meiosis and that of the production of elastin, for example, are relevant *not* directly in themselves, but because of their bearing on whether or not the traits currently play standard roles in the working of a given type of organism.

There is a recent modified version of the etiological view that is much closer to my own position than are standard etiological views, possessing superior resources than the latter for dealing with such cases as segregation distorters. This is the account offered by Griffiths, according to which biological functions are to be understood in relation to past contributions of the type of entity in question to the "fitness" of ancestral systems exhibiting it.[10] On this view, a case like the

---

[10] Griffiths (1993).

segregation distorter can be distinguished from more ordinary cases by distinguishing between the two relevant systems; indeed, he mentions this case explicitly, including it as a paradigm case of proper function in biology. Again, I think that an account of biological function, as this would ordinarily be understood, should really be stated primarily in terms of organism-types, with certain "renegade" sub-systems being treated as special, extended cases; but his appeal to systems is at least a significant improvement over standard etiological accounts.

Despite certain similarities, however, Griffiths' view is still importantly different from mine. This actually brings us to the second major difference between my account and etiological accounts generally, so I will first discuss the general issue and then return to Griffiths. Both etiological accounts and my account are *reductionistic* in a certain sense: Both involve an attempt to explicate teleological concepts in terms of non-teleological concepts, and to offer truth-conditions for teleological facts in non-teleological terms; teleological facts are held to supervene on various non-teleological facts (something also true of my account of functional systems in general). There are, however, important differences between my view and etiological views with regard to the nature of the reduction.

The crucial difference is that on etiological views, there is a *direct* reduction of teleological facts to certain facts about causal history, or to some conjunction of such facts with certain other facts, such as the fact that the item in question still has the effect that figured into its etiology (thus avoiding the problem of vestigials); that is, the fact that a given item has a certain function *just consists in* the fact that it has a certain kind of causal history involving that effect, or in this together with the fact that it still has that effect. Wright, for example, maintains that a trait's having a proper function just consists in its having a consequence-etiology (understood as implying both kinds of facts). Others, such as Neander, hold roughly that a trait's having a biological function just consists in its having a certain kind of causal history, such as a natural selection history, perhaps with some provision that the effect in question has contributed to the maintenance of the trait in the recent past.[11] And Griffiths argues in effect that a trait's having a biological function just consists in its having contributed in the

[11] Cf. also Millikan (1989, p. 293), who is explicit in claiming that having a certain kind of causal history is what *constitutes* having a function.

(relevantly recent) past to the "fitness" of ancestor systems (whether these are organisms, as in most cases, or subsystems, as in the case of the segregation distorter and its mechanism), such that this effect thereby figures into the etiology of the trait. This last view departs from the atomism of standard etiological views by bringing in explicitly the notion of contributions within a containing system, but it still involves a direct reduction to facts about causal history: Facts about present function are linked *directly* to facts about *past* contributions as such, with the former evidently just *consisting in* the latter.[12]

By contrast, on my view there is no *direct* reduction of teleological facts to facts about causal history. The function of a given type of item within a given type of working system is to be understood directly in terms of the *non-incidental role it presently plays in the working of such a system*; something's having a biological function, for example, consists directly in its presently playing a certain non-incidental role in the biological working of a certain type of organism (or in the case of a token entity, in its being a token of such a type of entity). These concepts are indeed further explicable in non-teleological terms, having in part to do with certain kinds of causal history; but the point is that any reduction of facts about function in such a context goes *through* the former concepts and facts, rather than being directly in terms of facts about causal history. To put it another way, facts about historical contributions in ancestor organisms are certainly *relevant* to facts about biological function, as I have argued: It is such facts that in large part determine the existence and shape of facts about entities presently playing non-incidental roles in the working of organisms. But facts about biological function must be understood directly in terms of these latter facts about present systems, rather than being directly reduced to facts about past contributions as such; the facts about past contributions are relevant to facts about function not *as such*, or simply because they provide an etiology of the entity in question, but rather only insofar as they bear on present facts about non-incidental roles played in the working of systems.

Two advantages of this intuitively plausible insistence on linking functions directly to certain facts about present functional systems have already been mentioned. First, it provides the most intuitive and straightforward way of distinguishing between vestiges and currently

---

[12] Griffiths (1993).

functional entities. While a type of vestigial organ and a type of currently functional organ, for example, may each have made historical contributions to the fitness of ancestor organisms, only the latter presently plays a role in the working of the type of organism in question. There are, of course, other possible ways of attempting to distinguish between them, such as by appealing directly to the requirement that at least some of the historical contributions be relatively recent if we are to attribute a present function to the entity in question; but the way offered above, in connection with a construal of functional facts as facts about roles currently played in the working of the relevant type of system, seems far more natural.[13]

Secondly, recall the hypothetical example of the sea turtle's tail discussed in chapter five, and mentioned at the beginning of this section. I argued that the tail could plausibly have the biological function of protecting the eggs, due to its regular and biologically functional use for this end, even if it was not selected—or even maintained against counter-selection pressures—for this effect. This intuitive claim is easily captured on the view I have defended, since the tail would indeed play a non-incidental role in the biological working of the sea turtle, in the relevant sense. By contrast, a view that just reduces facts about present function directly to facts about certain historical contributions to the fitness of containing systems obviously cannot capture the functional facts in such a case; and similarly with standard etiological views. Even if we suppose that the tail's egg-protection *did* in fact contribute to the maintenance of the tail, so that it could be allowed to have a function on such a view, the latter would still be missing the point: The tail has the egg-protecting biological function it does *not* because of anything about *past* contributions to turtle fitness, but simply because it currently plays a non-incidental role in the biological working of sea turtles.

---

[13] This is not to deny that there is still the problem of specifying *when* a given type of trait has ceased to play a role in the working of a given type of organism; clearly a sudden change of environment, for example, is not sufficient immediately to make a trait vestigial. I grant that an account of this matter will likely make appeal to some such notion as an "evolutionarily significant time period", and to the trait's having tended to make the relevant sort of contribution within such a period counting back from the present (Griffiths, 1993, p. 417). The point, however, is that such facts are relevant *not* directly, in themselves (as Griffiths suggests), but only insofar as they bear on the issue of whether the trait presently plays a role in the working of the type of organism, this being what is *directly* relevant to the distinction between vestiges and currently functional traits.

In addition to such considerations as these, there is a more general reason for rejecting direct reductions of facts about function to facts about causal history, which has to do with the issue of *explanation*. Appeals to proper functions are generally recognized to be explanatory *as such* of the items to which the functions are ascribed. But the only form of function-citing explanation for present functional items that is available given such direct reductions of functional facts to facts about causal history are consequence-etiological explanations of how the items with the functions in question came to be there (or something along these lines).[14] Now proponents of etiological views see this as unproblematic, since they take such explanations precisely to *be* functional teleological explanations, so that nothing is left out. If this view of functional teleological explanation is mistaken, however, as I shall argue in chapter eight, then there is a problem for etiological views of function—i.e due to their direct reduction of facts about function to facts about causal history, and the consequent lack of room for genuine functional teleological explanation. By contrast, on my view, while it is true that appeals to functions often provide the material for consequence-etiological explanations of the presence of the items in question, there is no need to identify such explanations as functional teleological explanations. Since there is more to functions than consequence-etiologies, there is room for another form of explanation that is characteristic of appeals to function *as such* and is genuinely *teleological* in nature, rather than being just a special case of efficient-causal explanation. That is, functional teleological explanations in the context of working systems can simply have to do with illuminating the non-incidental roles played by the items in question in the working of the system in question. Again, this is the topic of chapter eight and will be set aside until then. The present point is just that if the argument about the true nature of functional teleological explanation is successful, then this further supports the rejection of etiological reductions of facts about function in favor of the view I have defended.

I have allowed that on my own view, functional teleological facts pertaining to working systems supervene on various non-teleological facts, including facts about causal history; what I have rejected is just the attempt to make a direct reduction to the latter, as if they were relevant directly in themselves, or simply because they provide a

---

[14] This is discussed in detail in chapter eight.

consequence-etiology for the trait in question, as opposed to being relevant (as I maintain) only because of their bearing on complex facts about the present working of certain types of systems. The question arises, then: On my view, do facts about function at least *ultimately* consist in some combination of all these other facts—such as facts about causal history—that are relevant to determining the higher-level facts about working systems? The answer, I think, is that they do not: The "consisting in" relation seems to go only so far, plausibly reaching down to certain non-teleological facts, but not capturing all the particular facts that make the higher-level facts obtain.

An analogy may be helpful here. There are presumably various particular facts about the chemical composition of gasoline that make it the case that gasoline is flammable. But we should not, I think, say that the fact that gasoline is flammable *consists in* all these facts. The fact that gasoline is flammable *consists in* the fact that it is capable of being easily ignited: That is *what it is* for something to be flammable. If this in turn is to be said to *consist in* something further, this will be something further explicating what it is for gasoline to be capable of being easily ignited (e.g. its possessing a disposition to combine rapidly with oxygen), rather than the various particular *causes* of gasoline's being flammable; the latter are obviously relevant to, but not obviously constitutive of, the fact that gasoline is flammable. Similarly, there are many interesting facts—such as facts about remote causal history—that bear on the fact that a given type of part has a certain proper function within a given type of system, but we should not say that the latter fact *consists in* all those facts. The fact that this type of part has a certain proper function within such a system *consists in* its playing a non-incidental role in the working of such a system: That is *what it is* for something to have a proper function within a system. If this in turn is to be said to *consist in* something further, this will be something further explicating what it is for the item to play a non-incidental role in the working of the system, as spelled out in the account of working systems in chapter four, i.e. facts about the item's having certain kinds of effects that are involved in certain non-accidental relations, and so on; it will not, I think, be a description of the various particular facts, such as the particular facts about the relevant causal history, that make the above facts obtain. It seems, then, that an admission of reductionism does not entail any unlimited extension of the "consisting in" relation among the non-teleological facts on which teleological facts are said to supervene.

Finally, it is worth noting in connection with etiological views that my account does not sacrifice what is often regarded as one of the strongest virtues of at least some etiological accounts—namely that they preserve the unity that is to be found in the concept of function as it is applied across different contexts.[15] In particular, I have offered a unified account, at the *formal* level, of functions in connection with both living things and machines. The essence of the common formal account is this: A given type of entity has a function within a given type of system just in case the former plays some standard non-incidental role in the working of the latter; and tokens of that type of entity likewise as such have that function within corresponding token systems. Since etiological accounts are reductionistic, whether they can provide comparable unity at the formal level will depend on the sort of reduction that is proposed. On some versions, there is no attempt at all to provide a unified formal account of functions in machines and in biology. In the case of machines, facts about function are reduced directly to facts about intentional design, while in the case of organisms, facts about function are reduced directly to facts about natural selection; or one context is dealt with in this way while the other is simply ignored.[16] Others, however, do attempt to provide a unified account at the most formal level. Wright, for example, links functions in biology *not* directly to natural selection, and artificial functions *not* directly to intentions, but rather in both cases functions are linked directly to consequence-etiology; biological and artificial functions are then distinguished simply according to the way in which the consequence-etiology is realized, i.e. whether through natural selection or through intentions.

More recently, Kitcher has argued for an approach somewhere between Wright's and my own, linking functions directly to an extended notion of *design*, which again may be satisfied either by natural selection or by designers' intentions. Kitcher's view is thus more restricted than Wright's, since not every case of consequence-etiology can be regarded as a case of design, even in the extended sense in which Kitcher uses the term (though the use is less than entirely

---

[15] Cf. Kitcher (1993, p. 379).

[16] See Brandon (1981) and Neander (1991a).

clear); he therefore avoids certain counterexamples that have plagued Wright's account, while still maintaining a formal unity in the account of machine functions and biological functions.[17] What his appeal to "design" does not tell us, however, is whether or not the segregation distorter and its mechanism, for example, count as having been designed, and so possess paradigm biological functions; on the face of it, it certainly seems that they would, but then it is not clear what resources Kitcher has for distinguishing such cases from ordinary cases of function, as discussed above.

Also problematic is the fact—already noted in chapter four, section seven, in the discussion of crystal colonies—that there is a kind of cumulative natural selection to be found even in the inorganic world, which closely parallels biological natural selection. In fact, such selection among self-replicating silicates figures into one hypothesis concerning the origins of the more complex organic molecules leading to the emergence of life.[18] Now while some cases of such selection may get fairly complex, and *may* perhaps tempt us to speak of functions (though I doubt they generally will), there will surely be many cases that, as argued in the earlier discussion of such phenomena, are still far too simple for plausible talk of functional systems, even though they involve a cumulative selection process, whereby the crystals have over many "generations" acquired structures that make them better able to replicate. The question for Kitcher, then, is whether this cumulative selection background is sufficient for talk of *design* in his sense. If it is not, then more needs to be said about this crucial notion of "design"; and if it is, then it looks like his view captures too much, as he would be committed to speaking of functions in all such cases—e.g. the function of a certain feature of the crystal's structure, which makes it better able to replicate in its environment, this being the *end for the sake of which* it exists, and so on.

By contrast, there is nothing forcing me to apply teleological concepts to crystals shaped by cumulative selection. Since I link functions directly to non-incidental roles played in working systems, rather than to natural selection, the question for me is just whether

[17] Kitcher (1993, p. 384). He equally avoids the excessive liberality of Cummins' account (p. 390).

[18] The hypothesis is that of Cairns-Smith and is discussed by Dawkins (1987, ch.6). Bedau (1991) uses this as a counterexample to etiological accounts, even when they restrict the relevant consequence-etiology to natural selection etiology.

crystals or crystal colonies can naturally be spoken of as working systems—systems about which we could intelligibly ask *how they work*, and whether they are working *well* or *badly*, with parts and features making standard contributions to their working, satisfying certain non-arbitrary ends, and so on. Despite a superficial similarity to biological cases due to the possession of a natural selection history, the real analogue of a crystal is not an organism, or even a genetic replicator within an organism, but a lone genetic replicator; and the real analogue of a colony of crystals is not a set of mutually compatible genetic replicators in the nucleus of a cell, but a collection of many tokens of one type of genetic replicator clumped together. We are thus still very far away from what are primarily regarded as functional systems in biology, which are neither lone genes nor collections of many tokens of a single gene-type, nor even collections of many mutually compatible genes of different types, but complex hierarchically organized systems *constructed by* these sets of replicators through their interacting *phenotypic expressions*. It should be no surprise, therefore, that we do not rush to apply functional teleological concepts to a crystal clump as soon as we learn that it may possess a natural selection history with regard to certain of its features. Natural selection is relevant not directly in itself, but because of what it can give rise to; and what it gives rise to in the case of crystals may simply not be sufficient to constitute a teleological context.[19]

---

[19] To be fair to Kitcher, he does speak in places of contributions within hierarchical systems, so it looks as though he might be sympathetic to the sort of move I am making here, rather than treating the fact of "design" as a sufficient condition for functionality. But this appeal to systems should really be included explicitly in the account, as I have done, rather than putting the account primarily just in terms of "linking up with a source of design," which is too general to make the kinds of distinctions that need to be made—both in the case of the segregation distorter and in the case of crystals, for different reasons.

CHAPTER VIII

An Account of Functional Teleological Explanation: Historical Sensitivity Without Reduction

1. OVERVIEW

Appeals to proper functions or ends are generally recognized to be explanatory in some way. But what sort of explanation is typically involved here? This is a central question to be addressed by any account of function, and is the topic of the present chapter—focusing on functions and ends within functional systems, both artificial and biological. My aims are both critical and constructive. On the critical side, I seek primarily to undermine etiological accounts of functional teleological explanation, according to which this form of explanation is to be reduced to a kind of efficient-causal explanation. Such accounts have achieved the status of orthodoxy, but if I am right, they altogether fail to capture the form of explanation in question; consequence-etiological explanations may be perfectly legitimate and interesting, but they are not teleological explanations. This failure to come to terms with teleological explanation is significant in itself, but it is of particular importance insofar as the etiological view of teleological explanation has been a large part of the motivation for accepting etiological views of what functional facts consist in and of the truth-conditions for function ascriptions. If the reductionist, etiological view of teleological explanation is discredited, then as mentioned in chapter seven, this presents a problem for etiological accounts of functional facts, as well as for the atomistic understanding of the truth-conditions for function ascriptions that characterizes standard etiological views.

I will also examine the attempt, explored by Hempel and Nagel, to understand functional teleological explanation in terms of the "deductive-nomological" model of explanation. This is a very different approach from the etiological one, but the mistake involved is closely related. As Cummins has pointed out, both approaches begin with the assumption that what appeals to functions characteristically explain, if they explain anything, is "the presence of the item (organ, mechanism, process or whatever) that is functionally characterized".[1] Now unlike Cummins, I do not maintain that this assumption itself is the problem. Indeed, there is a way of understanding it according to which it is quite right, both for biological contexts and for artificial contexts: Functional teleological explanations certainly do purport to explain why a given structure, feature or act is present or occurs, in *some* sense. The problem is that both etiological accounts and deductive-nomological accounts get that sense wrong.

My positive aim is to identify the proper sense in which appeals to functions or ends as such are explanatory of functional items, and to elucidate the nature of such explanations in the context of functional systems. The view of such functional teleological explanation that I offer in section three is *non-reductionist* in the sense that it does not involve any reduction of teleological explanation to efficient-causal explanation. As will quickly become clear, it also fits naturally with the account of function developed in chapters four and five. It therefore provides further support for that account, showing how functions and ends on that view can indeed be cited to provide a special form of explanation, which latter most plausibly captures the sense of our everyday use of teleological language in connection with functional systems. That is, the challenge of showing how appeals to functions and ends can be explanatory is met, and in a way that makes better sense of teleological discourse than the alternatives do.

---

[1] Cummins (1975, p. 741).

2. THE ETIOLOGICAL ACCOUNT OF FUNCTIONAL TELEOLOGICAL EXPLANATION

The most popular approach to functional teleology over the last two decades has its roots in a natural solution to the puzzle of how an appeal to *effects* can possibly *explain* anything in the way that appeals to functions appear to—particularly where there are no intentions involved that might be thought to do the explanatory work as causal antecedents (with the effects coming in only as represented in the intentions). Consider, for example, the question:

"Why does the dry fruit of maples have a papery, wing-like structure?"

It has long been recognized that such a question can be taken in more than one way. On one interpretation, it is merely a request for a mechanistic explanation citing antecedent causes involved in the formation of the fruit. But there is another, equally familiar interpretation according to which it is a request instead for an account of the *function* or *purpose* of the structure, or what it is *for*—the assumption being that it is not incidental, but has some *biological point*. The appropriate answer to the question so understood involves an appeal to function, i.e. to certain *effects* of the thing in question: "The papery, wing-like structure slows the fruit's descent, increasing its chances of being blown free of the parent tree, thus promoting seed dispersal." But again, while such exchanges are familiar enough, they raise the question of exactly what sort of explanation is provided by such an appeal to effects.

The natural move that many have made here is predictable enough, closely paralleling the predominant move made with respect to psychological teleology. Just as psychological teleological explanation of action is widely construed as a species of efficient-causal explanation, so too functional teleological explanation has widely come to be construed as a species of efficient-causal explanation, concerned with how the structure or trait in question came to be present.[2] On this view, what sets functional teleological explanation of a structure or trait apart from ordinary causal explanation in terms of relatively proximate

---

[2] The classic paper in which *psychological* teleological explanation was argued to be a species of efficient-causal explanation is Davidson (1963).

antecedent conditions, is that functional teleological explanation involves an appeal to *remote* causes which include certain typical *effects* or *consequences* of the *type* of structure or trait in question—as illustrated in connection with natural selection just below. For such an appeal to something's typical effects in explaining its presence or occurrence is commonly taken to be the essence of teleological explanation.[3] Thus, continuing the parallel, just as in the psychological case the appeal to effects purportedly explains the occurrence of an action by pointing to certain antecedent causes (i.e. the agent's beliefs and desires wherein the effects are represented), so too in the functional case the appeal to effects purportedly explains the presence of the structure or trait by pointing to certain antecedent causes leading to the structure or trait—though in the latter case the antecedent causes are remote, and the effects in question needn't be mentally represented in any way.

It is easy to see how this works for biological adaptations—that is, phenotypic structures or traits of an organism resulting from the process of natural selection, where the structure or trait (-type) has played a causal role in the selection process. Consider a simple example involving a certain characteristic aspect of the coloration pattern on the flower petals of plants of a certain species F. Suppose that at some point in the past there was genetic variation at a certain genetic locus L affecting flower petal coloration; that is, there were a number of different alleles (in the gene pool of a given population of Fs), each of which originally appeared in some individual through random mutation. The different alleles tended, against the background of the genomic and external environments, to have different effects on flower petal coloration, which in turn had different effects on the success of

---

[3] This view of functional teleological explanation is famously articulated and defended by Wright (1976), and (in various forms) by many others, including Ayala (1970), Brandon (1981), Millikan (1989), and Neander (1991a). It is interesting to note that while Nagel (1977) offers a very different substantive account of teleological explanation (i.e. a deductive-nomological model), he agrees that its essence consists in the appeal to something's *effects* in explaining its presence. That is, he maintains that what are usually called "causal" explanations of a phenomenon account for it "in terms of antecedent conditions and the mechanisms that produce it," whereas "explanations [that] have traditionally been called 'teleological'" account for "the occurrence of a process or some other item in terms of certain effects these things have...." (pp. 277–8). According to the view presently being discussed, the two are not incompatible: Some remote *causal* explanations of the presence of a given structure or trait involve appeals to certain types of *effects* that the type of structure or trait in question has.

the plants in question at attracting pollinating insects. In particular, a certain gene G gave rise to a particular modification in flower petal coloration A, which proved especially effective at attracting pollinating insects, at least in comparison with the traits associated with other alleles. This in turn tended to improve the reproductive success of individuals carrying G and manifesting A, so that all else being equal, they enjoyed statistically greater reproductive success than individuals carrying the other alleles. As a result, individuals with G tended to pass on their genes—including G—to relatively more offspring, and so on with each generation, resulting in the spread of G relative to the other alleles throughout the population, and hence equally the spread of A relative to the traits produced by the other alleles; that is, the percentage of individuals carrying copies of G and hence manifesting A increased over the generations relative to those carrying copies of the other alleles and manifesting alternative traits. Eventually, A came to be a characteristic adaptation in such plants.[4]

Given this background, we can provide a certain explanation of the presence of instances of attribute-type A (a certain aspect of the coloration pattern on the flower petals) in current Fs (specimens of the plant species in question) in terms of its *effect*-type Z (attracting pollinating insects):

> Instances of (heritable) attribute-type A are present in current Fs *because* A has (or at least had) *effect*-type Z, instances of which contributed to the reproductive success of ancestor Fs who had instances of attribute-type A, causing A to spread by natural selection over the generations, hence coming to be increasingly manifested in descendent Fs, such that it is now a characteristic adaptation in Fs.

We cannot, of course, in this way explain the *initial* appearance of A in evolutionary history (i.e. before it was an adaptation), which *ex hypothesi* was due to a random mutation producing G. But given the initial appearance of A, there does seem to be a legitimate explanation here involving an appeal to typical effects. That is, we seem to have a

---

[4] This is, of course, an idealized case involving various simplifying assumptions, but it will suffice for the present purpose of illustration. For careful accounts of the many complications that arise in various cases—such as frequency-dependence in selection—see, for example, Dawkins (1982) and Kitcher and Sterelny (1988).

way of answering the question "why do the flower petals have that characteristic coloration pattern?" partly in terms of certain *effects* of the trait, i.e. "Because that helps to attract pollinating insects". And similarly in cases of more complex adaptations such as organs, though obviously the cumulative natural selection history required to fill out the explanation will be much more involved.

Notice that there is no threat here of any commitment to temporally backwards causation. It is not being claimed, for example, that we can explain how a particular instance of A came to be present in a particular F by appealing to *its* effects as such. Such a claim would indeed involve an appeal to temporally backwards causation; that is, it would involve the claim that "future events are active agents in their own realization", which would in no way be supported by anything having to do with natural selection.[5] As Wright clearly points out, however, the claim is rather that we can explain the presence of instances of attribute-type A in current Fs in terms of the *types* of effect produced by attribute-*type* A.[6] More precisely, since attribute-type A has advantageous effect-type Z, ancestor Fs who had instances of A benefitted from the instances of Z, and were thus better able to survive and reproduce; and according to the theory of natural selection, this is precisely what caused A to be selected over time, and so explains the presence of instances of A in current Fs—i.e. how they came to be there.[7]

---

[5] Ayala (1970). Of course, one could in principle appeal to the effects of a particular thing in "explaining its presence" if what is meant to be explained is not how it *came to be* present, but just how it has *remained* present; such an explanation would not have to involve any commitment to temporally backwards causation, so long as the effects appealed to are past effects of the thing in question. Consider a particular heart, for example. Part of the reason it is still present is the fact that *it* has successfully pumped blood (since if it hadn't, the body would have died and it would have ceased to exist). I shall ignore such explanations, however, since they do not represent the sort of explanation that is considered relevant to functional teleology by proponents of etiological views. It is generally held, for example, that a functional teleological explanation can be given for the presence of a token organ (such as a token ovary) even in a case where its continued existence is not explained in any way by *its* past token effects. The fact that in certain special cases, such as that of the heart, certain explanations of tokens can be given in terms of their particular effects, is not taken to be of any general significance for functional teleological explanation.

[6] Wright (1976, pp. 88, 90): "'It's there because it does that' is in a sense shorthand for, 'it's there because *things like it in the appropriate way* have that *sort* of property'" (my emphasis).

[7] A number of people, such as Achinstein (1983, p. 270), seem to miss this point. After noting that various things Wright says commit him to holding that a particular heart has a

Notice also that this sort of explanation is in no way incompatible with proximate causal explanations of adaptations. The manifestation of A in Fs can be explained both *proximately*, by citing the presence of G in Fs together with the ontogenetic mechanisms whereby this leads to A in an F, and *remotely*, by citing the effects Z relevant to the natural selection of (G, and hence of) A in Fs. Indeed, the effect-citing explanation in terms of remote causes itself *involves* proximate causal explanations, insofar as they enter into the explication of the evolutionary story, as brought out above. These two kinds of explanation are therefore not *rivals*, but just two different kinds of efficient causal explanation, both of interest in biology.[8] This compatibility also underscores the fact that the effect-citing sort of explanation involves nothing in the way of mysterious "life-forces". It appeals to nothing more than ordinary biological entities, causal processes, and natural selection.

Returning to the central point: Many have taken this sort of effect-citing explanation not only to be an interesting and legitimate form of remote, causal explanation, but to be nothing less than *functional teleological* explanation itself, since it accounts for the entities in question in terms of their effects. Thus, for example, in summarizing his account of functional teleological explanation—the form of answer appropriately given to a `what-for?' question—Brandon writes:

> Put abstractly, a what-for question asked of adaptation A is answered by citing the *effects* of past instances of A (or precursors of A) and showing how these *effects* increased the adaptedness of A's possessors (or the possessors of A's precursors) and so led to the evolution of A. . . . Put cryptically, trait A's existence is explained in

consequence-etiology, Achinstein objects: "Does the fact that my heart pumps blood causally explain how my heart came to exist...? How will natural selection lead causally from the fact that my heart does pump blood to the fact that it exists...?" But Wright makes it clear that even where what is being explained is the presence of a particular organ in a particular individual, the effects appealed to in a consequence-etiological account are not the actual effects of that particular organ, but the *types of effect* the organ-*type* has, past instances of which have led to the organ-type's being selected—i.e.to its being programmed for in the genetic makeup of the species, thus explaining the presence of instances of it in current specimens, including this particular individual.

[8] Explanation in terms of "ultimate [i.e. remote] causes" involving appeals to natural selection was identified by Mayr (1965), for example, as a necessary complement to explanation in terms of "proximate causes", both being vital on his view to complete understanding in biology.

> terms of what A *does*. More fully, A's existence is explained in terms of *effects* of *past* instances of A; but not just any effects: We cite only those effects relevant to evolutionary explanations.[9]

This is generally referred to as an "etiological" view of functional teleological explanation, following Wright's terminology. The claim is that functional teleological explanation is "consequence-etiological" explanation, i.e. remote causal explanation involving an appeal to certain effects or consequences of the type of thing in question, as described above, thus providing a "consequence-etiology" of the thing in question. As Wright puts it, "consequence-etiologies are the hallmark of teleology"; "the essence of teleology is the consequence-etiology", which in biology happens to be brought about through natural selection.[10] On this view, a functional teleological explanation can be put "cryptically" (as Brandon puts it) by speaking simply of what the trait in question *does*, i.e. what adaptively significant effect it has; but fleshed out, this is really just part of an evolutionary account of how the trait came to be present in current specimens, in terms of what past instances did.

Before going on to criticize this general view of functional teleological explanation, it is worth examining it in a little more detail, to bring out some interesting differences between Wright's and Brandon's versions, for example. First of all, Wright holds that biological function is *as a matter of fact* to be understood in terms of natural selection, but his general consequence-etiological account of function is deliberately formal, and so is perfectly compatible with some other mode of selection, such as conscious selection by a divine designer, which would equally provide a consequence-etiology.[11] His account thus differs from those of Ayala and Brandon. Ayala formulates his account directly in evolutionary terms, and Brandon

---

[9] Brandon (1981, pp. 102–3), my emphasis.

[10] Wright (1976, pp. 96, 116).

[11] Wright (1976, p. 97). Bennett (1976, p. 79) adopts this same formal approach to biological function, allowing for the possibility of grounding biological functional teleology in intelligent design, while holding that in fact it is grounded in evolution. One of Wright's motives for keeping his account of biological function formal in this way, rather than tying it essentially to evolutionary theory, is his desire to provide a unified theory of function, offering one formal consequence-etiological account for both artificial and biological functions. I say a little more about design-based consequence-etiology below.

explicitly takes his own account to be tied to neo-Darwinian evolutionary theory, disavowing any attempt to provide a more abstract, theory-independent etiological account of biological function.[12] So for Brandon (and at least implicitly for Ayala), biological function is essentially to be understood in terms of "adaptive significance", whereas for Wright this is only contingently so: In another possible world it may be understood in terms of divine purposes.

A second point concerns levels of explanation. Teleological explanations are, on Wright's view, consequence-etiological accounts of the presence of the item in question. But there are various levels of elaboration for such accounts, and it is a mistake to think that an account has explanatory value only if we can supply all the details of the etiology.[13] For example, we may be fairly certain that Z is the effect-type that explains why some trait A is there—and hence, on Wright's view, we would be fairly certain that Z is the function of A—though we are very much up in the air about the details of the particular selection background at work.[14] In such a situation, we might explain the presence of A by citing its function Z, without committing ourselves to any particular account of the underlying details. The consequence-etiological explanation at this level is equally consistent with a theological or an evolutionary account of the selection of A in connection with Z (i.e. an account of the conscious selection of A with a view to its doing Z, or an account of the natural selection of A due to its tendency to do Z); likewise, it is consistent with a variety of different theses about the precise workings of evolution, some of which can be expected to change as evolutionary theory progresses. This sort of explanation is what Wright calls a "function ascription-explanation" or a "functional explanation". Because an attribution of function offers "a consequence-etiological account of the existence or form of the thing

---

[12] Ayala (1970), Brandon (1981, p. 91).

[13] Wright (1976, p. 108). The same is true for non-teleological causal explanations: "Causes can be objectively determined without knowing much about the exact structure of causal laws governing the phenomenon. [For example,] Hertz had established the causal role of ultraviolet light in his photoelectric effect experiment long before the theoretical analysis of that effect was elaborated."

[14] On Wright's view, "the *function* of A is that particular consequence of its being where it is which explains why it is there"—in the sense of figuring into a consequence-etiological account of how A came to be there (1976, p. 78).

with the function," even if only at a certain general level, it is said to qualify as a teleological explanation.[15]

So according to Wright, we can and often do answer a teleological `why?' or `what for?' question with a *functional explanation*—something of the form "the function of A is to do Z", or just "A does Z" (where this is understood as a function ascription). The term "function" in the more explicit statement serves to indicate that the effect mentioned (Z) is being offered as the effect-type that explains how the thing in question (A) came to be there, by virtue of figuring into some consequence-etiological account—the details of which are left unspecified, just as the details underlying causal explanations are almost always left unspecified in everyday causal explanations. Similarly when the formulation of the explanation doesn't contain the word 'function', but makes use of a teleological connective, as in "A is there *in order to* do Z": The teleological connective simply serves as an indicator that Z is the effect-type that figures into a consequence etiological account of A. On Wright's view, there needn't be anything "cryptic" about such general explanations (as suggested by Brandon in the passage quoted above). Citing the relevant effect is perfectly acceptable as one level of teleological explanation. Indeed, Wright maintains that functional explanations are even preferable to fully specified etiological accounts to the extent that the former "are more certain than any underlying accounts of them because they are consistent with all such accounts".[16] Still, assuming that neo-Darwinian evolutionary theory is roughly correct, Wright is committed to the claim that evolutionary explanations are in fact to be understood as the fleshed out, most informative versions of the teleological explanations of functional items in biology.

---

[15] Wright (1976, p. 91). Establishing a functional explanation thus "merely indicates the presence of a selection background of some kind and it leaves an enormous amount of theoretical detail completely open. The exact physical mechanism, the precise details of the selection process, in principle even the type of selection (natural or conscious), are not determined merely by establishing a functional ascription-explanation." Wright (1976, p. 109).

[16] Wright (1976, p. 110). Note the desirable implication that functional explanations offered in pre-Darwinian times by those who had no knowledge of natural selection may very well have been more or less correct at this level. Rosenberg (1985, pp. 46–7) illustrates this fact nicely in a discussion of Harvey's discovery of the function of the heart, which was a well-founded improvement on previous theories.

A third point concerns the tense in which teleological explanations are cast. If a detailed evolutionary account is available, is there any reason to put the teleological explanation in terms of what A *does*, rather than simply putting it in the past tense, saying that "A is there because in the past As have done Z"?[17] Wright claims that there is, primarily because an evolutionary explanation cast wholly in the past tense, of the form Brandon offers, fails to discriminate between functional and vestigial organs. A consequence-etiological account, in a broad sense, can be given for both the kidneys and the appendix: Both are there because of something past kidneys or appendixes *did*; but only kidneys are there because of what they presently *do* (i.e. "because they do what they do"), and so according to Wright only they can be said to have *present functions*, and to be there *in order to* perform those functions.[18] Wright argues, therefore, that it is preferable in giving teleological explanations of things with present functions to put the etiological account in terms of what they *do*, rather than simply in terms of what their ancestors did. Instead of saying simply that A is present because ancestor As *did* Z, and Z had advantageous effects on reproduction, etc., we should say that A is present because As *do* Z, and Z had advantageous effects on reproduction, etc. The latter indicates that Z is the present function of A, i.e. that A is present *in order to* do Z, rather than being vestigial.[19]

This third point deserves some comment. First of all, it should be conceded that there is nothing objectionable about formulating consequence-etiological explanations in terms of what As *do*, as

---

[17] Wright (1976, p. 89).

[18] Ibid. As noted in chapter one, there is some unclarity in the use of the expression "consequence-etiology", and that earlier footnote bears repeating. In a broad sense, something might be said to have a consequence-etiology if its presence is causally explained in part by certain effects of that type of entity, whether or not the effects still occur; in a more restricted sense, something would be said to have a consequence-etiology only if that type of entity *still has* the effects that figure into the causal history leading to its presence. In the broad sense, which I employed in the text above, both the kidneys and the appendix have consequence-etiologies; in the more restricted sense, only the kidneys do, though at one time the appendix did as well. On Wright's view, it is only when something has a consequence-etiology in the more restricted sense that it can be said to have a present function.

[19] For example, the presence of chlorophyll is "explained functionally by pointing out the role chlorophyll plays in enabling plants to perform photosynthesis; it answers a 'why?' question by providing a perfectly respectable etiology; it provides the reason chlorophyll is there." Wright (1976, p. 91).

opposed to what past As *did*, so long as the claims about what As do are understood to imply corresponding claims about what past As did. Often this is perfectly clear from the context, and need not be at all "cryptic", as Brandon implies; we simply understand from the context that the claim that *As do Z* is elliptical *not* merely for "present As do Z", but for "As typically do Z", where this in turn is understood to imply *both* that present As do Z and that past As did Z. But it is important to recognize that such use of the present tense is of absolutely no significance to the *etiological explanation* except insofar as it implies the *past* tense claim, as indicated. It is obviously the latter that does the real work in this sort of historical-causal explanation. So while Wright may be *permitted* this use of the present tense, without being criticized for being unduly cryptic, it is surely misleading of him to insist on its importance, claiming, for example, that the kidneys—unlike the appendix—*are there because of what they do*. The only thing that is relevant to the kidneys' etiology, and hence to this 'because', is what past kidneys did—just as with the appendix—and the fact that kidneys still do it is entirely beside the point as far as the etiology is concerned.

There is, of course, a sense in which it can be said that the kidneys, unlike the appendix, are there because of what they do: The kidneys are there because of certain things past kidneys did, and these things are the same types of things that present kidneys do. But again, this does not in any way suggest that the fact that kidneys presently do these things is relevant to the etiological explanation; it is not, and that is why it is misleading to say that kidneys, unlike the appendix, are there because of what they *do*. Both are there because of what past instances did. So at best Wright's recommendation may be taken as a suggestion for a certain convenience: If we *can* put an etiological account in terms of what A does, we may as well, since in doing so we indicate (according to this view) that this is a present function of A. But nothing in fact hangs on this. The bottom line is that consequence-etiological explanations can be given for appendixes as well as for kidneys.[20] The difference, which is external to the etiology, lies only in whether the consequences appealed to are *still* associated with the things of the type in question; only if they are, on Wright's view, will the consequence-

[20] That is, in the *broad* sense of "consequence-etiological explanation" discussed in an earlier footnote.

etiological explanation be a functional explanation and legitimize claims about present function, or claims that the things are there *in order to* perform these functions. This, at any rate, is what Wright is committed to saying.

As for Wright's stronger claim that it would be positively *misleading* to put consequence-etiological accounts of things with present functions entirely in the past tense (as Brandon would), this rests on a confusion. He offers a parallel involving non-teleological causality: "The Titanic sank because when you tear a hole that size in the bow of a ship *it sinks*". His point is that it is perfectly appropriate to put the latter occurrence of the verb 'to sink' in the present tense, since the causal principle is one that presently holds; indeed, it would be misleading to put it in the past tense, as that "would imply that nowadays one could get away with tearing a hole that size in the bow of a ship without it sinking".[21] This is true, but the problem is that the parallel doesn't work for his purposes. First, the statement should be filled in to make the parallel more clear: "The Titanic sank because a big hole *tore* through the bow, and when you tear a hole that size in the bow of a ship *it sinks*". Notice that the italicized occurrence of the verb 'to tear' is in the *past* tense; it occurs here not as part of the statement of a timeless causal principle (as the second, non-italicized, occurrence does) but rather in the statement of the occurrence of a past event. Now the parallel for the etiological account of trait A would be this: "A was selected and so is present because past As *did* Z, and doing Z *is* reproductively advantageous for the organism. . . . etc." So Wright's parallel fails to show that we must speak in terms of what Xs *do* as opposed to what they *did*; all it shows is that if Z is still advantageous, we should perhaps put *that* statement in the present tense to avoid confusion.

More important than any of the above differences among etiological accounts, however, is what they all have in common. The essence of what I am calling "etiological accounts" of functional teleological explanation in biology is their construal of such explanation as nothing but a certain kind of historical-causal explanation of how instances of attribute-type A came to be present in current Fs—namely, a remote-causal explanation that offers a consequence-etiology. Proponents of etiological accounts generally

[21] Wright (1976, p. 90).

agree that in the actual world, at least, fleshed-out teleological explanations in biology are evolutionary explanations, accounting for adaptations in terms of natural selection. And the only causality such an explanation involves is *efficient-causality*—in particular, remote efficient causes, as becomes clear when the explanation is put in terms of the past token effects of past instances of the thing in question, and the token effects these in turn had on reproductive success.

The attraction of construing teleological explanation in biology in such a way that it is at bottom just evolutionary explanation is obvious. It involves no appeal to intentions, either in or behind biological phenomena; it is not committed to temporally backwards causes; it does not posit any non-empirical biological entities or forces; and it requires no commitment to any form of causation that is not already perfectly familiar and scientifically respectable: The 'because' (as in "A is present in Fs *because* A does Z") is just the 'because' of the sort of evolutionary account that has long since been accepted as fundamental to biology. The etiological account thus offers us a tempting vindication of teleological explanation in biology, achieved by reducing it to a scientifically respectable form of efficient-causal explanation.

3. A NON-REDUCTIONIST ALTERNATIVE ACCOUNT FOR FUNCTIONAL SYSTEMS

Let us grant for the sake of argument that proponents of the etiological view have indeed identified a legitimate and interesting form of question and causal explanation. The question, however, is whether it is really plausible to suppose that *this* is what *functional teleological* questions and explanations are all about. Those who think so assume that when we ask, for example, "why does the dry fruit of maples have a papery, wing-like structure?", we *must* ultimately be concerned either with:

(i) *how the dry fruit of maples comes to have such a structure*, i.e. how such a structure develops in the course of the life of a maple tree; or:

(ii) *how the dry fruit of maples came to have such a structure*, i.e. how such a structure has historically come to be characteristic of maple fruit.

That is, we might be asking either for a mechanistic explanation in terms of relatively proximate causes, or for a consequence-etiological explanation in terms of remote causes involving certain of the trait's typical effects, against some sort of selection background; and if it is the latter, then according to this view our question is the teleological one, equivalent to a "what for?" question and answerable in teleological language.

This, however, is hardly a natural interpretation of the "what for?" question. When someone asks "why does the dry fruit of maples have a papery, wing-like structure?", and means this in the *teleological* sense—where it is a way of asking "what for?", equivalent to "what function is served by the fruit's having a papery, wing-like structure?" or "what's its biological point?"—she does not seem to be asking to be told *how the trait came to be present* in maples. Indeed, she does not seem to be asking an *historical* question at all. Likewise, when someone provides a teleological answer to such a question, saying e.g. "it has this structure *for the sake of* slowing its descent when it falls, thereby aiding wind dispersal of the seeds", it is hardly natural to interpret this as being intended primarily to explain *how the trait came to be present* in maples.

It may, of course, be true that the effect that is appealed to in the teleological explanation, in response to the teleological question, *also* figures into an evolutionary explanation of how the trait came to be present; in fact, I concede that it does in such cases. It may even be true that there is some important connection between the role of this effect in the causal history of the trait and its ability to figure into a correct teleological explanation—something which will come up below. But even if this is so, it does not follow that this is its explanatory role in the context of the teleological explanation. It is still possible that the appeal to the effect in the teleological explanation, addressing the teleological question, has a distinct kind of explanatory force, not reducible to that of efficient-causal explanation.[22]

To see how this might be so, let's return to the possible interpretations of questions about adaptations. Is there another way of understanding the question "why does the dry fruit of maples have a papery, wing-like structure?" than as a request for some kind of efficient-causal explanation of how that trait comes or came to be

---

[22] Cf. Achinstein (1983, p. 289).

present in maple fruit—a way that plausibly captures the teleological sense of the question? Or again, to take a slightly different sort of example, is there such a way of understanding the question "why do we sweat when hot?", according to which what is being requested is neither an account of the mechanisms involved, nor an account (appealing to certain typical effects of sweating) of how the disposition to sweat when hot came to be manifested in human beings, but something else, that properly addresses the question understood as a question about what sweating is *for*?

I believe that there is, and that the key to understanding this is to approach a living organism in the manner proposed in chapter two, considering it as a system that can non-arbitrarily be said to *work* in a certain way, just as a machine can, with parts and features that make non-incidental contributions to that working. Let me set aside the biological case for a moment, and illustrate the points first in connection with artificial systems that have these same general properties. My central claim is that when we are dealing with such a system, we can raise legitimate and interesting questions about its parts, features and activities with a view simply to understanding their non-incidental roles in the working of the system. What we are after here is what might be called the *functional teleological significance* of the thing in question, as part of a broader understanding of *how the system works*, rather than any kind of causal account of how the thing came to be present or to occur.

Consider again the example of the engine, from chapter two. Someone who understands that a gasoline engine is such a system, but is unfamiliar with certain details of its working, might ask of the inlet valve "why does it move in and out periodically?" (referring to its opening during the induction stroke and then closing again). Let us suppose that he already understands what *causes* the valve to open and close—namely, the motion of the rocker arm, which in turn is caused by the pushrod's being moved by the valve-lifter, and so on. His question, he assures us, is different: He wants to know what *function* the opening of the valve serves—what the *point* of it is in the working of the engine, what it happens *for* in that context. And the natural answer to the question, so understood, is therefore something such as: "It opens *(in order) to* admit the fuel/air mixture into the cylinder, setting up the conditions for combustion." Admitting the fuel/air mixture is *what* the opening of the valve *does*, in the functional sense of the non-incidental role it plays in the working of the engine.[23]

Now again, it may be true that this effect of the opening of the inlet valve can *also* be appealed to in a remote causal explanation of how it is that inlet valves that open during the induction stroke *came to be present* in gasoline engines. That is, it was with this effect in mind that designers had inlet valves put in engines, set up in such a way as to open during the induction stroke. (This is an artificial analogue of consequence-etiological explanations in terms of natural selection. I will give two more examples, from Wright and Cummins, below.) But this is no reason to suppose that *this* is what the teleological question and explanation are directly about. Indeed, it seems plain that their concern is not with this historical matter as such, but with the *abstract present* matter of what non-incidental role the opening of the inlet valve plays in the working of the gasoline engine—even though this fact has been importantly shaped by the historical facts, as argued in chapter two. The effect, i.e. admitting the fuel/air mixture, sheds light on the matter addressed by the teleological question not by providing some sort of remote causal explanation of the phenomenon, but by

[23] Achinstein (1983, p. 288) has also argued for the existence of such a distinct form of question and explanation, not reducible to efficient-causal forms. My general view of functional teleological explanation is closest to his, though I focus on developing it for functional systems of the sort discussed in earlier chapters—i.e. systems, whether machines or living organisms, that can be said in a *non-arbitrary* sense to work in a certain way, with parts and features that play *non-incidental* roles in that working. Two views that might at first seem close to mine but are in fact very different are those of O'Grady (1986) and Hempel (1965). O'Grady rightly emphasizes that teleological "what for?" questions are non-historical, and even draws a parallel between an anatomist's functional approach to an organism and a mechanic's functional approach to an engine, as I am suggesting here. His view is very different from mine, however, inasmuch as he maintains that functional teleological claims in biology are merely metaphorical: They are just useful heuristic devices that essentially involve treating the organism "*as if* it were designed by an intelligent agent", so that teleological explanation in biology is "only an explanatory analogy" (1013). I have tried to show that such backing off from literal teleological discourse in biology is quite unnecessary and misguided; talk of design and intention in biology is indeed metaphorical, but talk of functions and ends (and the use of related teleological expressions) need not be. With regard to Hempel, while he does make an appeal to systems and their "proper working order", his account of function fails to distinguish incidental from non-incidental contributions to proper working, and so fails to provide the means for drawing the crucial function/accident distinction. He also fails to provide an account of what determines proper working in the case of a living organism, as I have done in earlier chapters. Most importantly for the present chapter, his "deductive-nomological" approach to functional explanation is completely different from the sort of account I am presenting here, as will become clear in section six.

contributing to a teleological understanding of it within the system in question, adding to our understanding of the abstract present matter of *how such a system works*. What we have here is a familiar form of puzzlement and understanding that seems irreducibly teleological. The attempt to analyze it by construing it in historical-causal terms results simply in turning it into something else—a different form of puzzlement and understanding that also has its place, but is not the distinctively functional teleological form we set out to investigate.[24]

If we remain sensitive to this form of question and explanation, it is not hard to hear even such a question as "why do gasoline engines have spark plugs?" as naturally asking a teleological question. That is, it can be construed as asking "what (non-incidental) role do spark plugs play in the working of the gasoline engine?", where this is not reducible to the historical question about agents' reasons for acting, i.e. "why were spark plugs put in engines—i.e. in view of what effect were they put there, explaining how they have come to be present in gasoline engines?" The same point holds for the question "what are spark plugs designed to do in gasoline engines?", and even for the question, asked of a particular set of spark plugs: "What were these parts designed to do?" For even the latter, which superficially appears to be an historical question about particulars, can just be a way of asking a question that is neither historical nor about particulars as such, but a teleological question about a certain type of thing. For example, after we identify the functional type of which the particular parts in question are instances, telling the questioner that "those are spark plugs", he might naturally rephrase his question in terms of what spark plug*s* are designed to do, or what *the* spark plug *does*, indicating that he was never concerned with the consequence-etiology of the *particulars*; perhaps it is still possible to interpret him as being concerned with the consequence-etiology of spark plugs collectively, but it is equally

---

[24] A certain qualification is necessary here. I conceded in chapter seven that my account of functions within working systems is reductionistic in a certain sense: Teleological concepts can be explicated in non-teleological terms, and teleological facts supervene on a complex of non-teleological facts. In a sense, then, a teleological explanation is in principle analyzable in terms of such a complex of non-teleological facts—though its distinctive explanatory force would likely be lost. In any case, my claims in the text pertaining to the irreducibility of functional teleological explanations, are concerned with a different issue. The point is that functional teleological explanations are not reducible to consequence-etiological explanations, or to any other form of efficient-causal explanation.

possible, and more natural, to interpret him instead as being concerned with a teleological understanding of such things in such a system.

These distinctions between etiological questions and explanations, on the one hand, and what I am claiming to be genuinely teleological questions and explanations, on the other hand, seem just to be overlooked by proponents of etiological accounts. Consider the following passage from Wright:

> The sweep hand of my watch may keep the numbers free of dust, but that is not *why it is there*: so that is not its function. It's function is making seconds easier to read: that is *the reason it is there*. And this sense of . . . 'why', and 'reason', this notion of *explanation*, is plainly etiological. It concerns *how* the thing with the function *came to be there*.[25]

Now it should be clear from what I have said in earlier chapters that I do not deny the relevance of etiology to function in artificial and biological teleology. Indeed, I have stressed that it has everything to do with the establishing of non-accidental relations among the parts and features of a working system and certain of their effects, which latter may thereby properly be regarded as relevantly non-incidental, which is crucial to their being genuinely functional. The difference between my view and Wright's is this: I acknowledge that etiology is relevant to determining present functional facts about working systems, but then recognize functional teleological questions and explanations as irreducibly teleological questions and explanations about systems and their functional parts and features—questions and explanations in the abstract present, that are clearly distinct from etiological questions about causal history. Wright, by contrast, does not appear to recognize that the sort of etiological background on which he focuses gives rise to

[25] Wright (1976, p. 78), my emphasis. His idea is that the sweep hand *came to be in the watch* because someone put it there for the purpose of making seconds easier to read, not for the purpose of keeping dust off the numbers; so the effect of making seconds easier to read comes into the explanation of *how it came to be there*—for it was by virtue of this effect that the sweep hand was *selected* for inclusion in the watch—whereas the other effect does not. (Again, this is an artificial analog of a consequence-etiological explanation in terms of natural selection.) Wright therefore identifies this effect with the function, and identifies the functional explanation of *why the sweep hand is there* with the historical-causal explanation of *how it came to be there* which makes appeal to the effect in question. It is the latter of these moves that is relevant to the present discussion.

such a form of question and explanation that is distinct from the etiological form of question and explanation; as the above passage makes clear, the only sense of the question "why do F's have A's?" that Wright seems to recognize is "*how* did F's *come to have* A's?", and similarly with the other related kinds of questions that I have claimed have distinct teleological interpretations.

Cummins shares Wright's tendency to hear functional teleological "why?" or "what for?" questions, i.e. questions that are answered by giving the thing's function, only as *etiological* questions—though unlike Wright he thinks such questions and explanations are properly restricted to artifacts.[26] He considers the following exchange: "Why is that thing there (pointing to the gnomon of a sundial)"—"Because it casts a shadow on the dial beneath, thereby indicating the time of day". On his view, this answer, which gives the thing's function,

> explains the presence of the gnomon because it rationalizes the action of the agent who put it there by supplying his *reason* for putting it there. In general, when we are dealing with the result of a deliberate action, we may explain the result by explaining the action, and we may explain a deliberate action by supplying the agent's reason for it. Thus, when we look at a sundial, we assume we know in a general way *how the gnomon came to be there*: Someone deliberately put it there. But we may wish to know *why it was put there*. Specifying the gnomon's function allows us to formulate what we suppose to be the unknown agent's reason for putting it there: He believed it would cast a shadow such that . . . , and so on. When we do this, we are elaborating on what we assume is the crucial causal factor in determining the gnomon's presence, namely a certain deliberate action. [The crucial assumption is] that the thing functionally characterized is there as a result of deliberate action.[27]

---

[26] Cummins (1975, p. 750) rejects the etiological account of *biological* functional teleology offered by Ayala, Brandon and Wright on the grounds that it "involves a subtle yet fundamental misunderstanding of evolutionary theory." In fact, his objection seems to me to be based on a misunderstanding of their position, and I do not find it persuasive. But this need not concern us here. My contention is that even if their position regarding the possibility of consequence-etiological explanation in biology is correct, it fails to account for teleological explanation.

[27] Cummins (1975, p. 746), my emphasis. This is another illustration of the artificial analogue of natural selection on Wright's view—i.e. "conscious selection".

Notice first of all the focus on a particular instance of the type of thing in question. There is nothing wrong with asking "why is it there?", pointing to a particular part; but as noted in the spark plug case considered above, if this is a functional teleological question, it must be understood as being not about this particular part *as* a particular, but as being about either this *type* of thing, or this particular *as* an instance of that functional type.[28] This is clear from the fact that, as in the spark plug case considered above, if the person asking the original teleological question "why is that thing there?" were told, as a start, that the thing he is pointing to is called a "gnomon", his question would be rephrased not as "what is this *particular* gnomon for?", or "what function is served by this *particular* gnomon?", but as "what is *the gnomon* for?" or "what is the function of *the gnomon* on *a* sundial?"

But now consider how far this question is from Cummins' question "*how* did this particular thing *come to be there*?", or more specifically "why was this particular thing put there?" It's not that such questions can't be asked. They can, of course, and his answer may be perfectly

---

[28] This is true, I believe, even where there is only one instance of the type of thing in question, as in the case of a singular invention. (Cf. Woodfield's (1976, p. 112) case of an inventor of a gadget who stipulates: "The function of that component is to vibrate the spindles".) As argued in chapter five, section eight, if we can really speak of a function or role played by something, the something in question is primarily a functional type of thing, and only derivatively the concrete item of that type, even if it is the only one in existence. For when we speak of *what it does*, or *the role it plays*, we are not making a claim about what the concrete particular actually does (as it may be defective and fail), but a claim about a functional *type* of thing, which could in principle be instantiated by other relevantly similar particulars. We are interested in what *the F* does—as opposed to what this particular concrete F does, even if this is the only F actually in existence—and this is what allows us to speak derivatively about what the concrete particular is *supposed* to do, or about its being *defective*, etc. Someone may argue against this that concrete particulars can possess functions even *as* concrete particulars, simply by virtue of an inventor's intentions for them (at least if they are moderately successful at doing what they were intended to do). But even if we grant the importance of the inventor's intentions, it is not obvious why the items should be thought to possess their functions *as* particulars. For a particular item falls within the domain of functional discourse not simply *as* e.g. a token bit of metal and silicone, but *as* a "Z-er" (for some function Z), of which others could in principle be made, even perhaps of different materials. And it is judged in functional terms according to how well it fills the role played by *such* a thing, not simply *as* a particular. It is thus plausible to suggest that by intending this part to have a certain function in the machine, the designer made it to be an instance of a certain *type* of thing (at least if he was moderately successful), and its function belongs to it *as* such a thing—whether or not the type happens to have any name. However this may be, it does not at any rate affect biological cases, where the problem of singular functions does not arise.

legitimate. But to say that something was put somewhere by someone *in order to* do something is to make a *psychological* teleological claim, not a functional one (regardless of how the two may be related). We are interested not in why someone in the past did something that resulted in the presence of a particular object, or even a set of particular objects (i.e. all the actual gnomons in all actual sundials), but in what purpose or function or end is served in a sundial by the gnomon. The answer to the original teleological question is not a historical story of the sort Cummins proposes, but simply this: "That thing is a gnomon, and the gnomon casts a shadow on the dial beneath, thereby indicating the time of day: That's how a sundial works". In other words, this is the gnomon's *function* in a sundial, what it is *for*: It casts a shadow on the dial beneath, *in order to* indicate the time of day, which is the *function* of a sundial. To reach beyond this for the consequence-etiological explanation of the presence of the particular gnomon is simply to miss the point of the teleological question. The teleological question has been answered once the functional teleological significance of the gnomon (and hence derivatively that of this particular gnomon as an instance of that functional type) has been made clear.

Let's return now to biological contexts. Given the preceding discussion, together with the arguments from earlier chapters elucidating the nature of organisms as working systems, analogous in important respects to machines such as the engine considered above, we should equally have available a non-reductionist account of functional teleological explanation in biology. And I think this again provides a more plausible construal of such discourse than that offered by the etiological view. Just as one can legitimately ask an abstract present, functional teleological "why?" or "what for?" question about a part, feature or activity of a machine, so too one can ask such a question of an organic part, feature or activity. What is sought here is not a causal explanation of how the entity comes or came to be present or to occur, but an account of the non-incidental role such an entity plays in the working of such an organism—an account of its biological function, which brings out its teleological significance within the system. Thus, for example, when we ask the question about maple fruit in the *teleological* sense, what we want to know is what proper function this feature of the fruit plays in the characteristic life of the maple. And this is naturally answered by citing the non-incidental role it plays in the working of such an organism: The fruit has such a wing-like structure *in order to* slow the descent of the fruit when it falls, thus aiding wind-dispersal of the seeds. This use of a teleological connective, such as "in order to", in such an explanatory context is perfectly natural, unlike its use in the context of a consequence-etiological (i.e. historical-causal) explanation, which as noted earlier just seems out of place.

Etiology is certainly relevant to biological teleology on this view, but its significance is very different from that accorded to it on standard etiological views. It is true that we can properly understand what working is for a given type of organism only by considering its natural selection history, which reveals the non-accidental relations that obtain among organic parts and features, their organization, and the hierarchy of effects they produce.[29] But functional teleological explanations in biology are not etiological explanations—i.e. remote efficient-causal explanations of how traits came to be present, any more than in artificial contexts. They are rather irreducibly teleological explanations that shed a different sort of light on functional entities within organisms understood as functional systems, just as in the case of machines; and they are no less respectable than efficient-causal explanations—no more involving appeals to backwards causation or mysterious "life forces" than consequence-etiological explanations.

Whereas etiological accounts of function are often motivated by the confusion of consequence-etiological explanation and functional teleological explanation, my earlier appeal to etiology in the account of working systems is therefore obviously not so motivated. In fact, that account lends itself perfectly to the above non-reductionist view of functional teleological explanation, unlike etiological accounts of function, so that if the present view of teleological explanation is correct, this provides further support for my account of function over etiological accounts, as discussed in chapter seven.

---

[29] This is, of course, related to *one* of Wright's (1976, p. 92) rationales for appealing to etiology, i.e. the fact that it seems crucial to a non-arbitrary distinction between functions and accidents, or ends and mere effects. On my view, however, it comes in less directly, insofar as it conditions distinct, present teleological facts about functional systems, in relation to which functional facts about parts and features are to be understood.

4. SHIFTING QUESTIONS: CUMMINS' "FUNCTIONAL ANALYSES"

It should be stressed that even though a teleological explanation, on my view, is not an explanation of *how* the part or feature *came to be present* in the system by citing its proper function, there is still a distinct sense in which, *as* a teleological explanation, it accounts for the presence of the part or feature in the system by citing its proper function. That is, it brings out the functional teleological significance of such a part's or feature's presence in such a system—its proper role in the working of such a system—and this is a perfectly respectable example of removing a natural form of puzzlement one may have about it given the assumption that this is a working system and that the part or feature in question has *some* proper function relevant to it. My disagreement with etiological accounts is not over *whether* appeals to functions are explanatory of functional items, but over the nature of the explanation they provide *as* functional teleological explanations.[30]

Notice, therefore, that while my focus is on the non-incidental role played by a part, feature or activity in the working of a system, I have not just abandoned the sort of question we began with in favor of a different sort of question about the containing system, as others who reject the etiological view have done. For example, I have not proposed any shift from questions such as "why do gasoline engines have spark plugs?" or "why does the plug make a spark?", to questions such as "how is combustion made possible in gasoline engines?", or "why do engines with spark plugs carry out combustion, whereas those without them do not?".[31] The latter are legitimate questions one could ask, but they are distinct from the questions that we originally set out to understand—i.e. "why?" or "what for?" questions about parts, features or activities that are answered by appeal to their proper functions. Rather than making such a shift to a different sort of question, I have simply given the original "why?" or "what for?" questions a different

---

[30] Cf. Achinstein (1983, pp. 285–90), who notes that an *end for the sake of which* something exists is "one type of reason that, or for which, something can exist."

[31] Cummins (1975) makes the first sort of shift in his account of functional explanation, which he calls "functional analysis"; I will say more about this just below. Mitchell (1993) criticizes Bigelow and Pargetter (1987) for making the (closely related) second sort of shift—moving, for example, from the question "why do peppered moths have dark coloration?" to the question "why do dark moths in that environment reproduce more successfully than light ones?".

construal from that given by proponents of the etiological view—a construal that better accords with our use of teleological language.

There are, of course, circumstances in which these other questions are closely related to the questions on which I have focused. For example, if we are dealing with a genuine working system (as in the above example), we can ask *how* certain effects are made possible or brought about as part of asking how the system works, and this will involve an appeal to functional parts and their functional effects as such. Thus, if such system-related "how?" questions are restricted to functional effects in genuine working systems, as in the case of combustion in an engine, they will at least be closely related to functional questions as I construe them. But if they are not, then they will be very far removed from functional questions as I understand them, and any account of functional explanation built around them will be very different from the sort of account I have offered.

The prime example of this is Cummins' account of functional explanation, which is often cited as a paradigm alternative to etiological accounts.[32] On his view, the characteristic way in which "functions" enter as such into explanations is in what he calls "functional analysis" or "function-analytical explanation." This has to do with explaining (or analyzing) the capacities or activities of a system in terms of the capacities or activities of its parts or properties, so that the relevant question is of the form: "How does system S manage to display capacity X or to carry out activity Y?", i.e. "By virtue of what other contributing properties and capacities is S able to manifest X or Y?" The capacities or activities to be analyzed can in principle be *any* capacities or activities the system happens to have or to carry out as a result of its possession of certain parts or properties that contribute to them, and there is no concern with whether or not the contributions

---

[32] Cummins (1975). Cf. also Prior (1985). As noted above, in the case of *artifacts* (such as a gnomon on a sundial), Cummins does not deny that something's presence may be explained etiologically by appeal to its function; but he does not take this to be the sort of explanation characteristic of uses of functional language, and in fact thinks that such appeals to functions in etiological explanations are quite foreign to biology. I will not go into his reasons for rejecting the possibility of function-citing etiological explanations in biology (p. 750), since as already mentioned, I believe his objections stem from a misunderstanding of the role of natural selection on the etiological view. What I am interested in here is just the way in which he shifts the characteristic question associated with function from "why?" (or "what for?") questions about functional parts or features to certain "how?" questions about higher-level capacities or effects of containing systems.

made by the parts or properties are anything more than incidental. That is, neither the capacities or activities being explained, nor the means whereby they are exercised, need have anything to do with the system's *working* in any non-arbitrary sense. The only qualification Cummins introduces is that:

> the *explanatory interest* of an analytical account is roughly proportional to (i) the extent to which the analyzing capacities are less sophisticated than the analyzed capacities, (ii) the extent to which the analyzing capacities are different in type from the analyzed capacities, and (iii) . . . the relative complexity of the organization of component parts/processes that is attributed to the system.[33]

Now the notion of function Cummins is operating with here—i.e. what something *functions to do* or *functions as* within a containing system, helping to explain higher level capacities or effects that we might happen to take an interest in analyzing—is something much weaker and broader than the notion of proper function, as discussed in chapter one. It is therefore not relevant to an understanding of *functional teleological* explanation, which has to do with appeals to *proper* functions, as such, in explanations.[34] As Kitcher points out, on Cummins' view there would be "functional analyses" associated even with what are ordinarily recognized to be obvious cases of *malfunctioning*. For example, there would be a Cummins-style "functional analysis" of the complex higher level capacity of a given organism to form tumors, in terms of the lower level capacities of a certain mutant DNA sequence, by virtue of which it "functions to" produce tumors.[35] Such a "functional analysis" would even satisfy the above three conditions and hence have full blooded explanatory interest according to Cummins. On his view it would thus allow us to ascribe—in a full blooded manner—*functions* to the mutant DNA in connection with their tumor causing capacities. Again, this clearly does not accord

---

[33] Cummins (1975, p. 764), my emphasis.

[34] This point has been made by Millikan (1989, p. 294). Talk of "function" in Cummins' broad sense applies in principle even to the clouds' producing rain within the water system, which helps to explain *how the soil is kept moist*—though Cummins would probably say that in any case such a "functional analysis" would have little "explanatory interest". The example that follows precludes even this response.

[35] Kitcher (1993, p. 390).

with the notion of proper function, and however interesting Cummins' "functional analyses" may be, they are not what we are looking for.

In summary, then, while I agree with Cummins in rejecting the etiological construal of teleological questions and explanations, my positive view is very different. First, he shifts the standard functional question to a different sort of question altogether, rather than just construing it in a different way, as I do; that is, on his view "what we can and do explain by appeal to what something does [i.e. its "function"] is the behavior of a containing system," rather than providing any sort of account of the presence of the thing to which the function is ascribed, as is done (in very different ways) on the etiological account and on my account.[36] Second, the sort of question and explanation he identifies applies much more broadly than to cases involving proper functions. What we purportedly explain, on his view, are any higher level effects we might be interested in, in terms of lower level properties that contribute to bringing them about, even if in a completely incidental way, having nothing to do with anything that could non-arbitrarily be considered the proper working of the system.

There is thus nothing in Cummins' view comparable to accounting for a part, feature or activity by shedding light on its functional teleological significance within a system—something which I have argued does indeed answer the teleological "why?" or "what for?" question about it. In terms of his general view of "functional analysis," an appeal to something's "functions" in his sense could never answer such questions, which is why he sticks to the broad "how?" questions about capacities or effects brought about within a containing system. He could, of course, address the question "what function does part or feature A serve in S?", but within his general framework of functional analysis this could only mean: "What does A function to do in S, thus contributing to capacities of S which we might take an interest in analyzing?" The answer to such a question will cite indefinitely many things, most of which will not qualify as proper functions even assuming that we are dealing with a genuine functional system, such as a machine or an organism. There is no connection here with teleological language—i.e. no place for a claim such as that A is *for the sake of* Y in S—and however interesting such an answer might be, it does not shed anything like the sort of light on A that is shed on entities

[36] Cummins (1975, p. 748).

with proper functions by *functional teleological* explanations as I have construed them.

5. QUALIFICATIONS

There are two important qualifications to make concerning the view put forth in section three. First, I have mentioned several different ways in which functional teleological questions may, on my view, be formulated: "*why* do F's have part or feature A?", (or "*why* does A occur in Fs?", or "*why* are A's G?" or "*why* do A's ϕ?"); "*what* is A *for*?"; "*what function* is served by A?" or "*what* does A *do*?"; and "*what* is A *designed* to do?" or "*what* are Fs *designed* to do?" While a given functional question can often be formulated in several of these ways, these formulations are not equivalent. There are important differences regarding built-in presuppositions. For example, asking what something is designed to do presupposes that there is indeed a design background, whereas asking what function something serves does not—though it has been argued by some that it does for an extended notion of design, not requiring conscious agency.[37] Similarly, asking why something is present, or is the way it is, *even in the teleological sense I have identified*, seems to impose the condition that what is cited in the answer must be etiologically relevant. That is, we could not correctly answer the question "why do Fs have A?", *even in the teleological sense*, by appealing to an effect that is etiologically irrelevant to A's presence—a point that Achinstein fails to appreciate, I think.[38] By contrast, it is less obvious that such a condition obtains for questions phrased in terms of what something is for, or what function it serves.

The situation here is actually a bit tricky. I have argued in earlier chapters that etiology is relevant to the establishing of non-incidental relations between parts or features and functions and ends within systems, so that etiological relevance is indeed usually a condition for the truth of teleological explanations in response to *any* of these questions—whether posed as "why?" questions or posed in terms of what something is for, etc. But there will be cases, involving changes in the *functional use* of parts, that plausibly allow for ascriptions of proper

[37] Kitcher (1993).
[38] Achinstein (1983, pp. 289–90). I am indebted to Robert Adams on this point.

functions to parts or talk of what they are (functionally used) *for*, but do not allow for teleological answers to "why?" questions about the presence of the parts themselves. The sort of case I have in mind here—shifting now to a biological example—is something like the sea turtle's tail, discussed in chapter five. If the tail's egg-protecting effect is *etiologically* irrelevant to its presence, then I do not think we can cite this effect even in a correct *teleological* explanation of *why* sea turtles have tails. But if the turtles' use of the tail to protect eggs is biologically *functional* use (e.g. if this behavioral trait is an adaptation, to take the simplest case), and if that use is sufficiently regular among sea turtles, then we may indeed ascribe an egg-protecting biological function to the tail itself; that is, regular functional use of the tail to protect eggs is sufficient to make it the case that it plays a non-incidental egg-protecting role in the life of sea turtles. In that case, we can plausibly cite egg protection as part of an answer to the question "*what function* does the tail serve in sea turtle life?" or "what is the tail *for*?", even though we cannot cite it in answer to the question "*why* do sea turtles have tails?"

The second qualification concerns the scope of the present account of functional teleological explanation in the context of functional systems. I have not here attempted to provide a *general* account of functional teleological explanation, but have limited my focus to functional systems, since I take this to be what is relevant to biological teleology, which is my primary interest. Nothing has been said, for example, about this form of explanation in connection with simple tools—i.e. the sort of explanation given in response to a question such as "why is the back side of a hammer head split and curved?". In fact, my account should not be taken even to cover every sort of functional teleological explanation that might be found in connection with functional systems. For example, besides having such parts as spark plugs, an engine might also have a handle on the top to make it easier to carry. I certainly don't want to claim that the function of the handle is to be understood in the same way as the function of the spark plugs, i.e. as making some contribution to the working of the engine. Its function is quite distinct from the working of the engine, having nothing to do with the conversion of fuel into constant rotational power. What we have here, I think, is just a compound entity, with different ranges of components that must be treated differently. While the functions of a wide range of parts are to be understood in relation to the working of the engine-system, there are also some parts, such as the handle, that must be understood differently.

I have not attempted to give an account of the functions of such things as handles or simple tools. However, I do think that the general moral applies to such cases as well. When we ask the "what for?" question about a tool—e.g. "what is a hammer (used) *for*?"—or about some feature of a tool, as in the above question about the hammer claw, or about something like a handle, we are asking a distinctly *functional teleological* question, which does not appear to be reducible to an historical-causal question, any more than in the cases on which I have focused. Likewise, the explanation should not be thought to be reducible to an efficient-causal explanation of how the thing or feature came to be present. The functional teleological question and explanation here will not be fleshed out in terms of roles played in the working of systems, but something loosely analogous to this seems plausible. For example, perhaps the proper function of a tool can be understood as the proper role it plays in a certain craft or other activity—such as the role the hammer plays in carpentry, or the role the shoe horn plays in getting dressed, in relation to which we can speak of particular hammers or shoe horns as being in good condition or defective, being used properly or misused, and so on. I shall not, however, pursue such cases here. My primary interest is in living organisms, and the relevant artificial comparison is a machine with a single hierarchical structure of functions and ends, all ultimately geared toward one relatively clearly definable end.[39]

6. ACCOUNTS BASED ON THE DEDUCTIVE-NOMOLOGICAL MODEL OF EXPLANATION: HEMPEL AND NAGEL

Before leaving the topic of teleological explanation, it is worth examining one other prominent approach that, while very different in substance from the etiological approach, involves a related misunderstanding of the way in which appeals to functions, as such, are typically explanatory of functional items. Proponents of this approach—which is associated primarily with Hempel and Nagel—

---

[39] Thanks to Stefano Predelli for an interesting example that brought to my attention the need for the second of the above two qualifications.

begin by construing questions of the form "why do Fs have feature A?" or "why are Fs A-ish?" as asking: "what accounts for the presence of A in Fs?". To this extent, it is similar to the etiological approach, and in fact, as remarked in section one, this initial move *needn't* be problematic; for on my own view there is a sense in which teleological questions *are* requests for "an account of the presence" of the item whose function is in question. The problem with the etiological account, I argued, is that it gets this sense wrong: Teleological questions are *not* requests to be told *how the item came to be present* (i.e. how that type of item came to be instantiated in Fs, or how instances of the type came to be present in Fs). What I want to show in this section is that the approach of Hempel and Nagel is equally misguided, again misconstruing the sense in which teleological explanations account for the presence of functional items.

According to this approach, an explanation consists of an argument with the *explanans* as premises and the *explanandum* as the conclusion. In both causal explanations and teleological explanations of a trait A, the *explanandum* is *the presence of a token occurrence of a certain trait A in a current token F*—or, if we like, the presence of token occurrences of A in current token Fs, considered collectively (assuming we are interested in something about Fs generally). What sets teleological explanations apart is that the premises have something to do with the *consequences* of the *explanandum*—i.e. the *consequences* of the presence of A in Fs (or of this instance of A in this F, if we choose to focus on a particular case).

Hempel's initial attempt to understand functional teleological explanations involves formulating them as "deductive nomological explanations" of the presence of the items in question. Deductive nomological explanation is "explanation by deductive subsumption under general laws"—"construed as an argument in which the explanandum is deduced from the explanans".[40] Again, so construed, functional explanations are structurally similar to ordinary_causal explanations as understood on the deductive nomological model—both being taken essentially to be deductive arguments; the difference lies just in the fact that in causal explanations, conclusions are derived from causal laws and antecedent conditions, whereas in functional explanations conclusions are derived in part from facts about the crucial

---

[40] Hempel (1965, p. 299).

role played by the *consequences* of the thing whose presence is to be explained.[41]

Hempel (1965, p. 310) thus offers the following schema as roughly the form that a functional analysis would have to take in order to have explanatory import, i.e. in order to be a functional *explanation*:

Hempel Schema (H):

(1) At t, a system S functions adequately in a setting of kind E (encompassing both internal and external conditions).

(2) S functions adequately in a setting of kind E *only if* a certain condition, N, is satisfied.

(3) N can be satisfied *only if* trait A is present in S.

---

(4) Therefore, at t, trait A is present in S.[42]

Hempel argues, however, that instances of 3 are typically false, so that such arguments will typically be unsound: "The assumption of functional indispensability for a given item is highly questionable on empirical grounds: In all concrete cases of application, there do seem to exist alternatives". As Nagel explains Hempel's point: "The presence of some specified item in an organism, which is to be explained in terms of its function, is in general not a *necessary condition* (or is not known to be a necessary condition) for the performance of that function"; there are always functionally equivalent alternatives, at least in principle.[43] To take a simple example offered by Cummins: "It is false . . . that the

---

[41] Cf. Nagel (1977, p. 299–300; 278): "Functional explanations can be shown to have the same structure as explanations in the physical sciences," or again, "explanations of both goal and function ascriptions are *structurally* similar to *causal* explanations in the physical sciences." Indeed, he maintains that explanations of *goal* ascriptions just *are* causal explanations in terms of antecedent conditions and causal laws, while explanations of *function* ascriptions should be distinguished as *teleological*, inasmuch as their premises are not in terms of causal laws and antecedent conditions, but are rather in terms of facts and lawlike statements which are not causal laws, but have to do with the *consequences* of the item in question. An example will be given below.

[42] Hempel (1965, p. 310). Notice that this paradigm involves a *token* system S, as such, with no mention of the type—an odd feature if my remarks about types and tokens in chapter five were on the right track.

[43] Nagel (1977, p. 292).

heart is a necessary condition for circulation in vertebrates, since artificial pumps can be, and are, used to maintain the flow of blood."[44]

We could try to escape this problem with 3 by replacing it with the claim that A is a *sufficient* condition, rather than a necessary condition, for N:

(3'): "If A is present in S, then N will be satisfied".

This would allow for the possibility of alternative ways of meeting N, and so would perhaps make for a more satisfactory functional analysis (though 3' would have to be qualified with a *ceteris paribus* clause to make it tenable). But as Hempel points out, this argument schema is no longer deductively valid, and so again fails to be sound. All that we can conclude from 1,2,3' is that A *or some functional equivalent of A* is present in S at t. So on the assumption that what we were out to explain was the presence of A in S at t *as opposed, among other things, to some other functional equivalent*, and that to explain this is to deduce it, clearly we have failed.[45]

Thus, the only general explanatory or predictive value of functional analysis according to Hempel is what little we get by deducing the weaker conclusion from the original two premises, i.e.:

(1) At t, a system S functions adequately in a setting of kind E (encompassing both internal and external conditions).

(2) S functions adequately in a setting of kind E *only if* a certain condition, N, is satisfied.

(4') Therefore, some item which, under conditions E, would bring about N, is present in S at t.

---

44 Cummins (1975, p. 743).

45 Hempel (1965, p. 310). Cf. Nagel (1961, p. 70). Hempel considers the possible move here of settling for the deductively invalid argument obtained by substituting 3' for 3, perhaps augmenting it with another premise stating that there are at any rate not many alternative functional equivalents of A and that their likelihood is relatively small, and viewing this as an good *inductive* argument for the presence of A in S at t. If that were successful, then perhaps functional explanation could be saved after all, construed as inductive-nomological explanation. But he rejects this move as unpromising because of the great difficulty in knowing and specifying the range of alternatives with any precision, and in assigning probabilities to the various cases, concluding that "the information typically provided by a functional analysis of an item [A] affords neither deductively nor inductively adequate grounds for expecting [A] rather than one of its alternatives" (313). I shall not go further into this, as it is not necessary for my purposes.

Such an inference, he points out, is generally rather trivial unless we happen to have further knowledge about which items would bring about N in E. In particular, if it could be established that the presence of A is the only physically possible way for condition N to be realized in S under E, then we would have a deductive argument (conforming to the original schema H above) for the conclusion that A is present in S at t—and hence, on Hempel's view, a non-trivial functional explanation. But again, since this is generally not the case, Hempel concludes that functional analysis has little if any explanatory value.

Before moving on to Nagel's attempt to save functional explanation within this framework, it is important to notice and query a key assumption made by both Hempel and Nagel—an assumption that was crucial to Hempel's case against the explanatory import of functional analysis: namely, that what we are out to explain in giving a functional teleological explanation is the presence of A in S at t *as opposed to some other functional equivalent*. The thought seems to be that "if there is nothing in the explanation itself which rules out all alternatives to the thing being explained . . . then [we cannot] legitimately say we have explained why specifically IT occurred [or is present], as opposed to something else."[46] But why think that teleological questions and explanations are generally concerned with this contrast, i.e. with accounting not only for the presence of the thing in question, but for why it rather than a functionally equivalent alternative is present? Wright, for example, explicitly rejects this assumption. Whether we have successfully explained something's presence or occurrence depends on the nature of the puzzlement we were addressing, and on Wright's view the puzzle generally associated with teleological questions is *not* "why A rather than any other functional equivalent?" but just "how actually did A come to be there?"[47]

---

[46] This is Wright's (1976, p. 102) formulation of Hempel's thought, which he rejects.

[47] Compare: "Why did Johnny come down with the measles?" As Wright points out, "there is no specific alternative to contrast [this] with, but that does not (fortunately!) imply a contrast with all alternatives. What [this question] principally requires is some intelligible way of avoiding the null hypothesis; and this our normal causal hypothesis provides...[e.g.] exposure to Jimmy." (Ibid. p. 104.) The fact that Jimmy may have several viruses, so that someone *might* want to know why Johnny got the measles rather than one of the others, does not imply that we *must* be concerned with this contrast in asking why Johnny came down with the measles; we may be satisfied just knowing that

I have rejected this etiological account of what teleological questions ask, but Wright's more general point here is well taken, and I agree with his negative claim about the puzzlement associated with teleological questions. On my view, what is at issue is not why A rather than any other functional equivalent is present, but just what functional role A actually plays in the life of such an organism, as this notion was fleshed out in chapter four. And like Wright's question, this can be answered without answering the question why A *rather than* any other member of a set of indefinitely many functional equivalents is present. So we already have good reason to be suspicious of Hempel's attempted construal of functional explanation: It seems to have the wrong sort of orientation if it can succeed only by accounting for the presence of A as opposed to any other functional equivalent.[48]

Let's set this aside for the moment, however, and look at Nagel's attempt to rehabilitate functional analysis on the deductive-nomological model so that it will have explanatory value after all. Nagel accepts Hempel's schematic account of what would be required for a teleological explanation (H above), but parts company with him over the interpretation of 3 and hence over the question of the explanatory import of functional analyses. Nagel argues that the premise:

(3) N can be satisfied *only if* trait A is present in S,

does in fact hold for all practical purposes in biological cases. That is, it holds with regard to "organisms in their natural state"—even though there are in principle alternatives which strictly speaking falsify it if it is taken strongly, i.e. as Hempel takes it; and this weaker, narrower interpretation of the necessity is enough, he thinks, to save teleological explanation:

> For example, a convincing case can be made for the claim that in normal human beings—that is, in human bodies having the organs for

---

he got the virus from Jimmy, so that citing his exposure to Jimmy is a perfectly satisfactory explanation.

48 Why does Hempel assume without question that the explanandum must be of this nature? This seems just to be the result of his assumptions (shared by Nagel) that what is to be explained in any given case is the actual occurrence of a particular item A in a particular system S (or a collection of such items in a collection of S's), and that to explain such a thing functionally is to argue from the existence of certain present effects in S to the conclusion that A is present as a causal condition for those effects. If what I've said above is right, both of these assumptions about functional explanation are radically mistaken.

which they are at present genetically programmed—the heart is necessary for circulating blood; for in normal human beings there are in fact no alternative mechanisms for effecting the blood's circulation.[49]

Nagel thus qualifies the modality in 3, so that what is relevant is not logical or abstract physical necessity but only necessity "in view of the limited capacities [organisms of the relevant kind] possess as a consequence of their *actual* mode of organization".[50] He argues, therefore, that in cases such as the heart's, 3 may be accepted as true in the relevant sense, making H a sound deductive argument for the presence of a heart in a given human body after all. And in other cases, where there are in a very ordinary sense genuine functional alternatives (e.g. neither ear by itself is necessary, even in Nagel's weak sense, for hearing), it is simply to be recognized that "functional explanations account for the presence in the system under discussion of a *set* of items".[51]

Nagel's view of the structure of functional explanation is roughly just a version of Hempel's general formulation (H). He gives the following example of a functional explanation:

> (i) During a stated period, a certain green plant is provided with water, carbon dioxide and sunlight, and performs photosynthesis.
>
> (ii) If during a given period a green plant is provided with water, carbon dioxide and sunlight, and performs photosynthesis, then the plant contains chlorophyll. (That is, a green plant will perform photosynthesis when provided with water, carbon dioxide and sunlight *only if* chlorophyll is present.)

---

[49] Nagel (1977, p. 292).

[50] Nagel (1961, p. 70). By "abstract physical necessity" I mean necessity such that if A is necessary for N, N cannot be attained in the absence of A without violating the laws of physics. The heart is obviously not necessary for the circulating of blood in *this* sense. Nagel's claim might be put by saying that the heart is *physiologically* necessary for the circulation of blood in human beings, meaning that circulation of blood in human beings in their natural state (e.g. without artificial hearts) can take place only if they have hearts.

[51] Nagel (1977, p. 293). This is Nagel's response to one of Cummins' objections (1975, p. 744).

(iii) Therefore, chlorophyll is present in the specified green plant during the stated period.[52]

Nagel points out that the first premise is an "instantial statement", the second is "lawlike", and the conclusion follows logically. But the second premise is not a *causal* law, since "the performance of photosynthesis is not an *antecedent* condition for the occurrence of chlorophyll"; it rather brings out the essential *contribution* chlorophyll makes to the system.[53] So what we have here, according to Nagel, is a *teleological*, not a causal, explanation—though because it does not contain any explicitly teleological language, i.e. "does not explicitly ascribe a function to chlorophyll", it is a *non-teleological formulation* of the teleological explanation.[54]

This is basically a simplified version of Hempel's formulation, leaving out the appeal to the plant's functioning normally, and the claim that photosynthesis is necessary to the normal functioning of the plant.[55] Here we start simply with the fact that a given plant *is* photosynthesizing in certain conditions, add the premise that a certain trait or item, such as chlorophyll, is necessary for this activity, and deduce that the trait or item is present. Nagel thus offers this as a perfectly good example of functional teleological explanation, so long

---

[52] Nagel (1977, pp. 299–300; 278). Again, notice the primary focus on the particular case, which is odd in a discussion of functions in biology.

[53] Ibid.

[54] Nagel (1961, p. 71). Nagel holds that "teleological (or functional) explanations....can be reformulated, without loss of asserted content, to take the form of nonteleological ones, so that in an important sense teleological and nonteleological explanations are equivalent" (p. 69). This last part is perhaps a bit misleading, as what he seems generally to mean is rather that teleological and nonteleological *formulations* of *teleological explanations* (i.e. explanations that involve not causal laws, but laws concerning an item's effects in a system) are equivalent in content. Thus, the above argument represents the non-teleological formulation of the teleological statement: "The function of chlorophyll in plants is to enable plants to perform photosynthesis (that is, to form starch from carbon dioxide and water in the presence of sunlight)" (p. 69).

[55] Actually, photosynthesis is not merely necessary to but *constitutive of* the normal functioning of the plant. Perhaps a better analogue for Hempel's 'N' in this case would therefore be the presence of starch, which is effected by photosynthesis; the point remains that N is satisfied only if chlorophyll (A) is present, and N is necessary to the proper functioning and maintenance of the plant.

as the modality in the second premise is understood in an appropriately weak sense.

There are several problems here, however. First of all, there is a clue that something is wrong in the fact that so much weight is placed on the necessity in the second premise. Nagel does indeed weaken it from the overly strict necessity Hempel had been working with, but it is still crucial to his account that functional items in biology are *necessary* for the performance of their functions; if it turns out that they often are not, and that there are genuine functional equivalents in biology—beyond the obvious cases, such as two ears, which he recognizes and accommodates—then his account will fail, just as Hempel argued. This leaves his account rather vulnerable, and while that is not in itself an argument against it, one wonders why an account of functional explanation should be predicated on such an apparently needless assumption. What reason is there, independent of a *prior* commitment to the deductive-nomological model of explanation, for supposing that Z is a function of A only if A is a *necessary condition* for Z?[56] We often apply teleological concepts and language with little or no knowledge of whether or not there are functional equivalents that could perform the function but don't for one reason or another, and I see no reason to suppose that we must be implicitly assuming that there aren't any. The issue just doesn't seem to have much to do with our application of teleological concepts, and this, I think, tells against an account which makes it absolutely central.[57]

Secondly, there are more general problems concerning the explanatory value of such deductive arguments. Even if Nagel is correct in claiming that the chlorophyll argument is sound, it is far from clear why it should be thought to capture the essence of functional teleological explanation. The only reason so far to suppose it does is that it purportedly explains something's presence not by appeal to causal laws and antecedent conditions, but by appeal to laws concerning effects of which the item in question is a causal condition.

---

[56] Cf. Woodfield (1976, p. 114), Wright (1976, p. 101), Beckner (1968, p. 113), and Ruse (1973, p. 183).

[57] It is interesting to note that we do not generally associate functionality with this sort of necessity in the case of *artifacts*. Nothing prevents our ascribing a function to one thing even though there is something else that can do it as well. For example, the function of a dining room table is to provide a surface for dining on, but there are many other things that provide surfaces suitable for dining on.

But it is not clear, first of all, that such an argument actually *explains* anything at all. And even if it did explain what it purported to, this would not suffice for it to constitute a functional teleological explanation.

The first point is well expressed by Cummins:

> Even if it were possible, as Nagel claimed, to *deduce* the presence of chlorophyll from the occurrence of photosynthesis, this would fail to explain the presence of chlorophyll in green plants in just the way deducing the presence and height of a building from the length of its shadow would fail to explain why the building is there and has the height it does.[58]

Unfortunately, he goes on to overstate the reason for this, once again betraying a failure to recognize any other sense of *why something is there* than the etiological one:

> To explain the presence of a naturally occurring structure or physical process—to explain why it is there, why such a thing exists in the place (system, context) it does—this does require specifying factors that causally determine the appearance of that structure or process.

Wright would of course be happy with this exclusive focus on etiology (causal origins) when it comes to answering "why?" questions about functional items. By contrast, I have argued that there is an alternative interpretation of such questions—which is in fact the teleological interpretation—according to which they are properly answered not by providing "factors that causally determine the appearance of that structure or process," but by explaining the functional roles such items play—the biological *ends* they non-incidentally serve, in the life (or biological working) of such organisms. But the myopia of the second passage above should not obscure the point of the first: Whether we are interested in etiological or genuine teleological explanations, merely deducing the presence of the item to be explained from certain effects of which it is a causal condition is not to *explain* its presence. We explain neither how the building came to be there, for example, nor what function it serves (e.g. to house corporate offices) simply by

[58] Cummins (1975, pp. 745–6).

deducing that a building of such a height must be present, given the presence of a shadow of certain dimensions.

The building example suggests another way of putting the problem. Nothing prevents us from formulating a parallel deductive argument for the presence of something in an organism where the effect mentioned in the premises is clearly not the function of the item in question. For example, someone might offer the following argument:

(i) A certain dog presently harbors fleas.

(ii) Dogs harbor fleas only if they have long hair.

(iii) Therefore, this dog presently has long hair.[59]

This shows first of all that to say that an effect Z is the *function* of an item A in a system S is obviously not just to say that A is a *causal condition* of Z in S, even if we restrict the scope of S to teleological ("goal-directed") systems. For it is certainly not the case that the proper function of this dog's hair (or of the hair of long-haired dogs generally) is to harbor fleas.[60] And it shows likewise that deductive arguments from effects to their causal conditions are not necessarily functional explanations—or explanations of any other sort, for that matter—of the presence of the causal conditions. Nagel recognizes this need to restrict function claims, and therefore adds the condition that "the function ascribed to an item *contributes* to the realization or maintenance of *some goal* for which the system is directively organized". He thus

---

[59] The example is from Ruse (1973, pp. 182 f.).

[60] Ruse took Nagel (1961) to be making this reductive claim, and so offered the above example as a refutation. Indeed, Nagel certainly did say things that justify this reading. He claimed, for example, that the teleological statement "the function of chlorophyll in plants is to enable the plants to perform photosynthesis" asserts "nothing that is not asserted by 'plants perform photosynthesis only if they contain chlorophyll', or alternatively by 'a necessary condition for the occurrence of photosynthesis in plants is the presence of chlorophyll'". That is to say, "when a function is ascribed to a constitutive element in an organism [or other goal-directed system], the content of the teleological statement is fully conveyed by another statement that is not explicitly teleological and that simply asserts a necessary (or possibly a necessary and sufficient) condition for the occurrence of a certain trait or activity of the organism." (Nagel 1961, p. 71.) These claims are later qualified, in response to Ruse, in Nagel (1977, pp. 295–7).

refers to his view of functions as the "goal supporting" view.[61] The addition of this condition for Z's being the function of trait A plausibly rules out Ruse's case and similar cases, since by hypothesis the harboring of flees "does not contribute to the maintenance of any goal for which dogs are goal-directed" (however exactly this is to be understood). But does it really suffice to make Nagel's view tenable? In particular, does it take us any closer to genuine *explanation*?

Nagel's use of the term "goal" is first of all very broad. It includes both biological ends and psychological goals, for example. He also fails to distinguish between incidental and non-incidental contributions to a "goal", and as already mentioned, focuses directly on particular systems at particular times, rather than appealing to the characteristic behavior of types of systems. All of this leads to serious problems. To take an example given by Woodfield, an animal's rumbling stomach might contribute to its goal of scaring away another animal, and the presence of the stomach may be a causal condition for such rumbling, but that does not make it the case that the rumbling was thus one of the stomach's functions.[62] Even if we grant that the first animal is "directively organized" toward the "goal" (i.e. biological end) of survival, and in a way that includes doing certain things to scare away other animals (which may or may not be a psychological goal of the animal), the convenient rumbling of the stomach is merely incidental to its proper functioning, and thus cannot be regarded as a proper function of the stomach.

Another problem is that any organ that contributes to the "goal" (i.e. biological end) of survival thereby contributes also to every goal or end of which survival is a condition.[63] The beating of a person's heart, for example, is plausibly a necessary condition not only for the circulation of blood, but for virtually all other life activities, which themselves contribute to further "goals" (i.e. to biological ends and to psychological goals). We could thus construct a variety of arguments

---

[61] Nagel (1977, p. 296). Such a move is naturally suggested by Hempel's (1965, p. 305) earlier discussion of the function/accident distinction: "Heart sounds are an effect of the heartbeat which is of no importance to *the functioning of the organism*; whereas the circulation of the blood effects the transportation of nutriment to, and the removal of waste from, various parts of the organism—a process that is indispensable if the organism is to remain in *proper working order*, and indeed if it is to stay alive."

[62] Woodfield (1976, pp. 126–7).

[63] Cf. Woodfield (1976, p. 125).

like the chlorophyll argument, deducing the existence and beating of a heart in someone, but where the first premise has to do with such things as his digesting, running, seeing, thinking, and even playing baseball or writing a paper. Since, according to Nagel, this "telescoped argument" is equivalent to the corresponding functional statement, we thus wind up with a number of very odd claims about the function of the heart, and some very strange "explanations" of why we have hearts.

This is true even if we restrict "goals" to biological ends, in an attempt to avoid being stuck with functional explanations of the heart in terms of baseball. We still have no principled way to distinguish between the proper function of an organ and all the other end-directed activities for which it is a necessary condition, but which do not constitute its proper function in any useful sense. For example, the presence of the heart is necessary to the operation of the lacrimal glands (which provide moisture for the eye), allowing us to construct a Nagelian deductive argument from the actual operation of the lacrimal glands to the presence of the heart; but the operation of the lacrimal glands is hardly constitutive of the function of the heart, and does not provide a functional explanation of why we have hearts. Or again, if the rumbling of an animal's stomach contributes to the biological end of survival by scaring away a predator, then we could formulate a Nagelian deductive argument from the rumbling of the stomach and the necessity of a stomach for such rumbling to the conclusion that the stomach is present, and we would thus be forced on this account to say that so rumbling was at least one of the stomach's *functions*, providing a functional explanation for the presence of the stomach.

In summary, then, Nagel's account of functional explanation is extremely problematic even on its own terms. A sound deductive argument from effects to the presence of a certain causal condition clearly fails generally to *explain* the presence of the thing in question in any sense—whether etiological or teleological. This is easily seen in cases where the effects are entirely irrelevant, as in the building case or the flea case. It is also illustrated by cases where the effects, while contributing to "goals", do so in a merely incidental fashion, as in the rumbling stomach case. The latter problem remains even when we focus on effects that contribute to biological ends, since the rumbling stomach might well contribute to the biological end of survival. Much more would obviously need to be said to distinguish between genuine functions and merely incidental contributions to a biological end in particular cases. It is not clear, however, how such considerations could be built into Nagel's scheme without changing it into a very different sort of account, more along the lines of the sort of account I have argued for. Finally, Nagel's account fails to provide any means for distinguishing an entity's proper function from all the many things for which it is a necessary condition and which either constitute or contribute to "goals" of the system. Even if we restrict relevant "goals" to biological ends, the difficulty remains; we hardly seem, for example, to provide a *functional explanation* of the heart simply by deducing its presence from the actual operation of the lacrimal glands.

The more one considers Nagel's deductive-nomological schema, the more puzzling it seems that it should ever have been thought to capture the structure or sense of familiar functional teleological explanations in biology, such as:

> The dry fruit of maples has a papery, wing-like structure *in order to* slow the fruit's descent, increasing its chances of being blown free of the parent tree, thus promoting seed dispersal; that's its *function*.

I have argued that such explanations are understood to be primarily about *types* of organisms and how they *work*, not about particular organisms on particular occasions, taken as such. To the extent that they apply to particular organisms, they apply to them not as particulars, but as instances of the relevant types. By contrast, on Nagel's account (following Hempel), functional explanation is taken to be merely the inference from the fact that (i) this token F actually did something on a particular occasion, which (ii) it is impossible for Fs to do without the presence of A, to the conclusion that (iii) a token of A was present in this F on that occasion—as we saw in the chlorophyll example.

It is really quite mysterious how such an argument is supposed to capture the content of an assertion like "the *function* of chlorophyll in green plants is to enable the plant to perform photosynthesis and produce starch" or "chlorophyll is present in green plants *for the sake of* photosynthesis." What does this sort of teleological claim possibly have to do with deducing the *token occurrences* of chlorophyll *on a case by case basis*, by appeal to what particular green plants *actually* did during particular periods of time, together with facts about certain necessary conditions for their doing so? Even setting aside the peculiar direct focus on the particular case, rather than on types of entities in types of organisms, this approach gets things completely wrong. As I've already stressed, we cannot arrive at the function of a token biological entity simply by looking at what it actually does on a given occasion. It is not sufficient for an instance of A to have function Z that it actually does Z and Z furthers some biological end, as the example of the rumbling stomach shows. And it is not necessary in order for an instance of A to have function Z that it actually does Z (or that Z actually furthers some biological end), as is clear from the fact that a defective organ that fails to do Z can still be said to have Z as its proper function simply by virtue of being an instance of a certain *type* of organ that has Z as its function. Nagel, however, does not seem able to account for such cases of functional but non-functioning token items (such as defective eyes), since in such a case the relevant deductive argument is unavailable.

Sometimes Nagel *seems* to be suggesting something closer to the sort of view I have defended, as when he says that teleological explanations "make evident one role some item plays in a given system".[64] But on his view, this can hardly mean a *non-incidental role* some item plays in the *non-arbitrary working* of a certain *type* of system, since all we get in his sort of "functional explanation" is a report of the token effects a token occurrence of some trait actually had in a certain token system on some occasion. Even if we add up a number of such arguments, this will not give us the idea of proper function, but just a report of what actual instances have done on a number of occasions. It will not allow us to say, for example, that even where a token organ fails to perform a certain operation, that operation is nonetheless its proper function, and the organ is therefore (all else being equal) *defective*; all we could say is that this organ does not in fact do what similar organs do, and so is unusual—though in fact even this may not be true, if there is a high percentage of cases of defect or malfunctioning in the sample. This indicates that Nagel has not really gotten to the heart of genuine functional teleology. Deducing something's presence as a causal condition of certain given effects is a red herring as far as functional explanation is concerned.

---

[64] Nagel (1977, p. 300).

7. CONCLUSION

It is widely agreed that appeals to functions can constitute explanations of certain phenomena. The dispute is over the nature of functional explanations: What exactly is meant to be explained by typical appeals to functions, and what sort of explanation is involved? According to both etiological views and the deductive-nomological view defended by Nagel, what is explained is the presence of the item to which the function is ascribed. They differ, however, in their accounts of the sort of explanation that is involved. According to etiological views, a functional teleological explanation tells us how the item to which the function is ascribed came to be present, by offering a consequence-etiology of it—i.e. a causal history involving, in part, the type of effect identified as that type of item's function, leading to the presence of the item's current instances. According to Nagel's deductive-nomological view, a functional teleological explanation amounts to a proof that the item to which the function is ascribed is present, where the premises contain an appeal to the effect identified as the item's function.

I have argued that neither of these views of functional teleological explanation is correct. The problem, however, is not with the very idea that appeals to functions are explanatory of functional items, as argued by Cummins against the two above views. We need not, therefore, follow him in abandoning the search for an account of functional teleological questions and explanations, traditionally understood. The problem lies rather in a misconstrual in each case of the sense in which such explanations account for the presence of the items to which the functions are ascribed. When we ask in a functional teleological sense *why the dry fruit of maples has a papery, wing-like structure*, for example, we are not asking for a historical-causal account of its presence, i.e. an account of how such a structure came to be there, involving an appeal to the consequences of that type of structure, thus providing a "consequence-etiology" of it. Nor are we asking for a proof that such a structure is present in a given maple (or set of maples), on the basis of certain other activities this maple (or set of maples) carries out and claims about the necessity of the structure to the performance of those activities. What we are asking is to be told the functional teleological significance of the structure's presence, in the context of the relevant type of functional system—i.e. what standard role the structure plays in the life (or biological working) of the maple. This, I have argued, is a perfectly familiar and legitimate sort of question in its own right, not reducible to a question about causal origins or to any other, non-teleological sort of question. Likewise, functional teleological explanation should be understood as an irreducibly teleological form of explanation—compatible with but not reducible to efficient-causal explanations or proofs of the presence of functional items. This view of functional teleological explanation fits naturally with the account of biological function I offered in chapter four, and thus provides further support for it: We can hold a system-oriented view of biological function not involving any direct reduction of facts about function to facts about causal history and still account for the explanatory aspect of functions; indeed, we can do the latter more naturally and satisfactorily than can proponents of etiological views of teleological explanation, thus removing what has been perceived to be one of the central attractions of the etiological view of biological function.

Finally, let me conclude by recalling the parallel I drew in section two between the etiological approach to functional teleological explanation and standard causalist approaches to psychological teleological explanation, along the lines of Davidson. If I am right that the former is a mistake, entirely failing to capture the nature of functional teleological explanation, then this raises the question whether the parallel move with regard to psychological teleological explanation might be implicated in a parallel mistake. What we have seen in the case of functional explanation is that even if the effect cited in the explanation (e.g. the wing-like structure's effect of slowing down the maple fruit's descent when it falls) can *also* be appealed to as part of a remote *efficient-causal* explanation of the presence of the functional trait, and even if this fact is indirectly important for the truth of the functional explanation, it does not follow that this is the nature of the explanatory force of the appeal to the effect in the context of the functional explanation. On the contrary, I have argued that its explanatory force in that context is irreducibly teleological, providing a distinct form of explanation that is not to be understood merely as a species of efficient-causal explanation.

This should at least make us suspect that something similar may be going on with the other sort of teleological explanation, where actions are explained by appeals to ends or reasons. *Even if* it is conceded that the beliefs and desires implicated by the citing of the end or reason could figure in an *efficient-causal* explanation of the occurrence of the action, and that this fact is somehow important to the truth of the teleological explanation, it does not obviously follow that this is the explanatory force of the appeal to the end or reason in the context of the teleological explanation. There may be a teleological sense of "why the agent ϕ-ed" that is no more reducible to "*what caused* the agent to ϕ" than the teleological sense of "why the maple fruit has a papery, wing-like structure" is reducible to "*what caused* the maple fruit to come to have a papery wing-like structure". I suspect—though I can't argue for it here—that this is indeed the case, and that the common reduction of psychological teleological explanation to efficient-causal explanation involving beliefs and desires is no less of a mistake than the etiologist's reduction of functional teleological explanation to consequence-etiological explanation.

CHAPTER IX

Biological Teleology in Human Life

1. SOCIOBIOLOGY AND BIOLOGICAL TELEOLOGY

What are the implications of my account of biological teleology for human life? Human beings are instances of a type of organism that has evolved by natural selection, as part of an evolutionary tree that extends back to the origins of life on earth. To a significant degree, then, it must be appropriate to view human beings in the way I have argued we must view all other organisms—namely, as complex systems of complementary genes and their well-integrated phenotypic expressions that tend to work as coherent units "designed" by the forces of natural selection primarily to bring about the replication of certain copies of these genes, as specified in chapters four and five. This should hardly be controversial as far as human physiology and morphology are concerned. Human organs, for example, are no less appropriate subjects for biological functional analyses than the organs of other species. Nor is there any reason to suppose that the overall framework will be different: The biological functions of human organs, just like the biological functions of organs in other species, will be understood ultimately in terms of non-incidental contributions to genetic replication.

Things get more difficult when we move to the realm of thought, desire, emotion and behavior. Can we likewise speak of biological function here? That will depend first of all on whether it is plausible to suppose that the influence of genes extends into this realm, and that there has in the past been genetic variation and selection pressures leading to natural selection among such genes by virtue of their phenotypic expressions in this realm. One need not buy into either genetic determinism or the exaggerated claims that have sometimes been made on behalf of sociobiology to think that this is indeed plausible.[1]

Consider first non-human animals. There is certainly nothing implausible in the very idea—at the heart of ethology—of genes having an influence on behavior. Indeed, quite apart from what is known about particular cases (e.g. "hygienic" vs. "non-hygienic" behavior in honey bees[2]), the idea is virtually forced upon any naturalist simply insofar as it is indispensable to the only remotely plausible naturalistic account of a wide range of animal behavior. It is no more plausible to regard complex and adaptive behavior patterns characteristic of a given animal species as merely fortuitous than it would be to regard complex and adaptive organs and organic processes as merely fortuitous. And if such behavior patterns are to be explained by appeal to natural selection, then that obviously involves the claim that there has been variation among alleles with phenotypic expressions in the behavioral realm, such that natural selection acting on such variation has gradually shaped the patterns of behavior, just as with other adaptations. As Dawkins remarks, "if we are so much as to discuss the possibility of a behavioral pattern's evolving by natural selection, we have to postulate genetic variation [at least in the past] with respect to the tendency or capacity to perform that behavior pattern," which obviously involves the idea of genes significantly influencing behavior.[3]

---

[1] A particularly awful example of such an exaggerated claim may be found on the back cover of the Bantam paperback edition of E.O. Wilson (1978): "Wilson shows how the qualities we value as most human—altruism, morality, religion, even love—are simply the survival strategies of our 'selfish' genes, biologically evolved through millions of years before we were human...and still operating today." I discuss such sociobiological excesses in the next chapter.

[2] See Gould (1982, pp. 312–13) for an account of Rothenbuhler's experiment involving the cross-breeding of hygienic and non-hygienic honey bees, and the resulting genetic model according to which differences in just two genes determine (in the context of the rest of the genome) whether a given bee will be fully hygienic, partially hygienic (in one of two ways), or non-hygienic.

[3] Dawkins (1982, pp. 18–19). Dawkins discusses the pit-digging behavior of antlions in this connection, which is generally accepted as being an adaptation even though no one has yet actually performed a genetic study of it. The point, again, is that we have independent reason to think that it is an adaptation (as it is too complex and clearly

The same can be said regarding appetites, drives, and emotions. It is surely in large part through affecting such things that genes ultimately affect behavior—for example, influencing an organism's eating habits through influencing its appetite, or affecting protective behavior through affecting tendencies to fear, or altering care-giving behavior or mating behavior through altering certain drives or emotional dispositions. These things, no less than the behaviors they tend to motivate, are generally plausible candidates for being adaptations, and hence likewise candidates for being entities with biological functions, on my view. Of course, this does not imply that any given particular emotional pattern, for example, is in fact an adaptation. The point is just that it is plausible that a great many are, particularly those that are clearly adaptive and cannot reasonably be explained in any other way.

Now it is a famously controversial matter how far this approach can be pushed in the case of human life. To what extent can typical human appetitive and emotional traits, and the behavior they tend to motivate, be understood in terms of our natural selection history—whether as adaptations, or vestiges, or modern forms or expressions of more general adaptations or vestiges? Both extreme positions on this matter seem clearly unacceptable.

On the one hand, the denial that human patterns of motivation, feeling and behavior reflect the pressures of our natural selection history at all runs up against the most obvious counterexamples, such as the existence of our basic appetites and drives for food, drink, sex and sleep, and of the pleasures connected with their satisfaction. To this may be added such things as the existence of instinctive fears, such as a common fear of snakes and spiders or a fear of dangerous heights, or the tendency to fear or mistrust intraspecific outsiders; the innate disposition to respond to the "baby scheme" (or "kewpie-doll scheme")—i.e. a certain set of traits characteristic of human infants—with nurturing feelings and behavior; the innate "desire for attachment and affiliation", especially clear in both infants and parents, for example; and the natural "sexual aversion. . . . between persons who

---

adaptive to be the result of chance), and this is sufficient to convince us that there must at least at one time have been variation among alleles having effects on such behavior. If genes didn't influence behavior, such behavioral adaptations could never evolve through biological natural selection.

have lived together when one or all grew to the age of six" (tending to discourage brother-sister incest).[4] Surely the most plausible naturalistic explanation of many such things with clear adaptive value is to be found within our natural selection history, just as with similar phenomena in many other animals. Moreover, quite apart from any particular examples, there is simply no good reason for supposing in general that our possession of sophisticated reflective capacities has the effect of wiping the human psychological slate clean of all the evolutionary conditioning so clearly found in other animals.

On the other hand, while these reflective capacities do not plausibly lift the appetitive, emotional and behavioral aspects of human life out of the range of genetic influence altogether, it would be equally absurd to proceed as if they made little difference. We are speaking here of advanced mental capacities by virtue of which we are able, for example, to reflect on and evaluate actions, desires and emotions, thereby regularly arriving at new ones that we are prepared to support with *reasons*. The presence of such capacities (which will be discussed in more detail below) makes it extremely implausible to suppose that human life is nonetheless as pervasively and heavily conditioned by genetic influences as ant or dog life is, the only real difference being that in the human case those influences are filtered through the medium of culture, leading to greater variation in particular forms of human behavior. Again, that thought runs into equally obvious counterexamples, where what is involved are not merely cultural variations on genetically-oriented themes, but radical departures from such themes altogether.[5]

Consider, for example, the choice to forego having children in favor of other valued pursuits that in no way tend to promote the spread of one's genes—as in Christ's urging of celibacy "for the sake of the

---

[4] See E.O. Wilson (1978, pp. 70–72, 37), Wilson and Ruse (1986, pp. 182–4), Eibl-Eibesfeldt (1977, p. 673), and J.Q. Wilson (1993, pp. 18, 125, 127). E.O. Wilson makes an interesting point about phobias: "It seems significant that they are most often evoked by snakes, spiders, rats, heights, close spaces, and other elements that were potentially dangerous in our ancient environment, but only rarely by modern artifacts such as knives, guns, and electrical outlets"—which latter are generally greater threats to us at present. J.Q. Wilson also notes that "young children can be conditioned to be fearful of caterpillars but not of opera glasses. Something in them prepares them to believe that creepy, furry things might be harmful, but odd, glass-and-plastic devices probably are not."

[5] Cf. Diamond (1992, p. 98).

kingdom of heaven"; or the choice to have fewer children than possible with a view to maximizing quality of life—by which I mean not merely the genetically adaptive choice to optimize the use of resources by having only as many offspring as will likely survive to maturity and reproduce (which would be expected in any case to result from naturally evolved dispositions, as discussed in the example of the swifts in chapter six), but the choice to have fewer still with a view to their living *more pleasant and fulfilling lives*; or the choice to adopt non-related children out of charity or love.[6] Such choices fly in the face of anything that can be attributed simply to psychological predispositions resulting from our natural selection history, even as filtered through culture (though there are some complications here that I will discuss below). Cultural influences may indeed contribute to such non-genetically-oriented choices, but not simply by giving particular forms to genetically-oriented psychological predispositions. What these examples indicate is that there are other influences involved that account for such clearly non-adaptive behavior—the obvious candidates being the reflective motivations people regularly cite in explaining and defending their choices. Moreover, quite apart from any particular examples—which will always be met with clever attempts to cast the speculative evolutionary net widely enough to capture them—the important point is that we should in any case fully *expect* to find various forms of independently motivated, non-adaptive behavior among creatures with sophisticated reflective capacities that can be given practical application. What would be truly surprising—and in need of explanation—would be the *absence* of such independently motivated behavior on the part of creatures with such capacities.

Someone who, ignoring the last point, insists on trying to explain everything in sociobiological terms might suggest that the above examples can be understood simply in terms of genetically-oriented psychological dispositions extending beyond their proper functional sphere of influence, rather than in terms of independent sources of motivation involving our advanced mental faculties. After all, a great deal of non-adaptive behavior in animals may plausibly be accounted for in this way, as discussed in chapter five (recall the examples of moths flying into flames or dogs chasing cars). For example, the

[6] The quote is from Matthew 19: 10–12. The text even mentions the extreme example of those "who have made themselves eunuchs for the sake of the kingdom of heaven."

feelings that may lead people to adopt non-related children, and the attachment they develop to adoptive children, may just be non-functional, non-adaptive extensions of psychological predispositions that were selected for their positive effects on natural children or close kin but do not perfectly discriminate between these and other children.[7]

I do not deny that there is often some truth in such claims. But there is no reason to think that they tell the whole story, in all cases, in a way that excludes independent sources of motivation. Even if they provide plausible accounts of certain feelings in at least some cases, they still fall short of explaining the kinds of choices mentioned above—for example, the choice to adopt *rather than* to give birth, particularly where the adoptive parents give *reasons* for their choice, appealing to such things as charity or a principled concern with the effects of overpopulation on the quality of human life. More problematic still would be the choice to forego having children altogether in favor of something like the concentrated pursuit of a career in advanced set theory or symphonic composition or science-fiction writing. It would really be stretching things to try to account for such things simply in terms of extensions of adaptive predispositions. And even where particular motivations may be accounted for in this way, there is still the problem of accounting for the choices people make among them—choices where someone gives more weight, for example, to his mathematical curiosity than to a desire for children. Again, it seems extremely likely that more is involved here than can be explained simply in ethological or sociobiological terms. In any case, as I have already remarked, the existence of biologically-independent

---

[7] See J.Q. Wilson (1993, p. 128, pp. 42–3). I call the extension "non-functional", in the sense of lying outside the scope of the proper function of the disposition, because such expressions of the disposition have nothing to do with the sort of genetic replication that constitutes the proper biological end served by the disposition. That is, the naturally selected disposition is functional, on the view I have argued for, in relation to its tendency to promote the inter-generational replication of copies of the genes contained in the individual's germ-line cells, in those of its offspring, and perhaps in those of other close kin as well—which it does insofar as it leads parents to care for their natural children and close kin. If it also leads people to care for children who are neither their own offspring nor close kin—or for that matter to dote over pets that share crucial features of the "baby scheme" and thus provoke similar feelings—then that is just a non-functional or dysfunctional side-effect of the functional disposition (that is, in terms of *biological teleology*, which is all I am addressing at the moment).

motivations is plausible quite apart from the force of such examples, just given the kinds of capacities we possess.[8]

For present purposes, I will make the plausible assumption that a *moderate* position concerning the application of sociobiology to human life is most reasonable, and go on to consider the implications regarding biological function. The moderate position, as I understand it, just involves the claim that *some* human psychological and behavioral phenomena must be understood *partly* in terms of certain genetically influenced psychological predispositions. This plausibly holds true even where cultural influences are involved, as such predispositions have plausibly had a strong influence (in conjunction with the given environment, as always) on the course of cultural history itself through influencing the behavior of the builders of culture, thus contributing to the shaping of specific conventions, such as conventional attitudes and practices surrounding aggression or sex, which in turn further condition individual and social behavior.[9]

Notice that nothing at all has yet been said about moral or rational *justification*. It is certainly not being suggested that pointing to evolutionary influences serves to provide a moral or rational justification for behavior that is so influenced. Secondly, there is no suggestion that genetic influence on behavior precludes the possibility of change. The above examples concerning family planning are sufficient to show that certain plausible genetic influences can be, and often are, overcome in favor of various goals, principles and independent motivations—though as E.O. Wilson stresses, there may well be limits to how radically human life can be changed in certain respects, and certain costs associated with various changes. Thirdly, there is no denial of the crucial role played by *cultural* evolution in shaping human behavior. Indeed, cultural evolution figures into typical sociobiological accounts themselves (though that does not exhaust its role, as there is plausibly a great deal of cultural evolution that is *not* driven ultimately by adaptational dispositions—something that is often underemphasized). As Wilson puts it: "The question of interest, then, is

---

8 These issues are further addressed in the next chapter.

9 See E.O. Wilson (1978) and Diamond (1992). Though I accept this moderate position as stated, in the next chapter I will criticize what I take to be significant excesses in Wilson's sociobiological approach to human nature.

the extent to which the hereditary qualities of hunter-gatherer existence have influenced the course of subsequent cultural evolution."[10]

A few examples will help to illustrate the sort of interplay between natural selection and culture that is typically suggested. Consider Wilson on aggression:

> The particular forms of organized violence are not inherited. No genes differentiate the practice of platform torture from pole and stake torture, headhunting from cannibalism, the dual of champions from genocide. Instead, there is an innate predisposition [resulting, predictably, from human natural selection history] to manufacture the cultural apparatus of aggression. Culture gives a particular form to the aggression and sanctifies the uniformity of its practice by all members of the tribe. . . .
>
> It is obvious that the specific conventions of aggression—for example, ambush as opposed to open warfare, and ornamental stone axes as opposed to bamboo spears—are heavily influenced by the materials at hand and the bits and pieces of past custom that can be conveniently adapted. . . . Only by considering the determinants of aggression at the three levels—the ultimate, biological predisposition, the requirements of the present environment, and the accidental details that contribute to cultural drift—can we fully comprehend its evolution in human societies.[11]

Similarly, Jared Diamond argues that just as a genetically influenced, general predisposition toward organized violence has arguably contributed to the evolution of various forms of aggressive behavior in various cultures, displaying both certain common elements and certain cultural peculiarities, certain general predispositions regarding sexual attitudes and behavior have given rise to various customs in various cultures—though he focuses on a striking common element in one particular type of convention. He begins by arguing that

---

[10] E.O. Wilson (1978, p. 91).

[11] E.O. Wilson (1978, pp. 117–119). Cf. also the following passage on altruism: "Let me grant at once that the form and intensity of altruistic acts are to a large extent culturally determined. Human social evolution is obviously more cultural than genetic. The point is that the underlying emotion, powerfully manifested in virtually all human societies, is what is considered to evolve through genes." (p. 160)

there is a "realistic biological basis for mens' widespread paranoia about being cuckolded"—a situation that has striking parallels in other species—and suggests that the influence of this predisposition may be seen in the nature of traditional adultery laws:

> Until recently, essentially all such laws —Hebraic, Egyptian, Roman, Aztec, Muslim, African, Chinese, Japanese, and others—were asymmetrical. They existed to secure a married man's confidence in his paternity of his children, and for no other purpose. Hence these laws define adultery by the marital status of the participating woman; that of the participating man is [considered] irrelevant. Extramarital sex by a married woman is considered an offense against her husband, who is commonly entitled to damages, often including violent revenge. . . . Extramarital sex by a married man is not considered an offense against his wife. Instead, if his partner in adultery is married, the offense is against her husband. . . . No criminal law against male infidelity even existed until a French law of 1810, and that law only forbade a married man to keep a concubine in his conjugal house against his wife's wishes. Viewed from the perspective of human history, the. . . . near-symmetry of modern western adultery laws is a novelty that appeared only in the last 150 years."[12]

Wilson offers a similarly structured account of another custom governing sexual behavior, which again is nearly universal, though it naturally takes on various particular, culturally conditioned forms—namely, the incest taboo. The prevailing sociobiological explanation of incest taboos locates the ultimate cause in

> the heavy physiological penalty imposed by inbreeding. . . . More than one hundred recessive genes have been discovered that cause hereditary disease in the undiluted, homozygous state, a condition vastly enhanced by inbreeding. . . . [Thus,] the elementary theory of population genetics predicts that any behavioral tendency to avoid

---

[12] Diamond (1992, pp. 94–98). Other cultural artifacts for which a similar account is given are traditional asymmetric codes of honor and shame common in Mediterranean countries, and (to take an extreme case) the practice of genital mutilation still found in parts of Africa and the Middle East.

> incest, however slight or devious, would long ago have spread through human populations. So powerful is the advantage of outbreeding that it can be expected to have carried cultural evolution along with it.[13]

I am not concerned with the truth of these particular claims, but only with the general implications of such ethological or sociobiological hypotheses about human life for biological teleology. It seems uncontroversial that *some* such claims are true, at least in the simplest cases of appetites or primal fears or basic parental instincts, and for the sake of argument I will suppose that the range of dispositions and behaviors that are plausibly genetically influenced extends beyond these, to include those involved in such things as aggression, territoriality, social organization, and sexual attitudes and behavior. Let us suppose, then, that there are certain general predispositions that are—or at least once were—adaptations, though the particular forms of emotion, attitude, and behavior to which they give rise have naturally been influenced also by particular environments and cultural histories.

Should such traits be considered to have *biological functions* in human life? On the view of biological function that I have developed, they should, provided that they *still* tend to promote gene replication of the relevant sort; if they do not, then at most they are vestiges, functionally akin to the appendix. So, for example, it may be that given current modes of social organization (as opposed to the hunter-gatherer and primitive agricultural societies in which the traits originally evolved), the general disposition toward organized violence has ceased to play any standard role in promoting inter-generational gene replication of the relevant sort, and may even tend to detract from it; it may be a behavioral and emotional analogue of the appendix, which no longer has a biological function in human life.

On the other hand, if we suppose, for example, that a disposition on the part of men to pursue what in other species would be called a "mixed reproductive strategy" (i.e. seeking additional sex outside a lasting pair bond) has its roots in natural selection history—as is clearly the case with many other species—this adaptational disposition might still plausibly have its original biological function. It may still

[13] Wilson (1978, pp. 38–9).

functionally contribute to mens' reproductive output, just as it originally did—and similarly for related attitudes and emotional dispositions, though of course all of this, and the particular behavior that results from it, will be conditioned by particular cultural histories.[14] Or to take a less controversial example, it seems clear that our basic appetites for food, drink and sex, and at least much of the behavior they normally motivate, still have their original biological functions. The same seems true of the natural disposition to avoid incest. These things still play functional roles in a human being's reproductive success (whether directly, or by contributing to his or her survival), just as they did in the past, and just as they do in other animals.

2. CAN LAWS AND CUSTOMS HAVE BIOLOGICAL FUNCTIONS?

The claim so far, then, is that there are plausibly some traits in the realm of human psychology and behavior that still have biological functions, along with others that were once functional but no longer are, as their effects no longer tend to promote the relevant sort of gene replication. It is a further, interesting question whether we should attribute biological functions also to related cultural artifacts. For example, should we attribute a biological function not only to the disposition to avoid incest, but also to the anti-incest laws or customs to which that disposition has given rise at the cultural level, and which in turn serve to reinforce that disposition? This is a tricky matter, involving issues similar to those discussed in chapter five. It is natural to attribute functions to such customs or laws, as in the case of other artifacts—the function of anti-incest laws, for example, being to discourage incest. And it is plausible that biological dispositions, such as the adaptive disposition against incest, were a major influence in the development of such laws. Thus, it is plausible to say that the functions of the laws are at any rate largely biologically-based functions, rather

[14] Cf. Diamond (1992, ch. 4). Again, note that I haven't yet said anything about moral or rational justification. I am here speaking simply of biological function, and if, as is generally the case in functional contexts, this claim about function supports certain related evaluative conclusions that seem clearly at odds with our considered judgments about *morality* and *practical rationality*, then this is just further support for what I have maintained from the beginning—namely, that it would be a hopeless mistake to try to understand facts about morality and practical rationality in terms of facts about biological teleology. I will say more about this below.

than stemming merely from independently motivated cultural practices.[15]

It is not so clear, however, whether the functions of the laws should be considered (at least partly) biological functions, right alongside the functions of the individual dispositions. An individual's disposition to avoid incest has a biological function insofar as it plays a role in his or her biological functioning, tending to make some contribution to the successful inter-generational replication of his or her genes—i.e. by avoiding wasting resources on making and caring for babies that are relatively unlikely to survive and one day produce healthy offspring themselves. By contrast, the function of an anti-incest law is *general*: the prevention of incest generally (i.e. among members of the society). That general end is not a biological end according to the view of biological teleology I have argued for, any more than *general* success at inter-generational gene replication is an ultimate biological end. The ultimate biological end for each individual organism is just the inter-generational replication of copies of *its* genes (whether residing in its own cells, or in those of close kin, in cases where kin selection is part of the natural selection history), and all other (subordinate) biological ends are such only insofar as they are instrumental to this.[16] On the other hand, *by* promoting the general end, the laws are thereby

---

[15] Cf. chapter five, section four.

[16] As mentioned in chapter five, section thirteen, there may indeed be subordinate biological ends at the level of a non-kin group—as, for example, where certain functional behaviors on the part of an individual serve the (subordinate) end of ensuring the survival of other members of a hunting pack, or the integrity of the pack, thus indirectly protecting the individual's own prospects for reproductive success (the ultimate end), since he depends on the others' cooperation to get sufficient food, and so on. In this way, then, the *survival* of useful partners, or the *integrity* of the cooperative group, may come in as a subordinate biological end, directed ultimately toward one's own reproductive success, i.e. one's own inter-generational gene replication. (Similarly with cases of symbiosis.) By contrast (and this is the point made above in the text), the inter-generational gene replication of other members of a non-kin group—i.e. other members of a society generally—is not among an individual's biological ends, nor is it a biological end in any other, more general sense; it is a biological end only insofar as it is the *ultimate* biological end *of each member*. One qualification: It is possible in principle for the reproductive success of others within a cooperative group to be among an individual's subordinate biological ends insofar as it might be instrumental to his own children's survival and reproductive success, since they too will need cooperative partners in the future. The point, however, is that just as in the other cases, the end here is subordinate to the individual's own success at genetic replication down the generations—*not* on a par with it, as if *general* inter-generational gene replication among members of the group were somehow an ultimate biological end.

promoting the biological ends of individuals (i.e. helping them to avoid incest), and helping to reinforce the original dispositions (i.e. providing cultural backing for them). Moreover, this promotion of biological ends is hardly accidental, since the laws are the direct result of adaptational dispositions directed toward those same ends. To this extent, then, it seems reasonable to speak of the laws as biologically functional, playing non-accidental roles in the promotion of biological ends.

These observations are meant just to illuminate the interesting issues involved in the attribution of biological functions to cultural artifacts. It would lead too far afield to go into this difficult matter in any more detail. I will therefore continue with a focus on human psychology and behavior.

3. HUMAN TRANSCENDENCE OF PROPER BIOLOGICAL FUNCTIONING

What is so interesting about the human case is precisely the extent to which we are capable of systematically *thwarting* the "designs" of natural selection, rebelling against many of the adaptational dispositions we have inherited—even those that have acquired strong cultural backing. I have already mentioned examples of this with regard to family planning. Another would be the development of and adherence to moral or religious codes condemning certain natural forms of violence, or uniformly forbidding extramarital sex, or praising unreciprocated altruism among non-kin. However pervasive biological teleology may be in our inherited psychological and behavioral dispositions, it is one of our distinctive features as human beings that we can and do systematically frustrate much of it in favor of our own values and designs for our lives.[17]

This ability to take significant charge of the course of our lives is rooted in the powers of reflection and rational thought mentioned above: the capacity to conceive of ourselves as agents, and to conceive of our actions as being subject not only to planning, but also, along

---

[17] Cf. Diamond (1992, p. 98): "In short, we have evolved, like other animals, to win at the contest of leaving as many descendants as possible. Much of the legacy of that game strategy is still with us. But we have also chosen to pursue ethical goals, which can conflict with the goals and methods of our reproductive contest. Having that choice among goals represents one of our most radical departures from other animals." A similar point is also stressed by Eibl-Eibesfeldt (1977, pp. 678–9).

with our desires and emotions, to justification and criticism. It is important to appreciate the depth of these practical mental capacities. Things would be much different if our powers were limited to the production of means/ends calculations culminating in behavior in the service of isolated specific ends generated by biological appetites and drives. But they are not. We are distinguished from other animals not merely by our advanced capacity for instrumental reasoning but by our capacity to conceive of and adopt what Anscombe calls "generic ends"—i.e. ends like happiness, glory, riches, and virtue—along with regulative ends such as principles of honesty and fairness. Moreover, we have the capacity to think *evaluatively* about such ends, as well as about more specific ends, and likewise to think evaluatively about *means* in light of such judgments about ends, and to take this into account in our practical reasoning, rather than just settling on whatever means might be calculated to further some given end in isolation.[18] When these capacities are sufficiently developed and exercised, we may eventually work out a sophisticated conception of the good life, which we then bring to bear on our everyday practical reasoning (though it will generally function primarily in the background). At this point, it is open to us to direct our lives at least in large part according to our own *reasoned conception of the good*, rather than being led about primarily by appetites and instincts directed ultimately at maximizing the inter-generational replication of our genes.

It is an interesting question where these sophisticated rational capacities came from. To some extent it is plausible to suppose that they are direct products of evolution. Gibbard, for example, argues that the general capacity for "linguistically infused normative motivation"—i.e. motivation resulting from the acceptance of norms—"evolved because of the advantages of coordination and planning through

[18] Cf. Anscombe (1981b), (1989, pp. 401 f.). The sense of "end" that I am employing here—in speaking of ends that we may be said to adopt and to pursue in the course of action—should not be confused with the sense of "end" I have been using in relation to functions, whether in animals, plants or artifacts. The former belongs not to functional teleology, but to *psychological teleology*. Such ends, which are spoken of as being adopted, held, pursued, etc., are attributed directly to individuals as such—indeed, they are attributed to them *qua* individual psychological agents—unlike function-related ends, as in biology; and nothing at all follows about such things as function and defect from the fact that an individual has certain ends of this type and performs certain actions with that end in view. There will not, therefore, be any direct move from facts about human psychological ends to facts about human functional teleology.

language."[19] The matter is less clear, however, when it comes to the most sophisticated dimensions of both our practical and theoretical intellectual capacities, such as our ability to think abstractly about the good life for human beings, or to learn to do advanced physics or philosophy; similarly with the kinds of advanced intellectual and emotional capacities involved in the composition of symphonies and poetry. One puzzle for the evolutionary view is that the human brain appears to have completed the bulk of its evolution by some point during hunter-gatherer times, thus possessing *even at that time* roughly the same general potentialities exhibited today. This would include potentialities for the sorts of advanced mental activities mentioned above—though obviously the actual carrying out of particular activities like calculus or symphonic composition is possible only given a certain background of cultural evolution and personal training.[20] If this is right, then the most natural account of such advanced intellectual capacities is not that they are themselves adaptations, since they do not seem relevant to promoting genetic fitness in the context of a hunter-gatherer lifestyle and environment.

It is hard to imagine, for example, how the enormous gap between the sorts of mental capacities plausibly relevant to hunting and gathering and the sorts of mental capacities exhibited in the development of quantum mechanics could be filled by a series of small steps (i.e. sufficiently small as to be attributable to mutations or changes resulting from recombination), such that each step toward the more advanced capacities provided evolutionarily significant advantage

---

[19] Gibbard (1990, p. 57). Cf. also J.Q. Wilson (1993).

[20] See Diamond (1992, ch. 2), who claims that "we were fully modern in anatomy and behavior and language by forty thousand years ago, and that a Cro-Magnon could have been taught to fly a jet airplane" (p. 56). The vast differences between modern human beings living in (or greatly influenced by) advanced industrialized societies and Cro-Magnons are attributed primarily to cultural evolution rather than to differences in genetically based potentialities, just as with the vast differences between members of modern societies and contemporary hunter-gatherers. (Cf. also Wilson, 1978, p. 35). Wilson is a bit more cautious: "Although we do not know how much mental evolution has actually occurred [in the last forty thousand years], it would be premature to assume that modern civilizations have been built entirely on the genetic capital accumulated during the long haul of the Ice Age. That capital is nevertheless very large. It seems safe to assume that the greater part of the changes that transpired in the interval from the hunter-gatherer life of forty thousand years ago to the first glimmering of civilization in the Sumerian city states, and virtually all of the changes from Sumer to Europe, were created by cultural rather than genetic evolution." (p. 91)

to the hunter-gatherers in question, thus leading to the evolution of increasingly sophisticated mental capacities right up to the capacity to learn quantum mechanics. A more plausible account of our highly advanced mental capacities is that they are not themselves adaptations, but are rather *concomitant* properties of the kind of brain that evolved under hunter-gatherer conditions because of the evolutionary advantages of more general and straightforwardly adaptive properties, such as the capacities for complex spoken language and for rational thought of the sort relevant to improving the reproductive output of hunter-gatherers. And, of course, in the particular forms they take—such as a developed capacity to do theoretical physics—they require certain forms of training and cultural background, which in turn rests on a long history of cultural evolution.

While this is in many ways an intuitively appealing explanatory move, however, it does seem to involve the acceptance of a very dramatic fact as a matter of faith. As Nagel remarks, if we endorse this account we must simply accept that "not only the physical but the mental capacity needed to make a stone axe automatically brings with it the capacity to take each of the steps that have led from there [via cultural evolution] to the construction of the hydrogen bomb," not to mention the composition of Beethoven's late string quartets or the writing of Kant's *Critique of Pure Reason*.[21] I do not find this an implausible hypothesis, as Nagel does, but he is certainly right to stress that it is far from obvious why things should be this way. Why should selection for relatively modest mental capacities equally generate such

---

[21] T. Nagel (1986, p. 80). His formulation of the point is perhaps a bit overstated, but the idea is clear enough. What is meant to be represented by "the mental capacity needed to make a stone axe" is presumably the more general sort of mental capacity that set Cro-Magnons (who were fully or nearly fully modern in their potentialities) apart from Neanderthals and even Middle Stone Age Africans. This sort of mental capacity was expressed in the making and use of tools with specialized functions, such as "standardized bone tools, bows and arrows, nets, fishhooks" and stone axes (where the stone is mounted on a handle, as opposed to merely being held in the hand), and in the sort of innovation that produced significant cultural variation for the first time. More to the point, the mental capacity possessed by Cro-Magnons included a capacity for complex spoken language. (Diamond, 1992, pp. 45 f.) The point, then, is that the common naturalistic account commits us to the claim that the evolution of a brain with such mental capacities—where these capacities were selected for because of their genetically advantageous contributions within the context of a hunter-gatherer way of life—automatically brings with it the potentialities, with the right cultural backing, for such things as theoretical physics and analytic philosophy.

impressive, sophisticated mental capacities—particularly capacities that are evidently so good at getting us to truths that go so far beyond anything that was relevant to hunter-gatherer gene propagation (such as the equations of quantum mechanics)? This is surely not something that should just be taken for granted. On the other hand, even if Nagel is right in suggesting that we cannot presently give a plausible answer to this question, this should not preclude our accepting the above naturalistic account of the origins of our advanced mental capacities, as far as it goes. We can consistently accept that account without falling into complacent disregard for the deep mysteries it leaves unexplained.

It is not necessary for my purposes to insist that this account is correct, but since it is the only naturalistic account I know of with any plausibility, I will assume it is basically right. What conclusions can now be drawn regarding biological teleology? Evidently, what has happened is that evolution has inadvertently provided us with the very tools that have enabled us systematically, though by no means entirely, to transcend its "designs", living lives that are directed at least in large part by our own reasoned conception of the good, rather than by forces "designed" to optimize the inter-generational replication of copies of our genes. Indeed, as already illustrated, our own designs for our lives may often conflict directly with the latter end, in favor of all manner of projects reflecting our own values and concerns.

To the extent, then, that our advanced mental capacities lead us to behavior that detracts from the sort of promotion of ultimate biological ends for which our mental capacities were in fact selected, those exercises of our advanced capacities and the resulting behavior must be considered *dysfunctional*—not in any absolute sense, certainly, but at least *in relation to the normative standards set by the biological teleological framework*. From the point of view of biological teleology, we are just gene replicating systems that have acquired minds of our own, so to speak—a development that often gets in the way of the functioning that we, like other organisms, have been "designed" by natural selection to carry out.

This might be thought an odd result, inasmuch as the sort of activity I'm calling "dysfunctional" from the point of view of biological teleology is so clearly a *natural* part of human life. If such activity is just the result of the natural operations of our faculties (in the context of cultures that are equally natural parts of human life, as natural human creations), shouldn't it just be considered part of natural biological functioning for human beings after all? The apparent conflict, however, just stems from an ambiguity in the notion of *natural biological functioning*. The objection employs a broad construal of this notion, according to which anything that a biological entity naturally does *as such*, i.e. where the activity is a natural manifestation of its being alive, counts as part of its natural biological functioning. So understood, the advanced operations of our brains that lead to systematically non-adaptive behavior, for example, are indeed part of the brain's natural biological functioning. By contrast, I have all along been employing the notion of *proper* functioning in relation to biological teleology. And it is consistent with what has just been said to maintain that such advanced operations of the brain are not part of its proper functioning from the point of view of biological teleology. All we have to recognize is that not everything that it is natural for a biological entity to do as such is part of its proper functioning in relation to biological teleology. As I have argued in earlier chapters, the standards for proper functioning set by biological teleology involve factors that go much deeper than facts about what it is natural for biological entities to do.

4. BIOLOGICAL FUNCTION IN CONNECTION WITH ADVANCED MENTAL CAPACITIES

It is compatible with the claims of the last section that in certain respects our advanced mental capacities may also prove to be useful in furthering our ultimate biological ends—for example, enabling us to develop drugs that improve our fertility or fight disease, or to conserve environmental resources so that we don't do ourselves in (never mind for the moment that we wouldn't be in such danger in the first place if it weren't for our own activities stemming from those same advanced capacities). The possibility of such positive contributions to biological ends raises the question: Can our advanced mental capacities thereby be accorded biological functions in connection with such contributions? Or is this just a fortuitous promotion of biological ends that cannot properly be regarded as biologically functional in nature? The answer will depend on whether, or to what extent, our use of these capacities in ways that tend to promote our biological ends can be assimilated to something like the sea turtle's use of its tail to protect its eggs, as discussed in chapter five. I said there that even though the tail did not evolve because of its usefulness in protecting eggs, the turtle's use of the tail for this purpose is plausibly biologically functional in nature, as it is plausibly either a behavioral adaptation or an improvised or learned way of trying to achieve a certain adaptational goal, i.e. protecting the eggs. And if this functional use of the tail for protecting the eggs is standard enough to allow us to speak of the tail as playing a standard egg-protecting role in the life of that type of organism, then there is no objection to speaking of egg protection as one of the biological functions of the tail. The question, then, is whether something similar can be said concerning our advanced mental capacities.

I am assuming that these capacities were not selected for the biologically useful purposes to which they may now be put by us (and they were certainly not selected for the many biologically useless, but distinctively human, purposes to which they are frequently put), just as the turtle's tail was not selected for egg protection. What we need to determine, then, is whether we can point to any disposition behind certain of our uses of our advanced mental faculties that is at least partly attributable to genetic influence, analogous to the turtle's adaptational disposition to protect its eggs. If we can, then such uses of our capacities to further certain biological ends are to that extent plausibly biologically functional in nature, like the turtle's use of its tail to protect its eggs.

On the other hand, to the extent that our uses of these capacities are motivated not by adaptational dispositions, but by reason-based motivations that cannot be reduced to such dispositions, it will not be correct to speak of these uses as biologically functional in nature. They will simply be non-functional uses (in biological teleological terms) of our capacities, which happen also to further our biological ends. Probably the most we are likely to get in the way of biologically functional use, where human beings are concerned and the use in question is improvised or learned, is a mixed case. Recall the example, also from chapter five, of our use of our noses to hold up eyeglasses: The motivating disposition may be partly attributable to genetic influences, but it is certainly also partly attributable to independent motivations.

Are there, then, biologically functional dispositions at least partly behind some of our uses of our advanced mental faculties, so that those uses are at least partly biologically functional?[22] I think that there plausibly are, but we have to be careful here. It is not obvious, for example, that any of the examples I gave earlier involves such an underlying disposition. Someone's requesting and taking a drug to increase his fertility might plausibly involve such an influence, as part of what is behind his desire to have children; but it is not clear that the researcher's use of her advanced mental faculties to develop the drug involves any significant influence of this sort. Or again, the environmentalist's concerns are generally too broad to be plausibly attributed to adaptational dispositions, which latter would presumably be centered primarily around his own survival and that of his kin. On the other hand, they might extend more broadly insofar as other things (such as the survival of a wider group or habitat) are perceived to be instrumental to this, and in this way they might after all be having a significant, though still only partial, influence. In any case, all we need to support the claim that some of our uses of our advanced mental faculties are biologically functional in nature are examples of such uses to promote some biological end of the agent, where this is at least partly motivated by an adaptational disposition to pursue such an end. These might include someone's studying the stock market in order better to provide for herself and her family, for example, or someone's performing intellectual feats to impress a potential mate.

It is a further question whether such biologically functional uses of our advanced mental faculties are sufficiently standard for us to go on to attribute biological functions to those faculties themselves, and if so, how specifically those functions should be described. Whether we can in fact make such function attributions depends, as in the case of the turtle's tail, on whether the functional use in question is standard enough for us to speak of the faculties' playing a standard role in human biological functioning insofar as they are used in these ways. (The same was said about the turtle's tail: While the use of it to protect eggs is biologically functional where it occurs, it is not right to count this as one of the tail's functions in sea turtle life if it is something that only a few turtles engage in, for example.)

It is illuminating in this connection to consider another closely related case of use, which often occurs in conjunction with our use of higher mental faculties—namely, our use of body parts, such as our

---

22 See also chapter five, section four, for a closely related discussion.

hands, for all kinds of purposes, some of which may further our biological ends (whether straightforwardly, as with eating, or in a less direct way involving our advanced mental faculties, as with operating a computer in the course of researching the stock market, in the example given above), and some of which may not (as in playing the guitar for fun).

There is first of all a general problem concerning levels of description. Hands, for example, are themselves adaptations, and it seems unproblematic to ascribe to them the biological function, say, of grasping. More generally, it is plausible to speak of biological function in connection with virtually the whole range of movements of which hands are normally capable, as the various elements of this dexterity have all plausibly played a part in the evolution of the human hand, and are still generally relevant to our biological functioning. We may even be able to speak in this connection somewhat more specifically of such things as crafting, grasping and carrying tools, which can still be said to have played a part in the evolutionary shaping of the hand, and so might be said straightforwardly to be among the hand's biological functions, so long as some such activities still play standard roles in our biological functioning.

When it comes to a specific, mentally advanced activity on the order of operating computers in the course of researching the stock market, however, we are no longer dealing with a level of description that allows for such straightforward attributions of biological function. At best it may be possible, in principle at least, to count such a thing as one of the biological functions of the hand in the way in which we may be able to count protecting eggs as one of the functions of the turtle's tail—i.e. by way of standard functional use rooted in a broader functional disposition. In fact, however, while it does seem plausible to regard such a use of the hands as at least partly a matter of biological functional use of something that was not specifically selected for such activities (as with the related use of advanced mental capacities), the description of the activity is too specific to be regarded as standard in the way that would allow us to speak of the hand as thereby playing a standard role in our biological functioning insofar as it is used in this way. That is, it seems implausible to say this about such a specific sort of modern activity, and I am thus inclined to stop short of such an attribution of biological function to the hand, while still recognizing that where such activities do occur they may be regarded as at least partly biologically functional in nature, as described above (though there will also be many activities that are not, as the example of using the fingers to fret guitar strings was meant to show).

With regard to advanced mental capacities themselves, we must be even more conservative. Again, whereas the hand is an adaptation, with certain straightforward biological functions (in terms of its basic movements), I have been supposing that the advanced mental capacities are just concomitants of an adaptation that was selected because of more primitive advantageous effects it had. These advanced capacities do not, therefore, have straightforward biological functions, as the hand does. (If I am wrong about this, then it should be clear from the discussion of the hand what should be said here.) The question that remains is whether they nonetheless have use-based biological functions, as discussed above. I have suggested that some of our use of our higher mental faculties can plausibly be regarded as at least partly biologically functional in nature—for example, using our mental capacities to increase our financial resources. Does this allow for attributions of biological functions to the capacities themselves? Again, the answer is the same as for the hand, and for the same reasons. It seems to me that if we can attribute biological functions to our higher mental capacities in this way, then they will have to be fairly general, so that these capacities can plausibly be regarded as playing standard roles—through being functionally used in these ways—in human biological functioning. Perhaps it can be said, for example, that our advanced mental capacities have as their biological function the working out of complex solutions to problems that arise in relation to the satisfaction of certain of our biological ends.

5. DEPARTURES FROM PROPER BIOLOGICAL FUNCTIONING IN ANIMALS

Let me return now to the claim that certain exercises of our advanced mental capacities and the resulting behavior—including many that are perfectly defensible from a rational or moral point of view—must be considered *dysfunctional* from the point of view of the biological teleological framework. It is worth noting that this sort of dysfunction seems to have parallels elsewhere in the animal kingdom, though there are important differences.

Consider, for example, the evolution of pleasure and pain as incentives to perform certain adaptive activities and to avoid various maladaptive ones. There is nothing in principle to rule out the possibility of animals sometimes pursuing the pleasure associated with a certain activity to a degree that outstrips its biological function (e.g. overeating), or in a manner that is not ordered to the proper end (e.g. masturbation). This may be less common than is the case with human beings, but it is not surprising that it should occur. Natural selection may be expected to give rise to adaptive mechanisms involving pleasure, even though they have such side-effects, so long as these mechanisms have on the whole served to promote the propagation of their coding genes over rival alleles. Such behavior needn't obviously be regarded as dysfunctional if it is just a harmless side-effect of the possession of the adaptive mechanisms in question. But if it gets to the point of interfering with the animal's proper biological functioning, then it seems natural to regard it as dysfunctional from the point of view of biological teleology. As with the human case, what has happened here is that natural selection has given rise to something that has a certain biological function, but that can also at least sometimes take on a life of its own, as it were, motivating behaviors contrary (either in kind or in degree) to those it was selected to motivate.

A rather different sort of example of a functional mechanism that can lead to dysfunctional types of behavior may be found in cases of brood parasitism.[23] The passerine birds that are parasitized by cuckoos, for example, have a natural, adaptive instinct to care for their own eggs and fledglings, which parasitic cuckoos exploit for their own advantage. In some cases, the host might simply have such crude powers of discrimination that it is unable to discriminate between its own eggs and various other objects placed in its nest (as with herring gulls), and thus cares for cuckoo eggs placed in its nest as if they were its own. In other cases, the host might have fairly keen powers of discrimination, but still be fooled due to the elaborate egg-mimicry that some cuckoos have evolved. Or again, a host may simply lack the power to discriminate between its own fledglings and cuckoo fledglings in its nest—even if it possesses fairly keen egg-discriminating abilities; for in the evolution of the species there may have been greater selection pressures leading to egg-discriminating ability than to fledgling-

[23] Dawkins (1982, pp. 67–70) and (1976, p. 110).

discriminating ability (since, for example, there is greater benefit in detecting the parasite sooner).[24] Or a host may be manipulated into feeding cuckoo fledglings by certain specific stimuli they produce, which provoke feeding behavior by the host—the production of such stimuli being a cuckoo adaptation, like egg-mimicry.

In any case, the result is that there is a misfiring of the maternal instinct, and the unwitting foster parents care for the cuckoo eggs and fledglings as if they were their own. From the point of view of the host's functional framework, such behavior is clearly *dysfunctional*—brought about and exploited by the cuckoo, saving it valuable time and resources by getting its eggs and fledglings taken care of for free. (From the perspective of the *cuckoo's* functional framework, the cuckoo's manipulative behavior is of course part of *its* proper functioning, and the host's consequent behavior is in effect the environmental result toward which the cuckoo's manipulative activities are functionally aimed. Indeed, from this perspective, the host's behavior can be said to play a functional role in the cuckoo's life—something which cannot be said from the perspective of the host's own functional framework, since the altruistic aiding of cuckoos is certainly not among the host's biological ends.)

Similar mistakes may occur also among organisms with much more sophisticated psychological faculties, where the misfiring instincts plausibly involve genuine emotions. In troops of monkeys, for example, an orphan will sometimes be adopted by a non-related female, especially if she has recently lost one of her own. As Dawkins remarks:

> In most cases, we should probably regard [this sort of] adoption, however touching it may seem, as a misfiring of a built-in rule [i.e. instinct]. This is because the generous female is doing her own genes no good by caring for the orphan. She is wasting time and energy which she could be investing in the lives of her own kin, particularly future children of her own. It is presumably a mistake which happens too seldom for natural selection to have 'bothered' to change the rule by making the maternal instinct more selective.[25]

---

[24] Dawkins (1982, p. 68).
[25] Dawkins (1976, p. 109).

Now it is admittedly debatable whether Dawkins' suggested take on this situation is in fact correct. Perhaps the social relations among monkeys are complex enough that such behavior is a natural result of a general and adaptational disposition to show solidarity with the troop. In that case, while adopting an orphan may not *directly* serve to promote the spread of the individual's genes, it may be part and parcel of the sort of behavior that does so in the long run, through maintaining the individual's connection with the group—without which she would be at a much more serious disadvantage, perhaps not surviving long enough ever to reproduce again.[26] If so, then the behavior needn't be thought to be dysfunctional at all. On the other hand, if Dawkins' characterization of the situation *is* correct for a given case, and what is happening really is just a "misfiring" of the maternal instinct, then such behavior is indeed dysfunctional from the point of view of biological teleology—and similarly if it is a result of some kind of relatively independent motivation, going beyond a mere displacement of instinct (i.e. the sort of thing referred to in the next chapter as "evolutionary spin off").

The above parallels between departures from proper biological functioning in non-human animals and departures in the human case should not be exaggerated, however. Not only is the departure much more widespread and pervasive in human beings than in other animals, but more importantly it has the potential to be (and often is) *systematic* in the human case, by virtue of the rational capacities involved. It is one thing to depart from standard biological functioning in certain arbitrary ways by being drawn to pursue certain isolated pleasures out of proportion to, or in a way that is irrelevant to, their biological function, or by having certain instincts misfire. It is quite another thing, however, to depart from standard biological functioning as a result of directing one's pursuits in accordance with a complex system of values and goals, in the course of trying to live a reasonably coherent and meaningful life. Even if the monkey's adoption of an unrelated infant stems from something more than a misfiring or overextended maternal instinct, it is doubtful whether monkeys have the intellectual resources to weave such behavior into a systematic, agent-directed departure from standard biological functioning. By contrast, even if some cases of human adoption of unrelated infants, or even cases of caring for pets,

---

[26] I am indebted to Robert Adams for pointing out this possibility.

do stem ultimately from misfiring or overextended natural instincts (as sociobiologists may claim, as mentioned earlier[27]), human beings may nonetheless assign value to these activities and relationships, and incorporate this into a scheme of values that they use in systematically directing their lives, departing from their biological functioning in a way not open to non-human animals.

The general difference between non-rational and rational agents to which I am appealing was formulated by Aquinas in this way:

> Those who have reason move themselves to an end, for they have dominion over their actions through their free judgment, which is a power of will and reason. Those lacking reason tend to an end by natural inclination, as moved by another and not by themselves, since they do not know an end *as* an end, and consequently cannot order anything to an end but are ordered to it by another.[28]

In relation to the present issue, we might say that even when a non-human animal is departing from proper biological functioning in the pursuit of a certain pleasure, for example, this is still just a matter of its being moved by a natural inclination. The activity does not stem from any rationally ordered scheme of ends the animal has set for itself—which would involve knowing ends *as* ends, as Aquinas puts it—and so departures of this kind do not add up to any *systematic* transcendence of the life the animal was "designed" by natural selection to lead. This is true even for hypothetical cases where the inclination somehow stems from the animal's faculties in a way that involves more than the misfiring or overextending of adaptational instincts. There is still no knowledge of ends *as* ends, no evaluation and ordering of ends into a scheme whereby the animal directs its life. Non-human animals may transcend their proper biological functioning in various isolated respects, but they do not have the ability to take general control of their lives and to set largely new courses for themselves. By contrast, we do possess and exercise that ability—evaluating, adopting and ordering both generic and specific ends for ourselves, as described earlier, so

---

[27] Cf. J.Q. Wilson (1993, pp. 127–8) on natural parental responses of attachment and the extension of these responses even to pets, particularly those resembling human infants in certain respects.

[28] Thomas Aquinas, *Summa Theologiae*, I-II, Q 1, art. 2 (trans. John A. Oesterle, in *Treatise on Happiness*, University of Notre Dame Press: 1983).

that the activities that stem from such a scheme may indeed add up to a systematic (though by no means complete) transcendence of the kind of life our natural selection history has "designed" for us.[29]

[29] By "non-human" animals, I mean other actual, known animals. I am not denying that there could in principle be animals that were both non-human and possessed of rational capacities comparable to ours in the relevant respects. If there were, then the distinction I am after would have to be put more generally in terms of rational vs. non-rational animals, where the former would include human beings and these other rational animals, while the latter would include all presently known non-human animals.

CHAPTER X

Human Nature and Morality

1. SOCIOBIOLOGICAL EXCESSES, ALTRUISM, AND THE TWO SIDES OF HUMAN NATURE

I have argued that *from the point of view of biological teleology*, much characteristic human activity—while perfectly natural—falls outside the realm of proper functioning and is sometimes positively dysfunctional. What this means is that *human nature*, properly understood, is a complex mixture of different elements. On the one hand, some of our natural tendencies are of the sort purportedly explained (at least in part) by evolutionary psychologists and sociobiologists; among these some must be regarded from a functional point of view as vestiges of once functional traits, while others may retain their biological functions to various degrees, and still others may never have been functional, being only side-effects of other traits. On the other hand, other natural human tendencies stem from our possession of sophisticated reflective capacities, as discussed in the previous chapter, and from the point of view of biological teleology may well lead regularly to behavior that falls outside the scope of proper biological functioning.

An example of the second aspect of human nature is the fact that human beings "are not merely concerned to bring about such circumstances as that object A be moved to point B...[but] want e.g. happiness, glory, riches, power".[1] Sociobiologists may be able to give partial accounts of the adoption of some such ends (and of the mental

[1] Anscombe (1989, p. 399). She refers to this as "an important fact of human nature."

capacities that make this possible), but the point is that the general tendency to adopt and to pursue such generic ends, along with the capacity and tendency for evaluating and ordering ends, may also lead to the adoption of certain ends and the performance of certain actions that go beyond what can be explained in sociobiological or biological functional terms. All of these elements need to be recognized in an adequate conception of human nature, which (*contra* E.O. Wilson) is much more than just "a hodgepodge of special genetic adaptations to an environment largely vanished, the world of Ice-Age hunter-gatherers," just as modern human life is (again, *contra* Wilson) more than just "a mosaic of cultural hypertrophies of the archaic behavioral adaptations."[2]

We should fully expect to find certain complex human traits that result from the combined influences of different aspects of human nature—traits that therefore cannot be adequately explained exclusively in sociobiological terms or exclusively in terms of independent exercises of higher faculties, but must be explained by appeal to both factors. A plausible example of such a trait is human altruism. Two equally shortsighted mistakes are common here.

On one side, someone understandably impressed with the differences between human beings and other animals might just assume that all manifestations of human altruism are simply attributable to arbitrary cultural influences or to reason-based emotions and convictions, for example, ignoring the possibility that to at least some degree, certain ranges of human altruistic feeling and behavior have their roots in evolved biological predispositions. On the other side, someone understandably impressed by the successes of sociobiology, particularly with regard to similar phenomena in other species, might just assume that the existence of human altruism requires some kind of Darwinian explanation, and indeed that such an explanation—along with an account of how exactly various cultures have contributed to the particular forms that the general altruistic tendencies have taken—is exhaustive. To someone in this position, the only real question, apart from the details of cultural shaping, will seem to be the nature of the

[2] Wilson (1978, p. 203). A *hypertrophy* is "the extreme development of a pre-existing structure." Wilson's general thesis is that "most kinds of human social behavior are hypertrophic forms of original, simpler responses that were of more direct adaptive advantage in hunter-gatherer and primitively agricultural societies" (226).

selection pressures involved and the characterization of the resulting adaptations. The discussion will center, for example, around the extent to which human altruism is plausibly to be traced to kin selection (which would explain what Wilson calls "hard-core" altruism toward kin), and the extent to which it is to be traced, perhaps with the help of game-theoretic models, to ordinary selection in non-zero sum contexts in which reciprocal altruism within a group is genetically more advantageous to members than rival, purely selfish "strategies" (which would explain what Wilson calls "soft-core" altruism toward other members of the group).[3]

I want to focus on the second of the above two mistakes for a moment—the sociobiological excess. What is overlooked in such an approach is that at least some human altruism may stem not from culturally shaped genetic predispositions, but from culturally shaped moral convictions that cannot simply be traced back to genetic predispositions. Such convictions or values may instead have their source in sophisticated reflection about what is important in human life and about how this bears on questions about how we should live. This possibility is nicely articulated by Sober, for example, who refers to the phenomenon as "evolutionary spin-off":

> Perhaps the ability to reason abstractly evolved because of its individual advantageousness. But once in place, this intelligence led human beings to see that rational considerations oblige them to take the interests of others as seriously as they take their own. If something like this is right, then vernacular altruism may find its pedigree not in evolutionary altruism, but in the sophisticated thoughts and feelings that a mind produced by individual selection was first able to formulate.[4]

I have suggested that there may actually be a *double* spin-off here, as even the advanced rational capacities appealed to here may be

---

[3] Wilson (1978, ch. 7). For a lucid discussion of "evolutionarily stable strategies" (or mixes of strategies) involving reciprocal altruism, based on Axelrod's game-theoretic approach, see Dawkins (1989, ch. 12).

[4] Sober (1988, p. 96), who makes reference in this connection to Singer (1981). It is irrelevant to the present point whether the moral convictions in question are true or false or even non-cognitive. The point is just that such convictions, and hence the behavior to which they give rise, would not be directly explicable in sociobiological terms.

largely *concomitants* of the evolved brain, rather than traits that were themselves selected for their adaptive effects. If this is right, then natural selection has given rise to certain traits that happen to bring certain concomitant traits with them, and the latter in turn may give rise to desires, feelings and behavior that go well beyond anything relevant to promoting our natural biological ends. This provides the possibility of a *naturalistic* but *non-reductionistic* way of accounting not only for at least a certain range of human altruistic behavior, but perhaps even for something like *agapeic* love. Naturalism, in the broad sense in which it contrasts with supernaturalism and non-naturalism, certainly does not commit us to the view that "the qualities we value as most human—altruism, morality, religion, even love—are simply the survival strategies of our 'selfish genes'."[5]

There would seem to be many traits that do not themselves invite Darwinian explanations, but are rather side-effects of the possession of certain adaptations or their concomitants in certain contexts (where it is only the adaptations that are directly subject to Darwinian explanations). Plausible examples might include the capacity to appreciate (or more generally, to learn to appreciate) classical symphonic music, the interest many people take in things like philosophy, and the ability to enjoy a good joke or to appreciate irony. Even certain general traits that are *universal* across cultures, and so may be considered *natural* to human beings generally, might be accounted for in this way: If they naturally stem from typical human mental capacities—as perhaps the general capacities for humor, or sarcasm or music do—then we may reasonably expect to find them universally manifested even though they are not themselves adaptations and may confer no biological advantage.

Thus, it is a mistake to claim, as James Q. Wilson does, that:

> If our willingness to help others or shun cruelty is in some sense natural, then this trait must have had some evolutionary value. Otherwise, it would have been extinguished, because its existence would have reduced the prospects of survival of those endowed with

---

[5] As stated in the advertisement on the back of Wilson (1978).

> it. . . . If I am to maintain that sympathy is natural, I must show that it has enabled the species to survive.[6]

Or again:

> Sympathy for persons who are not offspring and creatures that are not human is a characteristic of almost all humans. . . . *If sympathy is widespread, it must have been adaptive*. . . . [though] what was selected for was not a simple desire for reproductive success; what has been selected for is a generalized trait that both encourages reproductive fitness and stimulates sympathetic behavior. . . . [namely,] attachment or affiliative behavior.[7]

Again, these claims are false because it is at least *possible* for a trait to be natural and widespread even though it is *not* an adaptation; and if it is not an adaptation, then it need not itself be adaptive: It might have nothing to do with genetic fitness at all, being just a side-effect of adaptive traits, for example. Indeed, contrary to what is claimed above, it may even be *maladaptive* to a certain extent, provided that this is outweighed by the adaptive advantage of the trait of which it is a side-effect. So, for example, even if sympathy or altruism were maladaptive, it does not follow that they "would have been extinguished" by natural selection. If they are natural side-effects of the evolved human brain, for example, rather than direct phenotypic expressions of genes subject to selection pressures, then they should be expected to persist so long as the possession of such a brain is on the whole more adaptive than any alternative that arises. Unless brains appear that are more adaptive than this one and that do not give rise to such "spin off" traits as sympathy

---

[6] J. Q. Wilson (1993, pp. 40–1), my emphasis. Not only is the general claim here false, for the reasons given in the text, but the appeal to the *survival of the species* is bad biology. In other places too Wilson characterizes the adaptive value of a trait in terms of its tendency to promote "the perpetuation of the species" (p. 23), but as discussed earlier, the adaptive value of a trait is properly understood in terms of a tendency to promote personal reproductive success or that of kin (including cases that work by way of certain group dynamics, of the sort stressed by multilevel selection theorists), not in terms of the perpetuation or survival of the *species*—except perhaps in certain rare, special cases where a "species-selection" model is plausible. To be fair, there are also other places where he is more careful to speak in terms of reproductive success (e.g. p. 70), and even in the above quote he speaks also of the survival *of those endowed with the trait in question*, which is more to the point than talk of the perpetuation of the species.

[7] J.Q. Wilson (1993, p. 44), my emphasis.

and altruism, the latter traits may be expected to persist, even if they are somewhat maladaptive. (I'm not suggesting that they *are* generally maladaptive. The point is just that they could be and yet still persist.) Wilson's reasoning depends on the assumption that the traits in question are under direct selective control; if they are, then it is true that they will tend to be eliminated by natural selection if they are maladaptive. But if they are side-effects of other traits, then the simple argument fails.[8]

I have not, of course, shown that such traits as human sympathy and altruism *are* in fact "spin-off" traits—that is, that they stem not from our particular genetically influenced predispositions, but from independent exercises of our advanced mental capacities. Indeed, I would not myself argue that this is the whole story, though I think it is a crucial part of it. The point is just that this possibility is enough to undermine the quick leap to sociobiological conclusions and one-sided scientific investigations of such traits. The same holds for emotions and values often associated with the enforcement of reciprocal altruism or cooperation—the disdain for cheaters, the offense taken or guilt felt over failures to reciprocate, and so on. E.O. Wilson, for example, is quick to come to a sociobiological explanation of such things:

---

[8] There is a way that Wilson's argument might be pressed even if the traits turn out to be side-effects, but this would require two additional assumptions: (i) that the side-effect traits require a certain kind of parental influence in order to be manifested (i.e. they are *not* just *automatic* side-effects of the possession of certain universal adaptations, even given the right sort of cultural context, but additionally require particular inputs by parents to emerge), and (ii) that whether or not a parent provides that influence depends on whether or not the parent possesses the trait herself (i.e. altruistic parents provide the influence that allows for the manifestation of altruism in their children, selfish parents don't). If these conditions were satisfied, then differential reproductive success between possessors and non-possessors would lead to differential representation of the trait in succeeding generations, just as if the trait were under direct genetic influence. Thus, if the trait is maladaptive, leading to such differential reproduction, it will indeed tend to be selected against. (Cf. Sober [1988, p. 79], who points out that "the mechanism of inheritance does not matter [to the process of natural selection]; only the fact of heritability is essential.") These assumptions would certainly take some work to defend, however, particularly in light of the powerful effects that can be exerted by social forces beyond parental influence. In any case, it should be stressed that all I am attacking is the oversimplified argument Wilson gives in the quoted passages. I am not suggesting that his theses themselves are implausible: Sympathy may well be a generally adaptive trait, for example, arising at least in part from more general naturally selected psychological predispositions (as discussed further below).

> It seems probable that learning rules, based on innate, primary reinforcement, lead human beings to acquire these values [i.e. honor and loyalty] and not others with reference to members of their own group. The rules are the symmetrical counterparts to the canalized development of territoriality and xenophobia, which are the equally emotional attitudes directed toward members of other groups.[9]

This is certainly possible. But it is equally possible that such things are to be accounted for not in terms of evolved genetic predispositions and subsequent cultural shaping, but rather in terms of independent exercises of our rational faculties (conditioned, of course, by cultural influences), stemming, for example, from a recognition of the lack of any rational justification for one person's free-riding on others' obedience to certain useful rules. Or, as I have suggested, it is possible that for a wide range of cases some appeal must be made to both factors in arriving at a complete explanation.

E.O. Wilson is especially prone to the above kind of sociobiological excess. There may be a great deal of truth in his sociobiological hypotheses, even with regard to altruism, cooperation and related emotions. But he seems to take it for granted that the obvious sociobiological alternatives exhaust the possibilities: On the one hand, kin selection has given rise to certain "hard-core" altruistic tendencies toward kin—tendencies that can regularly involve *genuine net sacrifice* on the part of the individual, since the possession of such traits may still increase the individual's inclusive fitness due to the positive reproductive effects for kin (who share a relatively high percentage of the individual's genes); on the other hand ordinary selection in certain social contexts has given rise to a broader, "soft-core", *reciprocal* altruism, which ultimately serves to promote each individual's own reproductive success (as the actions of cooperating selfish individuals involved in an iterated prisoner's dilemma situation ultimately serve to benefit each individual), and does not involve net sacrifice on the part of the individual. Thus, his conclusion:

---

9 Wilson (1978, pp. 169–70). Cf. also Gibbard (1990, pp. 67–8), on guilt, resentment and gratitude. Gibbard writes: "Propensities well coordinated with the propensities of others would have been fitness-enhancing, and so we may view a vast array of human propensities as [naturally selected] coordinating devices. Our emotional propensities, I suggest, are largely the results of.... selection pressures [having to do with "evolutionary bargaining situations"], and so are our normative capacities."

> The most elaborate forms of social organization, despite their outward appearance, serve ultimately as the vehicle of individual welfare. Human altruism appears to be substantially hard-core when directed at closest relatives, although still to a much lesser degree than in the case of the social insects and the colonial invertebrates. The remainder of our altruism is essentially soft. The predicted result is a melange of ambivalence, deceit, and guilt that continuously troubles the individual mind.[10]

This is, first of all, an oversimplification even from the limited perspective of the sociobiological framework. Even if we consider only altruism arising from naturally selected psychological mechanisms (i.e. certain instincts and emotional dispositions), and grant that these mechanisms are functionally directed ultimately toward the promotion of reproductive success— i.e. one's own in the case of "soft core" altruism, and that of kin in the case of "hard-core" altruism—there may well in fact be plenty of altruistic behavior that fails to fit into either category. Recall, for example, the case of overextended maternal instincts and adoption discussed in the previous chapter. The result, which might be quite common, is altruism that is like "hard-core" kin altruism (and unlike "soft-core" reciprocal altruism) in that it can involve genuine personal net sacrifice, but is like "soft-core" reciprocal altruism (and unlike "hard-core" kin altruism) in that it is aimed at non-kin. By the same principle, there could be altruism stemming ultimately from (an overextension of) one or the other of these two sources that is aimed at things having nothing ultimately to do with promoting reproductive success at all—such as the happiness of strangers. So even in terms of sociobiologically based altruism, the actual situation is likely more complicated than is suggested by the simple dichotomy corresponding to the two biologically functional kinds of altruism.

This might, of course, make someone more confident than ever that all forms of human altruism can be accounted for simply within the framework of sociobiology, regarding biologically non-functional or dysfunctional forms (as in the adoption case) as overextensions of general biologically functional psychological mechanisms. I will say more about this below. For the moment, however, I want to return to

---

[10] Wilson (1978, pp. 165–6). What he means by "individual welfare" in the first sentence must be the individual's *inclusive fitness*.

E.O. Wilson's remarks about the two kinds of altruism he recognizes, both being products of natural selection. What is so puzzling is the assumption not only that these exhaust the forms of altruism that actually exist, but that they likewise exhaust our available *choices* when it comes to altruism. When he claims, for example, that genuinely other-serving ("hard core", kin-directed) altruism is relatively rare compared to ultimately self-serving ("soft-core", broadly-directed, reciprocal) altruism, he hastens to add that this is generally fortunate, since the former tends to be inimical to civilization, as the focus of cooperation is too narrow to produce wider harmony. Thus, he concludes that we would be wise to choose to cultivate the latter instead:

> My own estimate of the relative proportions of hard-core and soft-core altruism in human behavior is optimistic. Human beings appear to be sufficiently selfish and calculating to be capable of indefinitely greater harmony and social homeostasis. This statement is not contradictory. True selfishness, if obedient to the other constraints of mammalian biology, is the key to a more nearly perfect social contract.[11]

But why should we believe that these are our only two choices, so that the only hope for greater social harmony lies with enlightened selfishness? Wilson is right that kin-altruism is not the answer, but it doesn't follow that the only other alternative for promoting social harmony is to emphasize our other *evolutionarily bestowed* altruistic dispositions (which amount to enlightened selfishness). There is a rather obvious third possibility when it comes to altruism, which has much more to do with what most of us mean by genuine human altruism. That is, we can cultivate a principled altruism based on something more than biological instincts and cultural constructs shaped by them, which is both (i) *genuinely other-serving* (which isn't to say wildly imprudent or naively exploitable), unlike "soft-core" altruism, and (ii) *broad in scope*, both in the sense of extending as a matter of principle well beyond close relatives, unlike "hard-core" kin-altruism, and in the sense of pertaining quite generally, again as a matter of principle, to things having nothing in particular to do with the

---

[11] Wilson (1978, p. 164).

promotion of one's biological ends (or with any biological ends, for that matter)—things such as the happiness of a stranger. What I have in mind here is not just a matter of a biologically dysfunctional extension of certain normally adaptive psychological mechanisms, but a reflective, principled set of values that can deliberately be cultivated and put into practice, and that can result in everyone's benefitting.

More to the point with regard to the social contract, consider the possibility of cooperation based not on selfish calculation, or on the ultimately self-serving imperatives of genes, but on a *principled respect for others' rights, fair play, and social well-being*, as typically cultivated by a thoughtful moral education. Such an alternative is not available to honeybees or hyenas, who are stuck with the particular forms of natural selection-based, gene-serving "altruism" included in their genetic heritage, but it is bizarre to proceed as if human beings were equally limited in our social repertoire. Yet that is the implication of Wilson's remark: that despite human capacities for adopting and valuing such ends or principles as the respecting of rights, fair play and the common good, "the key to a more nearly perfect social contract" is nonetheless to be found among the *innate predispositions* we have as a direct result of our natural selection history (i.e. the evolved emotional and behavioral dispositions associated with kin altruism and reciprocal altruism). To be sure, our concern with such things as rights and fair play may enter into the picture on such a view, but Wilson's suggestion is clearly that this comes in only as part of the development of our innate predispositions, shaping these inherited "guides" in certain genetically useful ways, rather than providing the possibility of biologically *independent* guides of behavior along the (genuinely optimistic) lines I have suggested.

Wilson often talks this way, suggesting that our choice is limited to choice among various inherited predispositions (or various culturally developed forms of them), ignoring the possibility of new alternatives involving principles and commitments stemming from our advanced mental capacities. For example, he writes:

> Human emotional responses and the more general ethical practices based on them have been programmed to a substantial degree by natural selection over thousands of generations. . . . Which of the censors and motivators should be obeyed and which ones might better

> be curtailed or sublimated? These guides are the very core of our humanity. . . . At some time in the future. . . . we must consciously *choose among the alternative emotional guides we have inherited.*[12]

Or again:

> In order to search for a new morality based upon a more truthful definition of man, it is necessary to look inward, to dissect the machinery of the mind and to retrace its evolutionary history. But that effort, I predict, will uncover the second dilemma, which is *the choice that must be made among the ethical premises inherent in man's biological nature.*[13]

To be fair, Wilson does mention *reason* in a few places (p. 164). But he seems to associate this merely with "soft-core" altruism as opposed to "hard-core" altruism, which latter he labels "irrational" inasmuch as it is selfless (p. 162). Thus, for him, the altruism associated with reason just amounts to one of the two dispositions yielded by evolution, rather than being a creative alternative stemming from our advanced mental capacities. His occasional allusions to reason do not, then, add up to a recognition of the sort of possibility I have suggested. Where he does seem finally to recognize this possibility is in the very last paragraph of the discussion of altruism: "But to the extent that principles are chosen by knowledge and reason remote from biology, they can at least in theory be non-Darwinian".[14] But this idea is not seriously pursued, except to note that "the cultural evolution of higher ethical values. . . . will be constrained in accordance with their effects on the human gene pool," leading Wilson to the reminder that "morality has no other demonstrable ultimate function" than (what I have referred to as) the promotion of inter-generational genetic replication. Again, this altogether misses the point I am after: I concede that insofar as "morality" (in Wilson's sense, i.e. a certain range of emotional and behavioral dispositions) has any *biological* function it is ultimately the

---

[12] Wilson (1978, p. 7), my emphasis. Likewise, he speaks later on of "the conscious choices that must be made among our innate mental propensities" (p. 203).

[13] Wilson (1978, p. 5), my emphasis. By "ethical premises inherent in man's biological nature" he presumably means *basic moral judgments based on evolved emotional predispositions.*

[14] Wilson (1978, p. 174).

promotion of a certain range of inter-generational genetic replication, but I am interested precisely in the possibility of certain *non*-biologically-functional dispositions stemming from our advanced mental capacities, quite independently of our particular evolutionary dispositions.

It is not obvious that the cultivation of such dispositions will have "effects on the human gene pool" that will significantly limit them, or that it would have had such effects in the past, such that we should not presently expect to find such dispositions manifested. I am talking about possible dispositions that are not genetically based, but rely only on general advanced mental capacities and certain cultural influences, i.e. a sufficiently rich moral tradition and moral education. As already argued above, then, there is not the same problem with instability that would tend to arise for *genetically based* selfless altruism. It is also important to remember that the conditions of human natural selection include complex cultural contexts, which makes it still less obvious that the manifestation of independent values of the sort I am imagining would be severely constrained by natural selection. The short shrift given to the idea of non-Darwinian dispositions and principles is thus unwarranted, as is Wilson's remarkable conclusion that "the true humanization of altruism, in the sense of adding wisdom and insight to the social contract, can come only through a deeper scientific [i.e. sociobiological] examination of morality."[15]

Perhaps if Wilson were to concede that the theoretical possibility of non-Darwinian, reason-generated dispositions and behavior is not as limited as he had claimed, he would be more inclined to concede the existence of actual cases of it. As it is, however, he seems determined to construe virtually any purported example as being Darwinian in nature after all. Apparent cases of independently motivated altruism are dismissed as products of underlying biological predispositions after all,

---

[15] Wilson (1978, p. 173). Perhaps Wilson's argument about the limits of reason-generated values would work with regard to a kind of selection at the *cultural* level against fully *pacifistic* cultures, on the assumption that there is no effective global or regional body that would protect such cultures from assault by aggressive powers. It does seem true that in a relatively hostile environment such cultures would tend to be eliminated in favor of those that at least sanction violence in self-defense. But the qualities I have been speaking of need not be construed as involving anything as extreme and unstable as pacifism; and even pacifism could flourish in the right contexts—for example, as long as there are enough non-pacifists willing to defend pacifists against aggressors.

ultimately serving the same biological ends—the underlying principles cited by agents being dismissed as mere rationalizations. This too is unwarranted: Nothing in Wilson's arguments begins to justify such a strong position.

He asserts, for example, that sainthood is "cheerfully subordinate to the biological imperatives above which it is supposed to rise," but this startling conclusion is obviously not supported merely by remarking, for example, that while "Mother Theresa is an extraordinary person. . . . it should not be forgotten that she is secure in the service of Christ and the knowledge of her Church's immortality," or by observing that Christ taught that those who did not accept the Good News and receive baptism (i.e. "outsiders") would not receive salvation.[16] It is *possible* that Mother Theresa's behavior and Christ's teachings reflect certain deeper sociobiological influences—such as an innate desire for membership in a powerful group, or an innate tendency to divide humanity into friends and enemies and to focus altruism and good feeling on the former group. But it is equally possible that they do not. From the fact that Mother Theresa may have an innate human desire to be part of a powerful group it certainly does not follow that everything she does that might happen in principle to satisfy that desire can be *attributed* to it. Likewise, from the fact that Christ may have had, as part of his nature, an innate human tendency to divide people into friends and enemies and to focus altruism and good feeling on friends, it certainly does not follow that any particular conviction about the fate of outsiders can be attributed to it. And even if there *are* certain sociobiological influences at work in such cases, that certainly does not show that this is the whole story—that religious and ethical convictions and commitments can just be reduced in their entirety to "cultural hypertrophies of archaic behavioral adaptations"; indeed, the fact that Mother Theresa, for example, has embraced her commitments to the point of foregoing reproduction—and not merely to focus her efforts on aiding close kin—strongly suggests that there is more going on here than can plausibly be accounted for exclusively in sociobiological terms.

Similarly, even if it were true that "the highest forms of religious practice, when examined more closely, can be seen to confer biological advantage," this does not support the strong conclusions Wilson

[16] Wilson (1978, pp. 172–3).

draws.[17] The "biological advantage" supposedly conferred may, after all, be largely *accidental*—especially *qua* biological advantage (in the evolutionarily relevant sense)—in which case it would prove nothing about the practices in question. Wilson is hardly in a position to assert that "commitment [to various religious ideas, practices, and fellow believers] is pure tribalism enacted through emotional self-surrender," as if he had somehow succeeded in showing the impossibility of commitment out of reflective and intelligent belief in the truth of certain principles or doctrines, through some combination of reasoning and experience, or of commitment to others' welfare out of enlightened compassion likewise stemming largely from the exercise of reflective mental capacities. Similarly with his bold claim that "the true reason for the universal rights movement" is to be understood in terms of "raw biological causation," on the grounds that "we will accede to universal rights because power is too fluid in advanced technological societies to circumvent this mammalian imperative: The long-term consequences of inequity will always be visibly dangerous to its temporary beneficiaries."[18] Again, the biological facts to which he appeals—such as the existence of selection pressures favoring the development of psychological dispositions that might lead human beings to be wary of gross inequities, even if they temporarily benefit from them—do not begin to establish the strong conclusion he draws (though this is not to say that the former are not interesting). In particular, they do not show that the moral considerations, principles and arguments cited by advocates of the universal rights movement are in fact nothing but "rationalization[s] contrived by culture to reinforce and euphemize" certain biological imperatives, as opposed to being genuine motivating considerations derived from intelligent reflection on moral issues.

I said above that E.O. Wilson's sociobiological account of altruism seems excessively narrow in considering only the two standard evolutionary possibilities of "hard-core" kin altruism and "soft-core" altruism within larger groups. My response was to offer an obvious third possibility, which might be expected to co-exist with the other two, and to suggest that it could plausibly be rooted in human capacities for reasoning and evaluative judgment, quite independently of any particular biological predispositions. But not all sociobiological

---

[17] E.O. Wilson (1978, pp. 196–7).
[18] E.O. Wilson (1978, pp. 206–7).

accounts are as narrow as E.O. Wilson's, and some may purport to encompass the third possibility within the sociobiological perspective itself.

James Q. Wilson, for example, argues that what natural selection has given us in this area are quite *general* desires and emotional predispositions centered around attachment and affiliation (something highly adaptive to creatures who benefit crucially from various forms of social interaction). From this innate sociability there then naturally emerge—particularly in early social development in the context of the family—various related traits such as sympathy, a concern with fairness, a concern with others' well being, an inclination to generosity, a sense of duty, a capacity and tendency to make moral judgments and to apply them to oneself, and so on. These traits prove on the whole to be adaptive to individuals in social contexts, but they are not as finely tuned as is often suggested by talk of genetically influenced "rules" or "strategies". Indeed, they are general enough that they often motivate such non-adaptive behavior as cooperation even in circumstances where the individual could easily get away with cheating, or "hard-core" altruism even toward non-kin (and even non-human animals, such as pets), or altruism toward non-kin even in contexts where there is no reasonable expectation of reciprocation or other gain—all things that would not be expected if what natural selection gave us were more finely tuned dispositions or behavioral "strategies", as with lower animals.[19] The evolutionary *point* of these general dispositions, as with any other adaptation, is simply the contribution they tend to make to the individual's reproductive success (or to that of his or her kin, insofar as kin selection was involved), given the environment, which in this case includes a social context. But since the dispositions are very general, they may well lead to plenty of social behavior that fails to be adaptive—including behavior that we often associate with "higher values", such as cooperating even where one could get away with cheating, or adopting and caring for unrelated infants.

---

[19] J.Q. Wilson (1993, especially pp. 121–140, and pp. 43–46, 23, 99). Wilson refers to the sentiments involved with such things as sympathy, generosity, fairness and duty—which on his view tend to lead both to the making of moral judgments and to what is commonly considered "moral behavior" (i.e. as opposed to selfish, deceitful or aggressive behavior)—as constituting a natural, human *moral sense*. Cf. also Ruse (1988), who similarly argues that a general moral sense, as natural to human beings as hands, eyes and teeth, can be accounted for in terms of natural selection.

I shall not try to evaluate the merits of this broad, psychologically-oriented view of the sorts of dispositions that have resulted from our natural selection history, as against the view that natural selection has provided us with more specific dispositions, more tightly directed toward the relevant biological ends. What I do want to point out is that even if the former is correct, that still does not cast doubt on the existence of an independent form of motivation, not directly traceable to general biological predispositions—in particular, motivation stemming from independent exercises of our advanced rational faculties.[20]

First of all, even if certain forms of non-adaptive behavior may sometimes be explicable in sociobiological terms after all, it does not follow that they always are. For example, suppose that the tendency to cooperate even where one could benefit by cheating can in principle be explained on a broad sociobiological view as a natural and direct by-product of certain more general adaptive predispositions in certain contexts, and that some such behavior is indeed correctly explained in this way. It is still entirely possible that other cases of such behavior are to be explained in terms of principles or convictions that agents have arrived at independently of such biological predispositions—as by coming to the conclusion that there is no rational justification for cheating just because one can get away with it. And it makes no difference to the present point whether or not this conviction is *true*—an issue I have not yet addressed. It is enough that people can come to believe it and to be motivated by it.

The latter possibility is especially plausible in cases where the agents in question cite principles or arguments in defense of their choices, even if these are very simple—such as the common thought: "How can *I* legitimately take the liberty of cheating, when I myself depend on others not to cheat, and I'm no different in any relevant respect from them?" Perhaps such reflection would never have occurred were it not for certain general biological predispositions that incline us to think about such things. But that does not change the fact that such

---

[20] As noted earlier, it may be that there is an evolutionary explanation for our general capacity to be motivated by evaluative judgment or "guided by words" (Gibbard, 1990, ch. 4), but this is compatible with our making and being influenced by evaluative judgments, and by related emotions, in ways that go beyond anything attributable to biological influences. This is what I mean by an "independent form of motivation", which is a kind of "evolutionary spin off" (to use Sober's expression from before).

thinking goes beyond the biological predispositions themselves, providing an independent source of motivation that need not generally track the biological predispositions, and may well conflict with them (and certainly does often conflict with the biological ends to which the biological predispositions are functionally directed).[21]

Secondly, even in cases where behavior is motivated by the extension of a general biological predisposition beyond its proper functional scope, thus falling within the purview of sociobiological explanation, it is precisely the above sorts of exercises of our rational faculties that often plausibly *lead* to that extension. For example, someone may be led to take responsibility for a non-related orphan not simply by a brute extension of her maternal instincts, but by the recognition that it would be rationally indefensible to place the great value she does on the well being of her own children—something which may be partly attributable to her maternal instincts—while placing so little on that of non-related children as to allow an orphan to go uncared for. This consideration might lead her to value the well being of the orphan enough to take responsibility for it, and this in turn might pave the way for the further extension of her maternal instincts toward it as the bond deepens. So even cases that lend themselves in part to sociobiological explanation in terms of extended biological predispositions may very well also involve independent sources of motivation. Indeed, motivation (at least in part) by rational considerations seems crucial to the explanation of many of our choices, even where evolutionary factors might in principle account for certain relevant desires and emotions. In such a case we still need to account for particular ways of *structuring* values, which may lead to radically maladaptive choices, and it is not clear how we can do so except by appeal to the force of such independent motivating factors. Again, a

---

[21] The only general grounds for skepticism that I can think of with regard to this matter would be some reason for believing that our practical thinking itself is so tightly in the grip of biological influences that, far from enabling us to transcend our biology, as I have argued, it amounts to nothing more than a slavish rationalization for the biological forces that continue to run the show. But while it is plausible that *some* of our thinking is vulnerable to such a charge (e.g. a man's attempts to defend his philandering), there is no reason to suppose that this is true of our practical thinking *in general*, and the burden of proof lies with the skeptic to provide genuine cause for concern here. Again, I do not think he will have an easy time even with the most obvious examples, such as the altruism of Mother Theresa, or the teachings of the Buddhist monk and peace activist Thich Nhat Hanh.

good example to illustrate this point is the choice to forego having children out of a principled concern with overpopulation. Even if sociobiologists could plausibly claim to account for the motivation here simply in terms of overextended biological predispositions, the question would remain why the agents in question give such weight to the relevant considerations as to override the powerful biological predisposition to procreate; thus, we are once again led beyond sociobiology, to take seriously the force of independent reflection on human behavior.[22]

I do not think that J.Q. Wilson would (or at any rate should) disagree with any of this. While he does wish to establish an evolutionary background for a common human "moral sense"—i.e. he sees a certain set of moral sensibilities as growing largely out of certain basic, naturally selected desires for attachment and affiliation, in certain social contexts—he does not, like E.O. Wilson, appear to have any general agenda of finding a sociobiological explanation for virtually everything in human life. He does not, at any rate, deny the possibility

---

[22] The above points serve as an answer to Ruse (1988, p. 47), who questions Ayala's motivation for claiming that our moral faculties and sensibilities go beyond or "sit [partly] on top of" our biology. Ruse argues that the reason is "certainly not because morality (that is, morality at the substantive level) goes contrary to our biological interests. The whole point is that it does not. We are moral because, ultimately, natural selection found that it paid." But this is a gross oversimplification. While certain aspects of "morality" may indeed serve our biological ends, this is by no means always the case—nor should we expect it generally to be the case for creatures capable of genuine reflection, evaluation and choice (traits rightly stressed by Ayala). Indeed, it is within our capacity to create moral codes that *systematically* promote non-biological ends over biological ones—promoting virtue or happiness, for example, over the ultimate biological end (in each person's case) of optimizing inter-generational gene replication. Ruse later acknowledges that moral judgments may sometimes conflict with certain biological dispositions, frustrating the realization of certain biological ends, but he dismisses this as unproblematic, since such conflict is perfectly compatible with an evolutionary view of the moral sense behind such moral judgments; there may be good evolutionary reasons why men tend to be disposed to philander, for example, and equally good evolutionary reasons why they are also disposed to feel the force of moral objections appealing to fidelity. But this misses the point. Even if such moral judgments stem from an evolved moral sense, the fact remains that they will often thwart rather than serve biological ends—and even systematically so, where cultures have taken the raw materials of the moral sense and woven them into moral systems that reflect their own priorities. To that extent, then, they still go beyond biology to a significant degree, even though they are partially traceable to biological dispositions. And if at least some such moral judgments instead stem from altogether independent exercises of our advanced mental capacities—a possibility that Ruse has certainly not ruled out, or even given us real reason to doubt—then the transcendence of biology is even more clear.

of significant input from exercises of our rational faculties that go beyond anything directly traceable to innate psychological predispositions. On the other hand, he does tend to downplay the actual significance of such inputs. For example, he claims that

> when people act fairly or sympathetically it is rarely because they have engaged in much systematic reasoning. Much of the time our inclination toward fair play or our sympathy for the plight of others are immediate and instinctive, a *reflex of our emotions* more than an *act of our intellect*, and in those cases in which we do deliberate. . . . our deliberation begins, not with philosophical premises (much less with the justification for them), but with feelings—in short, with a moral sense.[23]

Perhaps this sort of claim is motivated by an exaggerated conception of what is involved in "acts of our intellect" (or "systematic reasoning" or "philosophical premises"), setting up a false dichotomy between untenable extremes. It is true that most of us do not run around making and defending our choices by appeal to complex philosophical arguments on the order of those advanced in Kant's *Groundwork*, for example. But it certainly does not follow that "most of us give unreflective allegiance to [certain natural] sentiments" (i.e. those sentiments Wilson identifies as making up the human "moral sense") or that "most of the time we do not reflect any more deeply than [criminals] do on our ethical inclinations".[24] Most of us who have progressed beyond the rudimentary stages of moral development characteristic of small children typically have more to say in defense of our choices than is suggested by characterizing them as "reflexes of emotions"—even if it is something quite modest, as for example in perfectly common and ordinary thoughts about the reasonability of certain rules, our own dependence on others' following them, and the lack of any relevant difference that would justify our making exceptions of ourselves just because we can get away with it.

---

[23] J.Q. Wilson (1993, pp. 7–8), my emphasis.

[24] J.Q Wilson (1993, pp. 9, 11). About criminals, Wilson remarks that they "do not ordinarily engage in much that we would call moral reasoning". His claim in the above quote, then, is that most of the time non-criminals do not either.

Sometimes the role of such thoughts may be overlooked due to the fact that we often accept appeals to emotions as explanations of action, without pressing further. But such acceptance does not imply that there is no moral reasoning behind the emotions. It may be simply that the background moral thoughts are taken for granted, or it may be that they are not deemed relevant to the explanation that is being sought. Consider, in this connection, the following passage:

> .... even in our daily lives, there will be innumerable cases of our obeying rules, resisting temptations, and doing the honorable thing *for no other reason* than that we instinctively act that way and reflexively feel guilty when we don't.[25]

If what is meant here is just that we often act simply on the basis of our convictions about what is right, *without ulterior motives*, then I have no objection; and since this passage occurs in the context of a discussion of conscience, that may have been all that was actually intended. But given Wilson's other remarks above, as well as his use of the terms "instinctively" and "reflexively" here, it is likely that he would also take this passage to be expressing a different purported truth: namely, that we quite often act simply on the basis of certain *feelings* of responsibility and guilt, *and for no other reason*, i.e. we act on the basis of these feelings *rather than* on the basis of moral reasoning.[26] But *this* claim is far from obvious. The fact that we might explain our action simply by appealing to the guilt we would feel if we did otherwise certainly does not imply that we acted from a sense of guilt *and for no other reason*, if this means that our sense of guilt was primitive or "reflexive" as opposed to being motivated by various intellectually apprehended moral considerations. Even if we did not have to deliberate at the time of action, our sense of responsibility or of guilt may well have been informed by moral considerations the force of

---

[25] J.Q. Wilson (1993, pp. 33–4), my emphasis.

[26] A similar ambiguity arises for his claim that "sometimes *sentiment alone*, unsupported by utility, motivates our actions, as when someone makes an anonymous benefaction or a lone bystander helps an endangered person" (p. 43), my emphasis. Does "sentiment alone" mean just "moral sentiment, apart from any thought of personal reward", or also the stronger "moral sentiment, apart from moral reflection or reasoning"? It is the first that is directly relevant to the discussion in which the statement appears, but again, other things he says (as quoted in the text above) support the stronger reading as well.

which may be clear to us only as a result of previous moral reasoning. Our feelings may thus be "reflexive" in the sense of not requiring deliberation at the time of action, without its being the case that they are independent of the influence of moral reasoning.

To put it another way, it is important to see that the explanation "because I would feel guilty if I didn't (keep my promise)" is *not* generally like the explanation "because I would get a headache if I didn't (take my medicine)". When we cite a feeling like guilt in explaining—and particularly in defending—our actions, the point is not generally that avoidance of the unpleasant guilt feelings is our *reason* for performing the action, as if we had nothing more to say about the guilt than we would about a headache. Rather, our reasons for performing the action are the same reasons we have for thinking the failure to perform it to be *wrong*, and hence for feeling that guilt would be an *appropriate* response to such a failure. At any rate, this seems true for the wide range of cases where we have more to say than *just* that failure to perform the action would make us feel guilty, even if we don't always need to go into these reasons for certain purposes of explanation.[27]

When Wilson remarks that "the feelings on which people act are often superior to the arguments that they employ", this has a ring of plausibility *if* the arguments to which he is referring are the complex and highly controversial arguments offered by philosophers in the course of doing moral philosophy. We are generally more confident in the moral soundness of ordinary feelings of the importance of fairness or self-control, for example, than in the soundness of complex Kantian or utilitarian arguments concerning them.[28] But in fact the arguments that people typically employ in arriving at and defending their choices are not those sorts of arguments, but much humbler ones—common moral thoughts like those mentioned above, which are often deeply bound up with the very feelings to which Wilson is appealing, making our moral inclinations more sophisticated and discriminating than the simpler inclinations of children. And about *these* arguments his remark has little plausibility. If such everyday arguments or pieces of moral reasoning inform the feelings themselves, then there is no dichotomy of the sort he is suggesting to begin with, and it makes no sense to speak

---

[27] Cf. T. Nagel (1970, ch. 5), and (1992, p. 17).
[28] Cf. Williams (1985).

of the feelings as being superior to the arguments. And if the feelings are something quite independent even of such ordinary moral reasoning, then it is not clear why they should be thought to be morally praiseworthy at all. If someone does something because she would feel guilty if she did not, for example, and yet she has nothing to say about *why* it would be *appropriate* to feel guilt for failing so to act (it is just a feeling she has), then it is not clear that there is anything at all morally praiseworthy about her feeling; it may be nothing but a psychological vestige to be targeted for elimination in psychotherapy.

The same holds for primitive feelings of sympathy, as opposed to feelings of sympathy informed and tempered by mature, moral thought. Just as with the sometimes sweet but dangerously naive and easily misguided feelings of children, it would surely be a mistake from a moral point of view to romanticize primitive feelings of sympathy, taking them to be generally superior to more mature, discriminating feelings and the ordinary moral reflection bound up with them. Apart from the fact that such immediate inclinations are easily misguided (as in Herman's example of the sympathetic person moved to help the stranger struggling late at night with a heavy burden at the back door of the Museum of Fine Arts), there is the more fundamental problem that, as Herman puts it, those who are motivated by such inclinations are to that extent "indifferent to morality".[29] She argues, on Kant's behalf (though the point does not depend on any general acceptance of Kantianism), that while immediate inclinations of the sort Wilson sometimes praises "may well lead to dutiful actions, and may do this with any degree of regularity desired. . . . the dutiful actions are [nonetheless] the product of a fortuitous alignment of motives and circumstances. People who act according to duty from such motives may nonetheless remain morally indifferent", moved by their sentiments quite independently of any interest in the moral rightness of the actions. Herman goes on to suggest that "for a motive to be a moral

[29] To be fair to Wilson, he does indeed have much to say about why sympathy is usually not in fact as easily misguided as in the example above, but this involves precisely an appeal to *moral judgment* and *reasoning*. It is because of such reasoning and judgment on our part, for example, that "a rich boy distraught at the loss of a penny arouses not sympathy but derision." (1993, pp. 32 f.) His remarks about the common connection between judgment and sympathy seem quite right, but I am presently concerned with another of his claims, which is in some tension with this: namely, that much of our sympathy is "immediate and instinctive, a *reflex of our emotions* more than an *act of our intellect*" (pp. 7–8), and that such feelings are "often superior" to our moral reasoning.

motive, it must provide the agent with an interest in the moral rightness of his actions. And when we say that an action has moral worth, we mean to indicate (at the very least) that the agent acted. . . . from an interest in the rightness of his action—an interest that therefore makes its being a right action the nonaccidental effect of the agent's concern".[30] This condition is not met in the case of sentiments uninformed by moral reflection, though it can be in the case of sentiments that are so informed. My point, then, is again that the former should not be romanticized from a moral point of view.

Returning to the more general issue, it is somewhat puzzling that Wilson downplays the motivational significance of independent inputs from exercises of our rational capacities, since his own arguments suggest that such activities will naturally be triggered by the same things that give rise to the "moral sense", and in a systematic way. On Wilson's view, the various aspects of the moral sense have their origins in human sociability—in a natural desire for attachment and affiliation. Such desires naturally give rise early on to such things as a sense of sympathy, a concern with fairness, a sense of duty, and a tendency to moralize—i.e. "to judge the actions and motives of others as worthy or unworthy," and, "as we grow from childhood to adulthood, [to] increasingly judge ourselves as we judge others."[31] Indeed, he explicitly argues that part of what emerges from this mix is a natural "desire not only to be praised, but to be *praiseworthy*"—part of what is involved in the possession of a conscience. Another way of putting this is that from our natural sociability, together with resulting tendencies to engage in moral judgment and argument, there emerges a natural *desire to be able to justify* one's behavior and motives, both to others and to oneself. But this sort of desire, together with the possession of advanced rational capacities, opens up an important new source of motivation.

The sort of reasoning that is involved in such justification is no longer merely in the service of the sentiments constituting the natural "moral sense", but instead takes on a motivational life of its own. It may have been triggered by certain natural desires, but its influence is potentially as broad and far reaching as the possible results of such reasoning: One may well arrive at motivation that would never have

---

30 Herman (1993, pp. 4–6).
31 J.Q. Wilson (1993, e.g. pp. 32–4, 70, 105, 108, 121, 127–8, 131, 141, 148).

emerged simply as the *direct* result of the natural psychological dispositions Wilson associates with sociability and the moral sense.[32] The point, then, is that Wilson's own views about the genesis of moral sensibilities suggest that there will naturally be significant motivational input not only from natural sentiments and beliefs directly stemming from them, but also from exercises of our rational faculties, particularly in our attempts to arrive at justifiable practical principles by which to live our lives. Just as human nature involves both crucial elements, so too our actual moral sensibilities may generally be expected to be informed by both elements, reflecting both our evolutionary heritage and the (potentially quite radical) results of our practical thinking.

Finally, returning to E.O. Wilson's remark about "the key to a more nearly perfect social contract," it should be stressed that even if genuinely other-directed, broad-scoped altruism or trustworthiness in cooperation could successfully be argued by sociobiologists to be derivable directly from elements of our genetic heritage, there is no reason to suppose that we need to *rely* on this fact in thinking practically about possibilities for human life. As beings endowed with reason, we need not search for the keys to a better life among the emotional dispositions bestowed on us by evolution. Perhaps it will be easier for us if our evolved nature is such that the above qualities come relatively naturally for us. But even if they did not, that would not condemn us to settle for what we can get from our evolutionarily shaped nature, as if we had no other resources. This is not to deny that knowledge of the sociobiological aspects of human nature is important in understanding the costs that accompany certain social choices, for example, as Wilson stresses. Such knowledge may be useful for this and for much else besides.[33] But it *is* to deny that our choices are limited to various particular elements of our evolutionary heritage.

In summary, I think it is reasonable to suppose that with regard to such things as altruism, cooperation and related emotions, there is some

---

[32] An extreme example that comes to mind is the choice not to have children, motivated by reasoning along the lines of that offered by Seana Shiffrin—the conclusion being that the choice to have children cannot ultimately be justified. I do not accept that reasoning or its conclusion, but the point is that someone might, and if she did, she would thus arrive at motivation that is only indirectly traceable to the sorts of natural desires Wilson focuses on, being directly traceable to exercises of rational capacities. The latter may be triggered by certain natural desires, but they clearly go well beyond them as sources of motivation.

[33] E.O. Wilson (1978, pp. 139 f., 154). I say a little more about this below.

truth in both sociobiological accounts and accounts appealing to independent factors stemming from the advanced mental capacities that set human beings apart from other animals. Sociobiology plausibly shed light particularly on some of the powerful, seemingly innate emotions that tend to be almost universally present even in children, or on relatively primitive emotions that appear to be associated more with the limbic system than with the neocortex, while non-sociobiological accounts might give a more plausible explanation of more refined emotions and values, for which sophisticated reasons are often given and principles often sought.[34] Or perhaps, as seems most likely, many phenomena contain a mixture of elements, requiring both kinds of account in order to arrive at a complete explanation.

I see nothing problematic in this, so long as we recognize that human nature is a complex matter that must be investigated from more than one angle. The "very core of our humanity" (to use E.O. Wilson's expression) is defined not merely by our specific biological predispositions together with our tendency to give particular cultural expressions to them, but equally by our advanced mental capacities and the tendencies to which they naturally give rise. Human nature is complex not merely insofar as it encompasses a variety of biological predispositions, ranging from the violent and exploitative to the altruistic and nurturing, but also insofar as it includes, in addition to such biological predispositions, advanced mental capacities and derivative tendencies, which plausibly operate with at least some degree of independence from evolutionary influences.[35]

2. HUMAN NATURE AND MORALITY

Discussions of human nature are often taken to shed light on the topic of *morality*, but it is important to distinguish two very different concerns here: (i) a concern with *explaining* certain empirical phenomena commonly regarded as falling under the general heading of "morality", on the one hand, and (ii) a concern with the existence and nature of genuine *truth conditions* for moral judgments or the

---

[34] On the former, cf. J.Q. Wilson (1993, p. 133).

[35] This moderate and inclusive position seems to me to capture both the truth to be found in Kitcher's critique of "pop sociobiology" and the truth to be found in Ruse's most basic sociobiological claims. See Ruse (1988, pp. 45–6).

possibility of genuine moral *justification* of certain types of behavior, on the other.

These are not, of course, unrelated issues: One way of trying to *explain* an agent's having certain moral beliefs or performing certain corresponding actions, for example, would be to claim that those beliefs (and hence, indirectly, the actions that they motivate) result from the recognition of certain moral *truths*. (Or alternatively, if one comes to the conclusion that moral beliefs can all plausibly be explained without any appeal to moral truths, this might lead to skepticism about the existence of such truths.[36]) But in the sciences, the pursuit of explanations for the empirical phenomena referred to as "morality" is generally carried out independently of any philosophical concern with the question of the existence of moral truths or of the possibility of objective justifications or condemnations of certain types of behavior. Those who approach "morality" from the perspective of psychology, sociology, evolutionary psychology, sociobiology, and so on, are concerned primarily with describing and explaining certain empirical phenomena in empirical terms. The general human capacity and tendency to make and to be influenced by moral judgments, for example, are explained by appeals to elements of our natural selection history, subsequent cultural evolution, and individual development. Likewise, such explanations are offered even for our tendency to possess a certain specifiable range of desires and sentiments commonly labelled as "moral", and to engage in certain forms of behavior likewise regarded as falling under the heading of "morality"—sentiments such as sympathy and generosity, and behavior such as altruism and cooperation, generally associated with "the better side of our nature" (as against the selfish and aggressive side, which is also explained largely in sociobiological terms).[37]

This is clearly the approach, for example, of E.O. Wilson (a sociobiologist) and James Q. Wilson (a social scientist)—though as discussed below, there are places where each seems to take himself somehow to be doing something more as well, bearing on philosophical issues of moral truth and justification. I have already suggested that even for purposes of scientific explanation these approaches may tend

---

[36] For the latter argument, see Harman (1977, ch. 1) and, from an evolutionary perspective, Gibbard (1990) and Ruse (1988), which latter is discussed further below.

[37] J.Q. Wilson (1993, p. xv).

to be too narrow in downplaying the contributions of independent exercises of our rational faculties. Even though they do recognize the role played by culture in the development of moral sensibilities, for example, there is often still unwarranted emphasis on the guiding of cultural evolution by biological influences, with a relative neglect (particularly in E.O. Wilson's case) of cultural evolution due to independent influences, such as the products of independent human thought. But let me set that matter aside here. I'm presently interested just in what is at issue in discussions of human nature and morality, and again, in the social sciences it appears to be primarily just the extent to which morality *understood as a certain range of empirical phenomena* can be understood as either constitutive of human nature or as a product of more basic aspects of human nature (such as human sociability), which is in turn subject to scientific *explanation*.

So, for example, when J.Q. Wilson speaks of "morality", claiming that "scientific findings provide substantial support for its existence and power," he is referring to a certain array of feelings, convictions and tendencies—feelings of sympathy, a sense of the importance of fairness, a tendency to cooperate with others in certain groups, a tendency to feel disapproval toward those who break social rules or guilt when one breaks them oneself, and so on. His concern is to show that "morality", so understood, is not merely an accident of particular cultures, but something deeply rooted in human nature—in what he calls a "common moral sense", the origins and effects of which can be investigated scientifically.[38] Similarly, when E.O. Wilson calls for "a deeper scientific examination of morality" (in connection with a discussion of altruism), he means "morality" to refer to a certain range of emotional and behavioral tendencies commonly deemed at least superficially "moral" in nature, which he argues belong to our evolutionary heritage, as discussed in the previous section.[39] Likewise, when he writes that "innate censors and motivators exist in the brain that deeply and unconsciously affect our ethical premises" and that "from these roots *morality* evolved as instinct," the term "morality" is meant to refer roughly to the range of feelings and behavior specified

---

[38] J.Q. Wilson (1993, p. xii). Hence the definite article in the title of the book (*The Moral Sense*) indicating a particular array of feelings or inclinations purportedly common to human beings generally—a result (purportedly) of their having had adaptive value in our evolutionary history. Ruse (1988) makes a similar claim.

[39] E.O. Wilson (1978, p. 173).

above—and similarly when it is said that "moral feelings" (or "the human moral sense") and "moral behavior" are evolved traits.[40]

Now it is important to see that not only are the two general concerns distinguished above very different, but discussions of the first issue by sociobiologists and social scientists do not have any direct *positive implications* with regard to the second issue, i.e. the philosophical issue of moral truth or justification. To show that certain feelings or tendencies are biological adaptations is not even to begin to provide a moral or rational *justification* for them, or to establish the existence of any moral *truths* about them.[41] What it shows is at most just that in our evolutionary past these traits have tended to contribute to the propagation of the genes that code for them more effectively than traits produced by rival alleles contributed to the propagation of those alleles (by contributing to the reproductive success of the possessors of the traits or of their kin), which obviously has nothing to do with morality as such.

Some caution is necessary here, however. Notice that I am not claiming *in general* that facts about etiology are irrelevant to evaluative facts. Indeed, I have argued that insofar as facts about etiology bear on facts about functional teleology, they will likewise bear on evaluative facts. For example, if certain traits are shown to be human adaptations, then on my view unless they are vestigial they will generally have biological functions; and if they have biological functions, then these functional facts will support certain evaluations, such as the claim that someone lacking the traits, or failing to exhibit them in the right way, is *defective from the point of view of human biological teleology*. So the point is not that natural selection etiology is generally irrelevant to

---

40 E.O. Wilson (1978, p. 5), my emphasis.

41 A point clearly recognized, for example, by Ruse (1988), and much earlier, for example, by T.H. Huxley. Gibbard (1990, ch. 4) makes use of such evolutionary considerations in his account of moral and rational justification, but this use does not involve any illegitimate move from something's being an adaptation to its being justified, or any attempt to derive moral facts from facts about biological function. Rather, he appeals to evolutionary theory to confirm the reality and centrality in human life of the "psychological state" of norm-acceptance, and to clarify the notion of accepting a norm. He then plugs this into a *non-cognitivist* account of moral judgment and normative judgment about reasons for acting. Further appeals to the evolutionary basis of various psychological propensities (such as patterns of emotion) are then meant only to shed light on the elements that in fact figure into the subjective activity of normative judgment typically carried out by human beings, rather than to provide some sort of objective justification for them.

evaluative facts, but something more specific: namely, that it is irrelevant to *moral* facts. If there are facts about human nature that are in some way relevant to moral facts (which I think is plausible), they are not facts about the natural selection etiology of human traits.

The only way natural selection etiology could be relevant as such to moral facts would be if moral facts were simply identified with a subset of evaluative facts connected with biological teleology as I have construed it. This would involve the claim that morality—meaning here not the manifestation of certain psychological and sociological phenomena, but rather the actual set of facts about the moral goodness and badness of human action and character—has ultimately to do with each individual's optimization of the inter-generational replication of copies of his genes (in his own germ-line cells, or in those of his close relatives). That is, the true standards of moral goodness with regard to human action and character, on such a view, would simply be those given us by natural selection, and thus would have ultimately to do *not* with such things as human good, happiness, freedom, rationality or salvation, but—as with any other species—with a certain kind of genetic replication!

To say that such a claim is a non-starter would be an understatement. First of all, such a view of morality would be radically revisionary in a totally implausible way, providing moral justification not merely for a certain degree of sympathy and cooperation, which might seem attractive, but equally for such things as selective infanticide, philandering, deception and murder—while at the same time condemning such things as genuine altruism toward non-kin.[42] But quite apart from such bizarre results, it would be ludicrous *in principle* to suppose that the standards for acting well for creatures capable of reasoning about the good and acting on such judgments should be set

---

[42] Though it might acknowledge that the latter is a natural, dysfunctional side-effect of certain general functional traits, as discussed earlier. Indeed, one side-effect of some of the traits constituting the "moral sense" may be that people tend to condemn even genetically advantageous selective infanticide, deception, etc., and to praise such things as genuine altruism toward non-kin. But that is irrelevant to the present issue, which is not about what people might think, but about the nature of the moral system in question. The point is that *the justificatory structure of a moral system based on biological teleology* would subordinate everything to the ultimate end of gene replication, and hence would justify such things as selective infanticide and deception (at least to the extent that this is motivated by biologically functional dispositions) and condemn such things as genuine altruism toward non-kin as dysfunctional. And that is absurd.

by the historical pressures of natural selection, as biological norms are, ordered ultimately around inter-generational germ-line gene replication. One might as well suppose that the standards for *thinking* well, even for creatures with our advanced mental capacities, are set by the historical pressures of natural selection, rather than by independent rules of logic—that thinking well, like digesting well, is to be understood in biological functional terms (so that the ultimate measure has to do with functional contributions to gene replication), *even* for creatures who are capable of conceiving of and taking an interest in truth and logic in abstraction from anything having to do with biological ends.[43]

Some writers explicitly recognize that their explanatory projects with respect to human nature have no positive implications for the issue of moral truth and justification. Ruse, for example, makes a point of arguing that to provide an evolutionary background for a trait—as he attempts to do with regard to the traits constituting the "moral sense"—is certainly not to provide a moral or rational justification for it.[44] Others, probably thinking the point too obvious to require stating, simply ignore the latter issue and are generally cautious to avoid drawing fallacious inferences with regard to it. Still, the issues are sometimes conflated in particular cases—or at the very least, the language in which they are discussed is sometimes misleading—and it is worth noting a few instances of this in the works I have discussed.

Consider, for example, E.O. Wilson's discussion of sexual morality, in which he attacks the position of the Catholic Church:

> The Church takes its authority from natural-law theory, which is based on the idea that immutable mandates are placed by God in human nature. This theory is in error. The laws it addresses are biological, were written by natural selection. . . . and have been erroneously interpreted by theologians writing in ignorance of biology. All that we can surmise of humankind's genetic history argues for a more liberal sexual morality, in which sexual practices are to be regarded first as bonding devices and only second as means for procreation.[45]

---

[43] Cf. T. Nagel (1979).
[44] Ruse (1988, pp. 40–1, 58–60).
[45] E.O. Wilson (1978, p. 147).

Even if one agrees with the conclusion that sexual morality should be understood in a more liberal way than it is by the Church (and I do), that should not lead one to overlook the fact that this is a lousy argument for it. Suppose Wilson is correct in asserting that human nature was not designed by God, but was shaped by the forces of natural selection. And suppose further that he is right in maintaining that the primary biological function of a great deal of sexual activity and of the pleasure connected with it is to facilitate bonding—something crucial to the reproductive success of creatures that depend so heavily on long-term bonds, without which they would have much more difficulty successfully rearing their young over a period of several years, for example. This may indeed undermine the Church's claim that the primary natural function of sexual behavior is the insemination of wives by husbands: There are no natural functions based on God's design in any case (we are supposing), and the natural functions that do exist point in this case largely toward bonding, not simply toward procreation (though of course the ultimate end promoted by all biological functions is procreation). But nothing *positive* follows about sexual morality. The fact that sexual activity beyond what is necessary for procreation has a certain biological function (as a "bonding device") no more provides a moral justification for it than the fact that the tendency toward male philandering has a certain biological function (assuming for the sake of argument that this is a fact—if not, other examples are easy to come by) provides a moral justification for male philandering.

"Humankind's genetic history" no more "argues for a more liberal sexual morality," in the direction of recognizing the moral legitimacy of forms of sexual behavior other than intercourse, than it argues for liberalizing our moral views about male philandering. I am not, of course, denying that there are important differences between the two cases—i.e. between sexual activities other than intercourse between a committed pair (or intercourse in excess of what is necessary for procreation), on the one hand, and male philandering, on the other. The point is just that it is precisely these differences that need to be considered in a moral argument about these things—for example, differences in how they bear on human happiness or misery—rather than facts about biological function as such. To be fair, Wilson *may* have meant nothing more than that his sociobiological claims undermine the "natural law" claims that have traditionally been taken to support a more conservative sexual morality, and in that weak sense provide "support" for a moral liberal morality, i.e. by eliminating one powerful obstacle to it. But in any case, that is far from making a positive case for the more liberal view.[46]

Another instructive example may be found in Wilson's discussion of universal human rights. As mentioned earlier, he thinks our belief in such a thing is just a matter of "raw biological causation", amounting to nothing but a rationalization for certain biologically pre-programmed emotions and tendencies. Nonetheless, he suggests that as part of what he calls a "biology of ethics" (to be fashioned by "the new ethicists", working from the enlightened perspective of sociobiology), such an idea "might *properly* be regarded as a. . . . primary *value*."[47] Now if his reason for this evaluative claim involved a rational appeal to the importance of the idea of universal rights for human well being, for example, then the claim might not be very surprising: Evolution gives us certain tendencies, he might reasonably say, which in turn influence certain ideas, and here is a case of an idea that is *worth* promoting because of its importance to human good.

That is not how his argument goes, however. What he actually says is that "we will want to give a primary status [to the idea of universal rights] not because it is a divine ordinance. . . . or through obedience to an abstract principle of unknown extraneous origin, but *because we are*

---

[46] One might, of course, well question whether it is even *necessary* to make a positive case for a moderately liberal view of sexual morality, since it is far from clear why there should be any general moral presumption against non-procreative sexual activities between committed and consenting adults in the first place. Why should we suppose, for example, that there is any *prima facie* moral presumption against homosexual behavior, so that some justification is called for? I suspect that the tendency to think that there must be a general moral issue concerning non-procreative sexual behavior is probably in large part a vestige of a conflation of the issues of sin and immorality, along with a certain interpretation of sin and sexuality. In any case, even if it is not necessary to make a positive case for a moderately liberal view of sexual morality, the fact remains that any attempt to do so directly from sociobiological facts about human nature is misguided, and Wilson's remarks suggest that he is making such an argument. Again, to be fair to Wilson, there are also places where he clearly recognizes that sociobiological conclusions do not amount to moral or rational justifications. For example, after arguing that female hypergamy and infanticide are to be explained as "inherited predispositions to maximize the number of offspring in competition with other members of society," he rightly points out that they nonetheless "do not recommend themselves as rational processes" (1978, p. 42).

[47] E.O. Wilson (1978, pp. 204–6), my emphasis.

mammals." He then goes on to reiterate the general biological imperatives for social mammals, as shaped by natural selection: roughly, the primary biological imperative to maximize personal reproductive success, the secondary imperative to promote the reproductive success of close kin, and the tertiary imperative to cooperate with others to the minimal extent necessary "to enjoy the benefits of group membership". His point is that these conditions will give rise, particularly in advanced technological societies, to a concern with equity, as mentioned earlier—indeed, they will do so, he thinks, in a way that provides deeper roots for that concern than do any of our intellectual "rationalizations". But how does any of *this* support Wilson's evaluative claim about the genuine *moral value* of equity (unless we assume that moral value is simply to be measured in terms of genetic fitness, which again would be absurd)?

The peculiarity of Wilson's thinking here becomes especially clear in the thought experiment he goes on to offer, involving the counterfactual situation in which hymenoptera (ants, bees, wasps) have developed the powers of rationality. "A rational ant", he claims, "would find [a mammalian social arrangement, with the three ordered imperatives listed above] biologically unsound and the very concept of individual freedom intrinsically evil." This claim is bizarre for two reasons. First, if the ants were really rational, then there is no reason to suppose that they would *care* about being "biologically unsound", any more than we care about it when we make innumerable choices that fail to maximize the replication of copies of our genes in the next generation—such as the choice to devote much of our energy to the study of sociobiology or the pursuit of philosophy. The only reason to suppose that the ants would think and feel this way would be if their thinking were so thoroughly under the control of their biology that it amounted to little more than a slavish rationalization for it. But there is no reason to suppose that this would have to be so, since it is evidently not so in our case (and if it were, then it is doubtful that the capacities in question would really be rational capacities after all).

Second, why does Wilson imagine his rational ants making the inference from "biologically unsound" to "intrinsically *evil*"? That is precisely the fallacy I have been discussing, and it is fair to say that his attribution of such reasoning to his hypothetical hymenoptera suggests that Wilson is mixed up in the fallacy himself. A philosophically careful rational ant would certainly not infer that individual freedom is *intrinsically evil* from the fact that for ants such freedom would fail to optimize genetic replication, just as philosophically careful human beings do not infer that adoption of non-kin is *intrinsically evil* from the fact that adoption fails to get one's genes replicated into the next generation. Indeed, if ants were to acquire the powers of reason, along with the sort of culture that would accompany it, there is every reason to suppose that they, like us, would come to value some degree of personal freedom for themselves. And if they were previously slave-making ants, for example, we might well imagine anti-slavery movements eventually arising among them, with certain groups pointing out (in the face of significant initial resistance) that the fact that a tendency to conduct slave-raids is part of their genetic heritage is no justification for such behavior.

Finally, let me mention some misleading remarks in the generally much more cautious work of J.Q. Wilson. Wilson makes a point of noting early on that his book "is not an effort to state or justify moral rules; that is, it is not a book of philosophy".[48] But while he never attempts to justify moral *rules*, he does say things that seem to suggest that he takes himself to be engaged not only in an explanatory project (his primary concern), but also to some degree in a justificatory project—if not of moral rules, then at least of certain general motivations or tendencies. Yet he has not provided the philosophical resources for such justification. The first hint of this comes in his discussion of morality and philosophy (under the curious heading "Morality *versus* Philosophy"). Wilson begins by criticizing much of late-nineteenth and twentieth century philosophy for failing to take seriously "the possibility that men and women are naturally endowed with anything remotely resembling a moral sense"—something that was traditionally recognized by philosophers from Aristotle to Adam Smith.[49] So far, so good: The question whether we are naturally endowed with something like a moral sense is an empirical issue, and Wilson devotes most of the book to providing scientific support for the traditional position that there is a "social side of human nature" as well as a selfish and aggressive side, and that the former naturally gives rise

---

[48] J.Q. Wilson (1993, p. xiii).
[49] J.Q. Wilson (1993, p. 3).

to certain sentiments and tendencies that have been called our natural "moral sense".

He does not, however, restrict his criticism to what he takes to be false empirical views about human nature. When he complains, for example, that "much of modern philosophy abandoned morality without even a hint [of empirical examination]," he does not appear to mean (as he should have) just that many modern philosophers have abandoned the notion of a natural, common human "moral sense" without conducting a proper empirical examination of the matter. For he goes on to criticize Ayer not only for "relegating moral judgments to the realm of *personal* preferences not much different from a taste for vanilla ice cream" (my emphasis), but also for maintaining that moral judgments are "nothing more than 'ejaculations or commands', 'pure expressions of feeling' that have 'no objective validity whatsoever'". The obvious suggestion is that in defending "morality" against those who have so abandoned it, Wilson will be helping to show that moral judgments somehow *do* have "objective validity".

That sort of project is a philosophical one, however, and is not accomplished simply by arguing that evolution has provided us with a common human nature from which a common "moral sense" naturally emerges. To establish the naturalness and universality of certain sentiments and the judgments to which (on Wilson's view, as on Ayer's) they give rise is not to establish the "objective validity" of such judgments—more precisely, it is not to establish the *truth* of the judgments, or any *objective moral justification* or *condemnation* of the behavior motivated by those sentiments. Even if Wilson's arguments from evolutionary psychology are fully accepted, they do not themselves provide any reason for thinking that moral judgments are anything more than false claims (as suggested in Mackie's error theory) resulting from an erroneous projection of moral sentiments onto the world—in Hume's words, from our "gilding or staining all natural objects with the colours borrowed from internal sentiment". Or alternatively, for all the arguments from evolutionary psychology show, moral judgments might well be nothing more than true claims about our emotional reactions to things (as suggested in certain passages of Hume), or even just "claims" that are neither true nor false because they are mere expressions of such reactions (as suggested in other passages of Hume).[50]

Hume, like Wilson, thought that there are natural "sentiments of morality" common to human beings across cultures; indeed, he thought they are rather more robust than Wilson himself seems to think, writing confidently that they are "so deeply rooted in our constitution and temper that, without entirely confounding the human mind by disease or madness, it is impossible to extirpate and destroy them."[51] Nonetheless, Hume is a paradigm moral *subjectivist.* Far from providing morality with any kind of "objective validity", he unequivocally linked moral judgment to contingent (even if universal) sentiments, rejecting the possibility that moral judgments might express truths about the objects of judgment themselves (i.e. as opposed merely to truths about our contingent emotional reactions to them). The new contributions in the area of evolutionary psychology stressed by Wilson provide a scientific account of the universality earlier recognized by Hume, but they do not change the fact that the "moral sense" he identifies is still a contingent product of our evolutionary history. Thus, inasmuch as Wilson still sees moral judgments as stemming from our "moral sense", he has not gone beyond a Humean linking of moral judgment to contingent sentiments.

The result is that Wilson, like Hume, remains stuck with the awkward question of what to say, for example, about the hypothetical (if unlikely) situation in which the natural human sentiments involved with moral judgment change radically, for whatever reason. In such a

---

[50] See Mackie (1977, ch. 1), and Hume, *A Treatise of Human Nature*, Bk III, part I, sec. I, and *An Enquiry Concerning the Principles of Morals*, section IX and Appendix I. The passage from Hume that most clearly suggests a *cognitivist subjectivist* view of moral judgment (i.e. that moral judgments are claims about certain of our emotional reactions to things, and so have determinate truth-values) is the following: "So that when you pronounce any action or character to be vicious, you mean nothing, but that from the constitution of your nature you have a feeling or sentiment of blame from the contemplation of it" (*Treatise*, III.I.I). By contrast, other passages suggest a *non-cognitivist subjectivist* view of moral judgment, closer to twentieth century emotivism. For example, "morality.... is more properly felt than judged of", and "we do not infer a character to be virtuous because it pleases; but in feeling that it pleases after such a particular manner we in effect feel that it is virtuous" (*Treatise*, III.I.II); or again: "language must....invent a peculiar set of terms in order to *express* those universal *sentiments* of censure or approbation which arise from humanity or from views of general usefulness and its contrary" (*Enquiry*, sec IX, part I, my emphasis).

[51] Hume, *Treatise*, III.I.II.

case, people would make radically different moral judgments based on their radically different "moral sense", perhaps looking back on our present concern with such things as fairness, compassion and self-control with ridicule and contempt. But would Wilson want to concede that their new judgments would be just as legitimate as our actual moral judgments—particularly those of which we are the most confident, and about which we feel the most strongly? If not, as proponents of moral objectivity would certainly maintain, why not?

The answer to that question requires a philosophical argument showing that our actual moral judgments (or at least some of them) have a claim to truth—and not simply to truth having to do with our actual, contingent emotional reactions—so that even if some of our emotional reactions were suddenly to change, the moral truths would not, or at least not in the straightforward way suggested by subjectivist views that take moral truths to be about the emotional reactions of appraisers; that is, if the moral truths did change, it would be because of the way in which the change in moral sense altered the possibilities for human social life, not simply because of the change in moral sentiments themselves and in the judgments to which they give rise. And even in that case, there would be limits to how radically the moral truths could change. For example, no changes in the structure of human social life resulting simply from changes in our moral sensibilities could plausibly make it *true* that it is morally acceptable to enslave or otherwise abuse certain groups of people—even if we possessed sentiments that made us feel positively about it. In any event, the point is that the sort of philosophical argument that would be required in order to provide an objective justificatory foundation for morality as we generally understand it would have to be something very different from any kind of explanatory project within evolutionary psychology.

Similar points may be made in relation to Wilson's discussion of *moral relativism*—something he takes himself to be attacking. There is first of all an important ambiguity in his conception of moral relativism. On the one hand, by "moral relativism" or "a relativistic conception of human morality" he often means just the view that there is no common "moral sense" rooted in the universal aspects of human nature; moral sensibilities are thoroughly relative to cultures, personal backgrounds, etc. That is, insofar as there is a universal human nature, it does not contain the elements necessary to tend to give rise to a universal moral sense; human moral sensibilities are to be understood simply in terms of cultural and personal invention and influence, rather than being traceable to universal aspects of human nature—with the result that there may well be a plurality of "moral senses" even at the most fundamental level. But this is not what is generally meant by "moral relativism", and it does not seem to be what he always means by it, judging from the sort of importance he accords to the issue.

Moral relativism is generally understood roughly as the view that there are *no universal objective standards* of moral goodness or rightness. By "objective standards" is meant the *actual conditions* for moral goodness or rightness, as opposed to *beliefs or feelings about* the conditions for moral goodness or rightness (i.e. "moral senses"). The claim is thus not merely that there are a variety of moral standards in the sense that there are a variety of beliefs or feelings about the conditions for moral goodness or rightness (i.e. a variety of "moral senses"), both across cultures and across times. It is rather that there is not at any rate any single set of objective conditions (or standards) for moral goodness or rightness that is universal for human beings.[52] Notice that the "universality" in question could in principle be construed differently: One might, for example, espouse a kind of moral relativism that denies merely that there is a single set of objective standards for moral goodness that is universal for *rational agents*—a kind of "species relativism", for example, rather than "cultural relativism". For purposes of the current discussion, however, I will confine the scope of relevant agents to human beings, as above.

Moral relativism, so understood, is consistent with the view that there are certain *local* objective moral standards—for example, the view that there is a set of objective moral standards for each culture with fairly well defined moral beliefs and customs. The claim is just

---

[52] Cf. Stace (1937), who helpfully distinguishes between two senses of "standards", as above. Such a distinction would have been especially helpful to Wilson's discussion. For example, when he considers the question whether murder is a "universal wrong", he fails to note that there are two very different questions here: (i) whether murder is universally *held* to be wrong, and (ii) whether murder is *actually* wrong universally. (1993, p. 17) (A similar ambiguity exists for the expression "universal moral rules", which he also uses without disambiguating.) Wilson mentions and discusses only the first question, which in itself is perfectly all right so long as it is clear that the scope of the discussion is indeed limited to this sociological issue. The problem occurs when such a discussion extends, without the necessary disambiguation, to cover the general topic of moral relativism, as I go on to discuss with examples below.

that there is no set of objective standards applicable to all human beings—the thought evidently being that somehow differences in beliefs and practices generally make for significant differences in the moral facts themselves, or if it is held that there are no moral facts to begin with, then the thought is that the various subjective standards—i.e. beliefs about moral goodness or rightness—are the only sorts of standards there are.[53] What is at issue here is not the existence of a single, natural moral sense (as with the first conception of "moral relativism" distinguished above), but the existence of a single set of actual conditions for moral goodness or rightness—a philosophical issue rather than a psychological or sociological one.[54]

Wilson appears to conflate the above two conceptions of moral relativism. He maintains, for example, that "the perils [of moral relativism] can be very great". This is a familiar enough claim with regard to *philosophical* moral relativism (the second of the two conceptions distinguished above). If people become convinced that there are no universal objective standards of right and wrong, but only a *variety of culturally arbitrary standards* with nothing placing any one set above any other (or perhaps no even locally objective standards at all, but just certain feelings and tendencies), then what reason is there to

---

53 My characterization differs on this point from that of Stace (1937), who without argument reduces moral relativism generally to moral subjectivism, rather than recognizing the two forms I have allowed, i.e. "local objectivist relativism" *and* "subjectivist relativism".

54 It is instructive to note how a view like Hume's fit into this characterization. In terms of the *psychological* conception of moral relativism (the first of the two conceptions distinguished above), Hume is clearly *not* a moral relativist. As already mentioned, he maintains that the moral sense is firmly rooted in human nature. On the other hand, he clearly rejects the existence of universal *objective* standards for moral goodness or rightness, so that technically he is a moral relativist in the philosophical sense I have defined (the second of the two conceptions distinguished above). But to call Hume a moral relativist in this negative sense, and to leave it at that, would be misleading. For the *point* of denying that there are universal objective moral standards is usually to make room for the assertion that there are instead a variety of equally legitimate standards, whether locally objective standards or just subjective standards, and Hume is not interested in either assertion. On his view, there are no objective standards, and the subjective standards that exist are, at least at a general level, universal. Hume's *positive* relativistic claim is rather best expressed by the claim that *morals (subjectively understood, i.e. moral convictions) are relative to sentiments*, and it so happens that there is a general universality of sentiments among human beings, due to our shared basic constitution (which, he might have added had he lived after Darwin, is in turn due to our common natural selection history). By putting it this way we can see exactly the ways in which he is, and the ways in which he is not, a "moral relativist".

take any one set very seriously and to abide by it when one is tempted by another?[55] But oddly, it is not the philosophical conception of relativism that Wilson discusses in this regard; rather, it is the *psychological* conception (the first of the two conceptions distinguished above). He writes:

> If the moral sense is the result of nothing more significant than a cultural or historical throw of the dice, then it will occur to some people who by reason of temperament or circumstances are weakly attached to their own moral senses that they are free to do whatever they can get away with by practicing indulgent self-absorption or embracing an angry ideology.[56]

This is an extremely odd passage, at least in light of his implication that his own *non*-culturally-relativistic thesis about human nature is supposed to help the situation. One can certainly understand the danger that might result from thinking that one's moral sensibilities are "the result of nothing more significant than a cultural or historical throw of the dice". But how is it supposed to *help* to be told that rather than being the result of cultural invention, our moral sensibilities are happily the result of the forces of *natural selection*?! As I have stressed, the fact that traits have been shaped by natural selection in general establishes no more than that they have been good at propagating their coding genes. This is as true of something noble, like social cooperation, as it is of something base, like male philandering. Thus, the fact (if it is a fact) that the moral sense has firm roots in human evolutionary history, and is not just the product of culture, doesn't in itself provide us with reason to take it seriously in our decision making. Wilson's defeat of

---

[55] This worry does not arise, I think, in the case of a broader "species relativism". For in that case, even though one rejects universal (i.e. covering all rational agents) objective standards of right and wrong, the objective standards that remain need not be suspected of being *arbitrary*: they still apply to human beings generally, and one might well think that they do so because of morally important facts about human life. By contrast, the worry with standard moral relativism, i.e. cultural relativism, is that in rejecting any universal human moral standards, the only standards that remain are arbitrary products of culture, which do not obviously merit our serious consideration and loyalty; indeed, this is precisely what is suggested by popular views of cultural relativism according to which whatever members of a given culture tend to *think* is right, and incorporate into their practices, thereby *is* "right for them" (meaning not just that it is "right according to them", which is trivial, but that it is *actually* right in their case).

[56] J.Q. Wilson (1993, pp. 8–9).

psychological "moral relativism" does not help to defeat philosophical moral relativism, or to defuse the danger of the thought that our moral convictions are in an important sense lacking any relevant foundation, so that we are "free to do whatever we can get away with"; natural selection provides no better a justificatory foundation for our moral convictions than do cultural and historical accidents.

Similarly, Wilson claims that "the moral relativism of the modern age has probably contributed to the increase in crime rates, especially the increases that occur during prosperous times".[57] Again, the only way this claim has any plausibility is if "moral relativism" is understood not in the psychological sense, i.e. not merely as the view that there is no common moral sense rooted in human nature, but in the philosophical sense, as the view that there are no universal objective standards for moral goodness or rightness. Why, after all, should the thought that there is *no common moral sense* rooted in human nature tend to lead one down the dark path of immorality? What leads to immorality, if anything along theoretical lines, is the thought that the moral sensibilities that would tend to discourage one from immorality *have no real foundation of the right sort*: There are no universal objective standards of human goodness with which they are aligned. One might, of course, suggest that this thought tends to follow upon psychological moral relativism, i.e. from the idea that moral sensibilities are just the result of cultural and historical accidents. The problem is that Wilson's alternative psychological *non*-relativism (appealing to common human evolutionary history) fares no better as far as justificatory moral foundations are concerned. The issue here is a philosophical one, and can't be settled by any scientific explanatory project. So again, there seems to be confusion in the discussion of moral relativism: The dangers Wilson cites tend to be dangers (or at any rate, plausible dangers) of *philosophical* moral relativism, while the view he actually succeeds in arguing against is just *psychological* "moral relativism", the dangers of which are rather different (such as giving rise to misguided methods of moral education based on false views of human psychology—something he also discusses, which is much more to his point, and is certainly important).

---

[57] J.Q. Wilson (1993, p. 10).

3. A SKEPTICAL CHALLENGE

I have considered one way in which the sociobiological aspect of human nature might legitimately be thought (and has often been argued, as by both Wilsons) to bear on morality, and one way in which it certainly does not. If by "morality" what is meant is just a certain range of empirical phenomena, such as the "moral sense" discussed by J.Q. Wilson and the behavior it tends to motivate, then morality may intelligibly and plausibly be argued to be part of human nature, or to be a product of certain more basic aspects of human nature. Often the appeal to human nature here is narrow, involving only the former of the two aspects identified above, probably out of a desire for relatively neat and straightforward scientific explanation. A more satisfactory approach would take both aspects of human nature into account, taking seriously the possibility that a significant range of "moral behavior" is motivated at least in part by considerations that gain their motivating force through our exercises of our higher faculties, independently of particular psychobiological influences. But if by "morality" is meant the actual set of facts about the conditions for human moral goodness (assuming that there are any), then no straightforward positive conclusions can be drawn about morality from an examination of the sociobiological aspect of human nature identified above. No examination of what is natural to human beings as a result of our evolutionary history will tell us how it is good or rational for us to live.

This is not, incidentally, to deny that sociobiological facts may bear *indirectly* on the moral facts (assuming that there are moral facts). They plausibly do so, for example, insofar as they bear on the probable benefits and costs or obstacles associated with various social arrangements, resulting from certain natural and deeply ingrained facts about human psychology. This is brought out by E.O. Wilson in connection with social experiments such as the Israeli kibbutzim or the American communes of the 1960's.[58] Our psychological nature, which seems to include such things as a need to care for our own children, as opposed to participating in collective parental care, for example, may very well place rather tight constraints on the range of social arrangements within which we can live sufficiently fulfilling and contented lives. Attempts to force ourselves into certain constructed

---

[58] E.O. Wilson (1978, pp. 137–142).

social molds such as those of the American communes—perhaps out of concern with some abstract principle—might have serious costs, the result being, for example, unhappy parents and maladjusted children. And those costs may plausibly be relevant to the moral situation, having at least some bearing on the question of the true moral legitimacy of the principles in question for human beings, and perhaps supporting others involving, for example, special duties and prerogatives pertaining to family.

In any case, if facts about the sociobiological aspect of human nature do not provide any direct and general *positive* philosophical conclusions about morality, the possibility remains that they nonetheless provide support for a *negative* philosophical conclusion about it: namely, that there are no genuine moral facts at all (unless they are just construed as facts about our emotional reactions to things). This is the move made by Ruse. He argues that "once we have the biological account of morality's origins [i.e. the origins of the "moral sense" and the behavioral tendencies to which it tends to give rise], we see that there can be no metaethical justification of substantive ethics"; for given the evolutionary origin of our "moral sense" and even, he thinks, of our inclination to objectify it and thus to believe in objective moral facts, it follows that "calls for justification (in the sense of foundations) are illicit." Once such a purely psychobiological explanation has been offered for our moral beliefs, requiring no appeal to moral facts to account for the beliefs, there is according to Ruse no longer any reason to take seriously the possibility of objective moral truths corresponding to those beliefs, just as there is no reason to take seriously the possibility of truths corresponding to beliefs in the magic powers of crystals once we are convinced that we have an adequate psychological-causal explanation of the superstitious beliefs in such powers.[59]

This is a familiar argument, and it raises a real skeptical challenge. But its force is plainly overstated. I have argued that while there is plausibly some truth in sociobiological accounts of the "moral sense", we are far from having good reason to suppose that all or even most of our moral beliefs can be accounted for simply in this way. Much more would have to be said to make a compelling case for the claim that our

---

59 Ruse (1988, pp. 41–3, 58–60). Cf. also Harman (1977, ch. 1), and for an illuminating discussion of this type of argument, Sturgeon (1992).

moral beliefs can quite generally be "explained away" as mere sociobiological phenomena, or that "the evolutionary process contains the basis for a *full and adequate understanding* of human moral nature." And even if the strong premises about our moral beliefs were accepted, it obviously does not follow that "there *can be no* justification or underpinning to morality," or that substantive ethics has been "explained away", at least if this means that there are no objective moral facts.[60] Even if many or even most of our actual moral beliefs could be explained simply in sociobiological terms, rather than in terms of the recognition of moral facts, it would not follow that the moral facts don't exist. At most it would follow that these moral beliefs are not justified, even if they are in fact true—though admittedly this would be a disturbing result in itself.

4. CONCLUSION

Ultimately, the best response to such a skeptical challenge, beyond pointing out its limited force, is to provide a positive account that both shows how there can be objective moral facts and explains how it is that many of our actual moral beliefs may plausibly be true and justified after all. This is exactly what Foot has attempted to do with her teleological naturalism, as described in chapter one. The idea behind such a project, again, is that if natural teleology can be construed according to a welfare-based account linking proper natural functioning for living things to the satisfaction of species-typical welfare-related needs, then perhaps moral normativity can be construed as a species of objective, natural teleological normativity; particular human actions might be objectively evaluated in relation to species-typical natural teleological norms having ultimately to do with human flourishing. If my arguments in chapters three, four and six have been sound, however, then this project is hopeless: The welfare-based account of natural teleology, on which it relies, is just false—and the alternative account I have argued for and developed reveals that natural teleology is plainly irrelevant to moral normativity.[61]

---

[60] Ruse (1988, pp. 27, 60), my emphasis.

[61] There are, of course, other problems as well. As argued in chapter six, for example, even if a suitable welfare-based view of natural teleology along Foot's lines could be defended, it would provide virtually no guidance to understanding the structure of morality.

Where does this leave us? My own hope is that the skeptical challenge can still be met, through developing a more satisfactory account of the objectivity of morality. In fact, I am optimistic that such a project can still be carried out within what is still a largely Aristotelian framework. What has to be jettisoned is any appeal to natural teleology in trying to make sense of moral normativity—for example, Foot's and Wallace's appeals to objective, species-relative evaluations of organs, plants and non-human animals in trying to reveal something deep about life itself, which can then be applied in a morally illuminating way to human action. If such mistakes are avoided, however, there may still be hope for a view that seeks to understand moral facts as objectively grounded at least largely in facts about human needs and the conditions for their fulfillment, given a variety of facts about human capacities and the circumstances of human life.[62] We just have to abandon any hope of shedding some deeper light on the connection between moral normativity and human flourishing by appealing to natural teleology, i.e. to the normative nature of life itself.

Where *should* we look to understand this connection? Let me close with some very brief and forward-looking remarks, intended not as a complete argument, but only as an outline of the direction in which I believe a neo-Aristotelian might legitimately proceed. I have argued in this and the previous chapter that human nature must be understood not simply in terms of our possession of a set of innate psychological and behavioral predispositions traceable to our evolutionary history (as emphasized in sociobiology and evolutionary psychology), but also in terms of our possession of advanced rational capacities, which may well be only indirectly traceable to our evolutionary history and in any case plausibly give rise to "spin off" capacities and tendencies extending well beyond the scope of proper biological functioning. It is to the latter, sophisticated reflective capacities—in conjunction with quite ordinary facts about human goods and harms, among other things—that I believe we should look in seeking to understand the nature of moral facts and their connection to such things as human flourishing (rather than to genetic propagation, as with natural teleological evaluations).

---

[62] This, in fact, is precisely the theme of Foot's earlier work, which is not affected by my criticisms of her more recent teleological naturalism. See Foot (1978d,e).

I've already argued that this aspect of human nature makes it possible for us in large measure to transcend our proper biological functioning, and to do so in a systematic way that ultimately makes our lives *our own* in an important sense. That is to say, our advanced mental capacities do not merely provide new ways of realizing old Darwinian ends, making us simply into more interesting gene propagating machines. Instead, they provide the possibility of *setting our own ends*—such as the pursuit of moral goodness, or happiness, or love, or excellence in poetry or philosophy or physics for its own sake—and organizing our lives around them, even at the expense of what we were ultimately "designed" by nature to do. Nor is that all. What is perhaps most striking is that our advanced mental capacities make it possible for us to approach the designing of our own lives not merely in a piecemeal fashion, but in a coherent and unified way through practical reflection at the most formal level, putting to ourselves such questions as: *What is best and most worthwhile in life*? And *how should we pursue these things, so that we treat ourselves, each other, and the rest of the living and non-living aspects of our environment appropriately and responsibly*?

The possession of such capacities obviously does not guarantee that they will be exercised. Someone might well neglect them and live unreflectively, or limit his reflection to broadly instrumental matters, thus giving his life over to the desires and impulses with which he finds himself, largely as a result of his evolutionary and cultural heritage (along with the particular quirks of his nature). In so doing, he would in an important sense be forfeiting his opportunity to make his life truly his own, settling instead for living to a great extent in the ultimate service of his genes and of the cultural imperatives that may have conditioned him. The question, however, is whether he can escape objective criticism for such a choice. It is one thing, we might think, for an animal that does not and can not know any better to live primarily in the unreflective service of its genes; it is quite another for an adult human being to live that way. With the added ability plausibly comes accountability to higher standards that would, among other things, condemn behavior that causes pointless suffering or unnecessarily precludes the realization and enjoyment of important goods—as philandering, for example, even if part of proper biological functioning, obviously hurts people and arguably precludes the enjoyment of the goods bound up with deep and lasting intimacy, in favor of comparatively fleeting and hollow pleasures. If something along these lines is correct, then it seems to be a normative fact rooted in facts about human capacities: Because we are *able* to live our lives in such a way as to direct them toward what is good and worthwhile (at least to the best of our ability to determine this), it seems equally true that this is only *fitting* for us, as opposed to following passively along the path of our biological "designs" or cultural influences; there seems to be something like a moral parallel of childishness—a similar, though more serious, failure to live in a way that is appropriate to our complex nature—in failing to exercise our full practical abilities and living either unreflectively, or with only piecemeal reflection, failing to do our best to put together coherent, defensible and flourishing lives for ourselves.[63] This point might even be put in teleological terms, by speaking of practical reason as having a proper function or end having to do with choosing action that is, all things considered, good—this being understood, of course, not as its *biological* function or end (which would ultimately be genetically oriented, as with any other natural faculty, according to my account), but as a distinct kind of function or end it has by virtue of its special nature as a faculty capable of independent, comprehensive practical reflection on how it is best to live.[64]

There is of course a great deal of argument that would have to be forthcoming to fill in the above moves, to show that there are sufficient constraints on what can count as success in this broad reflective project to justify speaking—even in a sensible, appropriately flexible manner—of anything like a single, objective morality, and to elucidate its structure.[65] My purpose in these brief remarks has been simply to indicate what I take to be a promising general course to pursue, which is naturally suggested by the preceding discussion of the two sides of human nature. However that may turn out, my principal aim has been to show that in any case teleological naturalism in ethics must be

---

[63] Cf. Aristotle, *Eudemian Ethics*, bk. I, ch. 2.

[64] An example of this may be found in Quinn's (1993, ch. 12) neo-Aristotelian defense of the connection between practical reason (or reasons for acting) and goodness.

[65] I take up the first of these projects in a manuscript entitled "Normativity, Goodness and Realism".

abandoned, given what we have seen about the true form of the teleological norms of nature.

Bibliography

Achinstein, Peter

1983 *The Nature of Explanation*. Oxford: Oxford University Press.

Anscombe, G.E.M.

1963 *Intention*, 2nd ed. Ithaca: Cornell University Press.

1981a "Thought and Action in Aristotle," reprinted in *From Parmenides to Wittgenstein: Collected Philosophical Papers, Vol I*. Minneapolis: University of Minnesota Press.

1981b "Authority in Morals," reprinted in *Ethics, Religion and Politics: Collected Philosophical Papers, Vol III*. Minneapolis: University of Minnesota Press.

1981c "Modern Moral Philosophy," reprinted in *Ethics, Religion and Politics: Collected Philosophical Papers, Vol III*. Minneapolis: University of Minnesota Press.

1981d "On Promising and its Justice," reprinted in *Ethics, Religion and Politics: Collected Philosophical Papers, Vol III*. Minneapolis: University of Minnesota Press.

1995 "Practical Inference," in Hursthouse, R., Lawrence G., and Quinn W.S. eds., *Virtues and Reasons*. Oxford: Oxford University Press.

Aquinas, Thomas

Summa Theologiae, excerpted in Pegis, Anton C. trans., *Introduction to St. Thomas Aquinas*. New York: Random House, 1948.

Summa Theologiae, I-II, Q 1–21, in Oesterle, John A. trans., *Treatise on Happiness*. University of Notre Dame Press: 1983.

Aristotle

De Anima. Trans. Hamlyn, D.W. Clarendon Aristotle Series, 1968.

Nicomachean Ethics. Trans. trans. Ross W.D. and Urmson, J.O., revised Barnes, J. (Revised Oxford Aristotle, 1984), in *A New Aristotle Reader*, edited by Ackrill, J.L. Princeton: Princeton University Press, 1987.

Ayala, Francisco

1970 "Teleological Explanation in Evolutionary Biology." *Philosophy of Science* 37: 1–15.

Bedau, Mark

1991 "Can Biological Teleology be Naturalized?" *Journal of Philosophy* 88: 647–55

1992a "Goal-Directed Systems and The Good." *The Monist*, vol 75, no. 1.

1992b "Where's the Good in Teleology?" *Philosophy and Phenomenological Research*, vol 52, no. 4, pp. 781–806.

Bigelow, J. and Pargetter, R.

1987 "Functions." *The Journal of Philosophy* 84: 181–196.

Boorse, Christopher

1976 "Wright on Functions." *The Philosophical Review* 85: 70–86.

Brandon, Robert

1981 "Biological Teleology: Questions and Explanations." *Studies in the History and Philosophy of Science* 12, no. 2: 91–105.

Broadie, Sarah Waterlow
1987 "Nature, Craft and Phronesis in Aristotle." *Philosophical Topics* 15, no. 2: 35–50.

Cummins, Robert
1975 "Functional Analysis." *The Journal of Philosophy* 72: 741–765.

Davidson, Donald
1963 "Actions, Reasons and Causes." *The Journal of Philosophy* 60.

Dawkins, Richard
1976 *The Selfish Gene*. Oxford: Oxford University Press.
1982 *The Extended Phenotype*. New York: Oxford University Press.
1982a "Replicators and Vehicles," in *Current Problems in Sociobiology,* edited by The King's College Sociobiology Group. Cambridge University Press.
1987 *The Blind Watchmaker*. New York: W.W. Norton and Co.
1989 *The Selfish Gene*. 2nd ed., expanded. Oxford: Oxford University Press.

Diamond, Jared
1992 *The Third Chimpanzee: The Evolution and Future of the Human Animal*. New York: HarperCollins.

Eibl-Eibesfeldt, I.
1977 "Phylogenetic Adaptations in Human Behavior," in Grzimek, Bernhard ed. *Encyclopedia of Ethology*. Van Nostrand Reinhold Co.

Foot, Philippa
1978a "Goodness and Choice," in *Virtues and Vices*. Berkeley: University of California Press.
1978b "Euthanasia," reprinted in *Virtues and Vices*. Berkeley: U.C. Press.

1978c "Virtues and Vices," in *Virtues and Vices*. Berkeley: U.C. Press.

1978d "Moral Arguments," reprinted in *Virtues and Vices*. Berkeley: U.C. Press.

1978e "Moral Beliefs," reprinted in *Virtues and Vices*. Berkeley: U.C. Press.

1994 "Rationality and Virtue," in *Norms, Values and Society*, edited by Herlinde Pauer-Studer. Dordrecht: Kluwer Academic Publishers.

1995 "Does Moral Subjectivism Rest on a Mistake?" *Oxford Journal of Legal Studies*, Vol 15, No. 1.

Gaut, Berys

1997 "The Structure of Practical Reason," in Gaut, Berys and Cullity, Garrett eds., *Ethics and Practical Reason*. Oxford: Oxford University Press.

Gauthier, David

1986 *Morals By Agreement*. Oxford: Oxford University Press.

Geach, Peter

1977 *The Virtues*. Cambridge: Cambridge University Press.

Gibbard, Alan

1990 *Wise Choices, Apt Feelings*. Cambridge: Harvard University Press.

Gould, James L.

1982 *Ethology: The Mechanisms and Evolution of Behavior*. New York: W.W. Norton and Company.

Gould, Stephen Jay

1980 "Caring Groups and Selfish Genes," in *The Panda's Thumb*. New York, W.W. Norton.

Grafen, Alan

1984 "Natural Selection, Kin Selection and Group Selection," in Krebs, J.R. and Davies, N.B. eds., *Behavioural Ecology: An*

Evolutionary Approach. Oxford: Basil Scientific Publications.

Griffiths, Paul E.
1993 "Functional Analysis and Proper Functions." *British Journal for the Philosophy of Science* 44: 409–422.

Hampe, M. and Morgan, S.R.
1988 "Two Consequences of Richard Dawkins' View of Genes and Organisms." *Studies in the History and Philosophy of Science*, Vol 19, No. 1, pp. 119–138.

Harman, Gilbert
1977 *The Nature of Morality*. New York: Oxford University Press.

Herman, Barbara
1993 "On the Value of Acting from the Motive of Duty," reprinted in *The Practice of Moral Judgment*. Cambridge: Harvard University Press.

Hempel, Carl
1965 "The Logic of Functional Analysis," in *Aspects of Scientific Explanation*, pp. 297–330. New York: The Free Press.

Hsu, Andrew
"On the Natural History of Artifacts." Manuscript.

Hull, D.L.
1982 "Philosophy and Biology," in Floistad, Guttorm ed., *Contemporary Philosophy: A New Survey*, Vol. 2. The Hague: Martinus Nijhoff Pubs.

Hume, David
A Treatise of Human Nature, 2nd ed., edited by L.A. Selby-Bigge. Oxford: Oxford University Press, 1978.

An Enquiry Concerning the Principles of Morals. Reprinted in Aiken, Henry Ed. *Hume: Moral and Political Philosophy*. New York: Hafner Publishing Co., 1948.

Kant, Immanuel

Grounding for the Metaphysics of Morals, 3rd ed., trans. by Ellington, James W. Indianapolis: Hackett Publishing Co., Inc., 1993.

Kitcher, Philip

1993 "Function and Design." *Midwest Studies in Philosophy* 18: 379–97.

Kitcher, Philip and Sterelny, Kim

1988 "The Return of the Gene." *The Journal of Philosophy* 85: 339–61.

Korsgaard, Christine M.

1996 *The Sources of Normativity*. Cambridge: Cambridge University Press.

Kripke, Saul A.

1982 *Wittgenstein on Rules and Private Language*. Cambridge: Harvard University press.

Lack, David

1954 *The Natural Regulation of Animal Numbers*. Oxford: Clarendon Press.

1966 *Population Studies of Birds*. Oxford: Clarendon Press.

Lawrence, Gavin

1993 "Reflection, Practice and Ethical Skepticism." Pacific Philosophical Quarterly, Vol 74, No.4.

1995 "The Rationality of Morality," in Hursthouse, R., Lawrence G., and Quinn W.S. eds., *Virtues and Reasons*. Oxford: Oxford University Press.

MacIntyre, Alastair
1984 *After Virtue: A Study in Moral Theory*, 2nd ed. Notre Dame: University of Notre Dame Press.

Mackie, J.L.
1977 *Ethics: Inventing Right and Wrong*. Penguin Books.

Mayr, Ernst
1988 *Toward a New Philosophy of Biology*. Cambridge: Harvard University Press.

McDowell, John
1980 "The Role of Eudaimonia in Aristotle's Ethics," in Rorty, Amelie O., *Essays on Aristotle's Ethics*. Berkeley: University of California Press.
1995 "Two Sorts of Naturalism," in Hursthouse, R., Lawrence G., and Quinn W.S. eds., *Virtues and Reasons*. Oxford: Oxford University Press.

Millikan, Ruth
1984 *Language, Thought and Other Biological Categories*. Cambridge: MIT Press.
1989 "In Defense of Proper Functions." *Philosophy of Science* 56: 288–302.

Mitchell, Sandra
1993 "Dispositions or Etiologies? A Comment on Bigelow and Pargetter." *Journal of Philosophy*, pp. 249–59.

Mulder, Raoul A.
1994 "Faithful Philanderers." *Natural History*, November, pp. 57–62.

Nagel, Ernest
1961 "Teleological Explanation," from *The Structure of Science*. New York: Harcourt, Brace and World. Reprinted in J. Canfield, ed., *Purpose in Nature*. Englewood Cliffs, N.J.: Prentice Hall, Inc.

1977 "Teleology Revisited." *Journal of Philosophy,* vol. 74, no. 5.

Nagel, Thomas
1970 *The Possibility of Altruism.* Princeton: Princeton University Press.
1979 "Ethics Without Biology," in *Mortal Questions*. Cambridge: Cambridge University Press.
1986 *The View From Nowhere*. Oxford: Oxford University Press.
1992 "The Value of Inviolability." *Revue de Metaphysique et de Morale*, 99 (1994), 149–66.

Neander, Karen
1991a "Functions as Selected Effects: The Conceptual Analyst's Defense." *Philosophy of Science* 58: 168–184.
1991b "The Teleological Notion of 'Function'." *Australasian Journal of Philosophy* 69: 454–468.

Nussbaum, Martha
1988 "Non-Relative Virtues: An Aristotelian Approach," in Midwest Studies in Philosophy, XIII.
1995 "Aristotle on Human Nature and the Foundations of Ethics," in *World, Mind and Ethics: Essays in the Ethical Philosophy of Bernard Williams*, edited by Altham, J.E.J. and Harrison, Ross. Cambridge: Cambridge University Press.

O'Grady, Richard T.
1986 "Historical Processes, Evolutionary Explanations, and Problems with Teleology." *Canadian Journal of Zoology*, Apr-Jun.

Parfit, Derek
1984 *Reasons and Persons*. Oxford: Clarendon Press.

Plantinga, Alvin
1993 *Warrant and Proper Function.* Oxford: Oxford University Press.

Prior, Elizabeth
1985 "What is Wrong with Etiological Accounts of Biological Function?" *Pacific Philosophical Quarterly* 66: 310–28.

Quinn, Warren
1993 "Putting Rationality in its Place," in *Morality and Action*. New York: Cambridge University Press.

Ruse, Michael
1988 "Evolutionary Ethics: Healthy Prospect or Last Infirmity?" *Canadian Journal of Philosophy*, Supplementary Volume 14, pp. 27–73.

Searle, John R.
1992 *The Rediscovery of the Mind*. Cambridge: MIT Press.

Sober, Elliott
1988 "What is Evolutionary Altruism?" *Canadian Journal of Philosophy*, Supplementary Volume 14, pp. 75–99.
1993 *Philosophy of Biology*. Boulder: Westview Press.

Sober, Elliott and Wilson, David Sloan
1998 *Unto Others: The Evolution and Psychology of Unselfish Behavior*. Cambridge: Harvard University Press.

Sorabji, Richard.
1964 "Function." *Philosophical Quarterly*, vol. 14, no. 57.

Stace, W.T.
1937 *The Concept of Morals*. Macmillan Publishing Company.

Sturgeon, Nicholas
1992 "Nonmoral Explanations," in *Philosophical Perspectives*, 6, ed. Tomberlin, James.

Thompson, Michael

1995 "The Representation of Life," in Hursthouse, R., Lawrence G., and Quinn W.S. eds., *Virtues and Reasons*. Oxford: Oxford University Press.

Villee, C.A. and Dethier, V.G.
1971 *Biological Principles and Processes*. Philadelphia: W.B. Saunders Co.

Wallace, James D.
1978 *Virtues and Vices*. Ithaca: Cornell University Press.

Williams, Bernard
1981 "Internal and External Reasons," reprinted in *Moral Luck*. Cambridge: Cambridge University Press.
1985 *Ethics and the Limits of Philosophy*. Cambridge: Harvard University Press.

Wilson, Edward O.
1978 *On Human Nature*. Harvard University Press. (Page references are to the 1979 Bantam edition.)

Wilson, Edward O. and Ruse, Michael
1986 "Moral Philosophy as Applied Science." *Philosophy* 61: 173–192.

Wilson, James Q.
1993 *The Moral Sense*. New York: The Free Press.

Woodfield, Andrew
1976 *Teleology*. Cambridge: Cambridge University Press.

Wright, Larry
1976 *Teleological Explanations: An Etiological Analysis of Goals and Functions*. Berkeley: University of California Press.

Index

Kolibri 360 GmbH
Ferdinand-Jühlke-Straße 7
99095 Erfurt, Deutschland
produktsicherheit@kolibri360.de